Proceedings of the
International Conference

Semigroups and Formal Languages

In honour of the 65th birthday of Donald B. McAlister

Proceedings of the
International Conference

Semigroups and Formal Languages

In honour of the 65th birthday of Donald B. McAlister

Centro de Álgebra da Universidade de Lisboa (CAUL), Portugal
12 –15 July 2005

Organised with special support from
Centro Internacional de Matemática (CIM)

Editors

Jorge M. André and **Vítor H. Fernandes**
CAUL and Universidade Nova de Lisboa, Portugal

Mário J. J. Branco and **Gracinda M. S. Gomes**
CAUL and Universidade de Lisboa, Portugal

John Fountain
University of York, UK

John C. Meakin
University of Nebraska-Lincoln, USA

World Scientific

NEW JERSEY • LONDON • SINGAPORE • BEIJING • SHANGHAI • HONG KONG • TAIPEI • CHENNAI

Published by

World Scientific Publishing Co. Pte. Ltd.

5 Toh Tuck Link, Singapore 596224

USA office: 27 Warren Street, Suite 401-402, Hackensack, NJ 07601

UK office: 57 Shelton Street, Covent Garden, London WC2H 9HE

British Library Cataloguing-in-Publication Data
A catalogue record for this book is available from the British Library.

SEMIGROUPS AND FORMAL LANGUAGES

ISBN-13 978-981-270-738-3
ISBN-10 981-270-738-7

Printed in Singapore by World Scientific Printers (S) Pte Ltd

Preface

The International Conference on Semigroups and Languages took place in July 2005 at the Center of Algebra of the University of Lisbon (CAUL) and it was jointly organized by CAUL and by the International Center for Mathematics (CIM). It aimed to bring together researchers in topics on or related to Semigroups and Languages, and in this way to acknowledge the enormous importance of Donald McAlister's work in the area. Don's research has made him one of the world's leading figures in the theory of semigroups. In particular, he is renowned for his work on inverse and regular semigroups which combines deep insights with great ingenuity and which has set many of the directions of research in these and related areas. That his contributions continue to shape the theory of inverse and related semigroups is evident from several of the contributions in these proceedings. The conference took place on the occasion of his 65th birthday and it was a tribute to him.

The main speakers were J. Almeida (University of Porto, Portugal), R. Gilman (Stevens Institute of Technology, U.S.A.), M. Lawson (Heriot-Watt University, U.K.), D. McAlister (Northern Illinois University, U.S.A.), D. Munn (University of Glasgow, U.K.), F. Otto (University of Kassel, Germany), J.-E. Pin (University Paris VII, France), P. Silva (University of Porto, Portugal), B. Steinberg (Carleton University, Canada), M. Szendrei (University of Szeged, Hungary), D. Thérien (McGill University, Canada), M. Volkov (Ural State University, Russia) and P. Weil (University Bordeaux I, France) and there were also 31 short talks. Accounts of several of the talks are provided by the articles in this volume, all of which have been refereed in the usual way.

It was an extremely successful conference both on the mathematical and on the human side, attended by 82 people from 16 different countries.

We would like to thank all the participants without whom the conference would not have had this success and all the sponsors whose financial and logistic support made it possible, namely the Center of Algebra of the University of Lisbon (CAUL), the International Center for Mathematics (CIM), the Interdisciplinary Complex of the University of Lisbon (CIUL), the Portuguese Council of University Rectors (CRUP), the Faculty of Science of the University of Lisbon (FCUL), the Foundation of the University of Lisbon (FUL), the Luso-American Foundation (FLAD), the Portuguese Foundation for Science and Technology (FCT), the International Relations Unit of the Portuguese Ministry for Science and Higher Education (GRICES), as well as, the City Hall of Lisbon and the Blue Coast Tourism Board.

Special thanks are due to Patrícia Paraíba for her permanent support and enthusiasm during the preparation of the conference and of these proceedings.

Lisbon, January 2007

Jorge M. André
Mário J.J. Branco
Vitor H. Fernandes
John Fountain
Gracinda M.S. Gomes
John Meakin

Conference sponsors

Project POCTI/0143/2003 "Fundamental and Applied Algebra" of CAUL, financed by FCT and FEDER

Contents

A NOTE ON FINITELY GENERATED SEMIGROUPS OF REGULAR LANGUAGES*

SERGEY AFONIN AND ELENA KHAZOVA

Lomonosov Moscow State University, Institute of Mechanics
Michurinskij pr., 1, Moscow, 119192, Russian Federation
E-mail: serg@msu.ru

Let $\mathcal{E} = \{E_1, \dots, E_k\}$ be a set of regular languages over a finite alphabet Σ. Consider morphism $\varphi : \Delta^+ \to (\mathcal{S}, \cdot)$ where Δ^+ is the semigroup over a finite set Δ and $(\mathcal{S}, \cdot) = \langle \mathcal{E} \rangle$ is the finitely generated semigroup with $\mathcal{E}$ as the set of generators and language concatenation as a product. We prove that the membership problem of the semigroup $\mathcal{S}$, the set $[u] = \{v \in \Delta^+ \mid \varphi(v) = \varphi(u)\}$, is a regular language over Δ, while the set $\mathrm{Ker}(\varphi) = \{(u, v) \mid u, v \in \Delta^+ \ \varphi(u) = \varphi(v)\}$ need not to be regular. It is conjectured however that every semigroup of regular languages is automatic.

1. Introduction

Regular languages play an important role in both theoretical and practical aspects of computer science. As regular languages are closed under basic language operation it is interesting to ask whether one language may be represented in terms of other regular languages. The language factorization problem, i.e. the problem of representing a given regular language as a concatenation of other (regular) languages is a special case of such a representation. If the set of factors is fixed then language factorization may be considered as the membership problem for a finitely generated semigroup: a regular language $R \subseteq \Sigma^*$ belongs to the semigroup $\mathcal{S} = \langle E_1, \dots, E_k \rangle$ if and only if there exists a sequence $i_1, i_2, \dots, i_n$ of integers such that $1 \leqslant i_p \leqslant k$ $(p = 1 \dots n)$ and

$$R = E_{i_1} E_{i_2} \dots E_{i_n}.$$

The membership problem for a finitely generated semigroup of regular languages was shown to be decidable in [5]. This solution is based on reduc-

*This research was supported in part by the grant no. 10002-251 of the Russian Academy of Sciences Presidium program no. 17 "Parallel computations on multiprocessor systems"

tion to the so-called limitedness property of distance automata and gives an answer to the question whether a given regular language R belongs to $\mathcal{S}$ or not. Once we know that the answer is positive, a solution can be found by exhaustive search.

In [1] the authors considered finiteness conditions for semigroups of regular languages. It was found that a finitely generated semigroup $\mathcal{S} = \langle E_1, \ldots, E_k \rangle$ is finite if and only if for every set of non-repeating indices $\{i_1, \ldots, i_m\}$ $(m \leqslant k)$ there exists a natural number p such that $(E_{i_1} \ldots E_{i_m})^p = (E_{i_1} \ldots E_{i_m})^{p+1}$. In contrast, the semigroup given by the presentation $S = \langle \Delta \mid x^3 = x^2 \text{ for all } x \in \Delta^* \rangle$ is infinite [2] in the case of $|\Delta| \geqslant 2$. Actually, it is not surprising that semigroups of regular languages have a more complicated structure, because the structure of regular languages induces some additional relations between generators of a semigroup. In this paper we study the membership and word problems of the finitely generated semigroups of regular languages.

The layout of the paper is as follows. In section 2 we present basic definitions and briefly describe some useful results. In section 3 the regularity of the membership problem and the non-regularity of the word problem are proved. In the last section we discuss a connection between semigroups of regular languages and automatic semigroups.

2. Preliminaries

We assume familiarity with formal languages and finite automata, but recall, in order to fix the notation, the basic definitions. An *alphabet* is a finite non-empty set of *symbols*. A finite sequence of symbols from an alphabet Σ is called a *word* over Σ. The *empty* word is denoted by ε. A word $u = a_1 a_2 \ldots a_k$ is called a *scattered subword* of a word v (denoted as $u \sqsubseteq v$), if there exist $w_1, \ldots, w_{k+1} \in \Sigma^*$ such that $v = w_1 a_1 w_2 a_2 \ldots w_k a_k w_{k+1}$.

Any set of words is called a *language* over Σ. Σ^* denotes the set of all finite words (including the empty word), Σ^+ denotes the set of all non-empty words over Σ, $\varnothing$ is the empty language (containing no words), and 2^{Σ^*} is the set of all languages over Σ. The *union* of languages L_1 and L_2 is denoted by $L_1 + L_2$, concatenation by $L_1 L_2$, and iteration (or Kleene star) by L^*.

A *(non-deterministic) automaton* is a tuple $\mathcal{A} = \langle \Delta, Q, \rho, q_\circ, F \rangle$, where Q is a finite set of states, Δ is an input alphabet, $\rho \subseteq Q \times \Delta \times Q$ is a set of transitions, $q_\circ \in Q$ is the initial state and $F \subseteq Q$ is a set of final states. A successful path π in $\mathcal{A}$ is a sequence $(q_1, \delta_1, q_2)(q_2, \delta_2, q_3) \ldots (q_{m-1}, \delta_m, q_m)$

of transitions such that $q_1 = q_\circ$ and $q_m \in F$. We call the word $\delta_1\delta_2\dots\delta_m$ the *label* of π. The language $L(\mathcal{A})$ of the automaton $\mathcal{A}$ is the set of all words w such that there exists a successful path π labeled by w. The language L is called *regular* if it is recognized by a finite automaton. The class of regular languages over Σ is denoted by $\mathrm{Reg}(\Sigma)$.

The set X of pairs of words (u, v), where $u = a_1 \dots a_m$ and $v = b_1 \dots b_m$ ($a_i, b_i \in \Delta$) is called regular if there exists a finite automaton over $\Delta \times \Delta$ that recognizes X. This definition may be extended to the case when length of words u and v differs by introducing the *padding symbol* \$, and the mapping $\mu : \Delta^+ \times \Delta^+ \to (\Delta \cup \{\$\})^+ \times (\Delta \cup \{\$\})^+$ that appends the minimum number of padding symbols at the end of the shorter word in order to equalize their lengths.

Let Σ be an alphabet and $\mathcal{E} = \langle E_1, \dots, E_k\rangle$ be a set of regular languages over Σ. The concatenation of regular languages is a regular language and we write $(\mathcal{S}, \cdot) = \langle \mathcal{E} \rangle$ for the finitely generated semigroup, generated by $\mathcal{E}$ with concatenation as a semigroup product. The homomorphism $\varphi : \Delta^+ \to (\mathrm{Reg}(\Sigma), \cdot)$ between the free semigroup Δ^+ and the semigroup of regular languages with concatenation is called a *regular language substitution.* With every semigroup $(\mathcal{S}, \cdot) = \langle E_1, \dots, E_k\rangle$ of regular languages we can associate a language substitution $\varphi : \{\delta_1, \dots, \delta_k\} \to \mathcal{S}$, defined by the rule $\delta_i \mapsto E_i$. Conversely, every regular language substitution $\varphi : \Delta^+ \to \mathrm{Reg}(\Sigma)$ generates the semigroup $\mathcal{S}_\varphi = \langle\{\varphi(\delta) \mid \delta \in \Delta)\}\rangle$. For a regular language $L \subseteq \Delta^+$ by $\varphi(L)$ we mean the set $\varphi(L) = \{\varphi(w) \mid w \in L\}$ of regular languages over Σ.

Definition 2.1. Let $\varphi : \Delta^+ \to \mathrm{Reg}(\Sigma)$ be a regular language substitution. The *maximal rewriting* of a regular language $R \subseteq \Sigma^*$ with respect to φ is the set

$$M_\varphi(R) = \{w \in \Delta^+ \mid \varphi(w) \subseteq R\}.$$

The following theorem is due to Calvanese *et al.* [3]

Theorem 2.1. *Let $\varphi : \Delta^+ \to \mathrm{Reg}(\Sigma)$ be a regular language substitution. For any regular language $R \subseteq \Sigma^*$ the maximal rewriting $M_\varphi(R)$ is a regular language over Δ.*

3. Membership and word problems

The regularity of the membership problem for the semigroup of regular languages is a simple corollary from Theorem 2.1 and Higman's lemma, which may be stated as follows.

Lemma 3.1. *In every infinite sequence* $\{u_i\}_{i\geqslant 1}$ *of words over a finite alphabet there exist indices* i *and* j*, such that* $u_i \sqsubseteq u_j$.

Let $w = \delta_{i_1} \dots \delta_{i_m}$ be a word over Δ and $A \subseteq \Delta$. We shall call the language $w \Uparrow A^* = A^*\delta_{i_1}A^*\delta_{i_2} \dots A^*\delta_{i_m}A^*$ the *shuffle extension* of w. By $E(w, A)$ denote the language $(w \Uparrow A^*) \cap M_\varphi(\varphi(w))$. Clearly, $E(w, A)$ is a regular language for all $w \in \Delta^+$.

Proposition 3.1. *Let* $\varphi : \Delta^+ \to \mathrm{Reg}(\Sigma)$ *be a regular language substitution,* $u \in \Delta^+$*, and* $\Delta_0 = \{\delta \in \Delta \mid \varepsilon \in \varphi(\delta)\}$*. For every* $v \in E(u, \Delta_0)$ *we have*

$$\varphi(u) = \varphi(v).$$

Proof. Let $\delta \in \Delta_0$. Consider the word $v = u_1\delta u_2$, where $u_1, u_2 \in \Delta^*$ and $u = u_1u_2$. We have

$$\varphi(u) \subseteq \varphi(v) \subseteq \varphi(u).$$

The first inclusion is due $\varepsilon \in \varphi(\delta)$ while the second one follows from the definition of the language $E(u, \Delta_0)$. □

Theorem 3.1. *Let* $\varphi : \Delta^+ \to \mathrm{Reg}(\Sigma)$ *be a regular language substitution and* w *be a word in* Δ^+*. The membership problem for the semigroup* $\mathcal{S}_\varphi$

$$[w] = \{u \in \Delta^+ \mid \varphi(u) = \varphi(w)\}$$

is a regular language over Δ.

Proof. By Theorem 2.1 the set $M_\varphi(w) = \{u \in \Delta^+ \mid \varphi(u) \subseteq \varphi(w)\}$ is regular. Clearly $[w] \subseteq M_\varphi(w)$. Let $[w]$ be an infinite language. We prove now that there exists a finite subset $F \subseteq [w]$ satisfying

$$[w] = \bigcup_{u \in F} E(u, \Delta_0).$$

Note that $u \sqsubseteq v$ implies $E(v, \Delta_0) \subseteq E(u, \Delta_0)$, so without loss of generality we may assume that if the language F contains a word v then it does not contain subwords of v. The finiteness of F follows immediately from Lemma 3.1, and thus $[w]$ may be represented as a finite union of regular languages. □

This proof is not constructive because Higman's lemma does not provide an algorithm for the construction of the set F. In [1] the authors provide the algorithm that for a given regular language substitution $\varphi : \Delta^+ \to \mathrm{Reg}(\Sigma)$, a regular language $K \subseteq \Delta^+$, and a regular language $R \in \Sigma^*$ checks whether or not R belongs to the *rational set of regular languages* $\mathcal{R}(K,\varphi) = \{\varphi(w) \mid w \in K\}$. On the basis of this result the set F may be constructed by the following procedure (the input of this algorithm is a word $w \in \Delta^+$):

- Assign $K = \Delta^+$, $F = \varnothing$
- Repeat the following steps until $\varphi(w) \in \mathcal{R}(K,\varphi)$
  - Find the shortest word $v \in K$ such that $\varphi(v) = \varphi(w)$
  - Add v to F
  - Assign $K = K \setminus E(v, \Delta_0)$

Proposition 3.2. *The algorithm is correct.*

Proof.

(1) The algorithm always terminates by Lemma 3.1 — elements of the set F are not subwords of each other.
(2) On each round of the algorithm we have that F does not contain subwords of K.
(3) On each round $E(v, \Delta_0) \subseteq [w]$ by Proposition 3.1. □

The result of Theorem 3.1 is nonextensible in the following sense.

Theorem 3.2. *Let $\varphi : \Delta^+ \to \mathrm{Reg}(\Sigma)$ be a regular language substitution. The set*

$$\mathrm{Ker}(\varphi) = \{(u,v) \mid u,v \in \Delta^+ \ \varphi(u) = \varphi(v)\}$$

need not be regular.

Proof. We show that if the set $\mathrm{Ker}(\varphi)$ is regular then the equivalence problem for a rational set of regular languages, i.e. the problem to decide whether or not two given rational sets $\mathcal{R}_1 = (K_1,\varphi)$ and $\mathcal{R}_2 = (K_2,\varphi)$ are equal as sets of languages over Σ, is decidable. Then we reduce the later problem to the finite substitutions equivalence problem, that is known to be undecidable.

For a given regular language $K \in \Delta^*$ by $\overline{K}$ denote the closure of K with respect to φ:

$$\overline{K} = \varphi^{-1}(\varphi(K)) = \{u \in \Delta^+ \mid \exists v \in K \ \varphi(u) = \varphi(v)\}.$$

Let $\mathcal{R}_1 = (K_1,\varphi)$ and $\mathcal{R}_2 = (K_2,\varphi)$ be rational sets of regular languages. We have $\mathcal{R}_1 = \mathcal{R}_2$ if and only if $\overline{K_1} = \overline{K_2}$. Now, suppose that the set

$\mathrm{Ker}(\varphi)$ is regular, i.e. there exists a finite automaton M that recognizes this language. By standard direct product construction of automata M and K we construct the automaton that recognizes the language $\{v \in \Delta^+ \mid \exists u \in K\ (u,v) \in \mathrm{Ker}(\varphi)\}$. Thus, if $\mathrm{Ker}(\varphi)$ is regular then the so is $\overline{K}$ for every regular language $K \in \Delta^*$.

Let φ_1 and φ_2 be finite substitutions, i.e. homomorphisms between Δ^+ and a semigroup of finite languages. The equivalence problem of finite substitutions on a regular language L

$$\varphi_1(w) = \varphi(w) \text{ for all } w \in L$$

is known to be undecidable [6] for $L = xy^*z$. Let $\Delta = \{x_1, y_1, z_1, x_2, y_2, z_2\}$, φ be a finite substitution, and rational sets $\mathcal{R}_1 = (K_1, \varphi)$ and $\mathcal{R}_2 = (K_2, \varphi)$ are given by languages $K_1 = x_1 y_1^* z_1$ and $K_2 = x_2 y_2^* z_2$. By considering the length of the longest word in the image $\varphi(w)$ we have that $\mathcal{R}_1$ and $\mathcal{R}_2$ are equal if and only if finite substitutions φ_1 and φ_2 (induced by φ) are equal on the language xy^*z. We have a contradiction, so the set $\mathrm{Ker}(\varphi)$ is not regular in general. □

4. Connection with automatic semigroups

Let us recall the definition of automatic semigroups [4]. Let S be a semigroup, A be a finite set, L be a regular language over A, and $\psi : A^+ \to S$ be a homomorphism with $\psi(L) = S$. The pair (A, L) is called an *automatic structure* for S if

(1) $L_= = \{(u,v) \mid u, v \in L, \psi(u) = \psi(v)\}$ is regular;
(2) $L_a = \{(u,v) \mid u, v \in L, \psi(ua) = \psi(v)\}$ is regular for each $a \in A$.

If a semigroup S has an automatic structure (A, L) for some A and L, then S is called *automatic*.

Automatic semigroups include many naturally appearing semigroups, e.g. finite semigroups and finitely generated subsemigroups of a free semigroup are automatic [4]. An attractive property of automatic semigroups is the solvability of the word problem in quadratic time.

Example 4.1. Let the semigroup $\mathcal{S}$ be generated by languages

$$x = (a+b)^*a, \quad y = \varepsilon + a + b, \text{ and } z = b^*$$

over $\Sigma = \{a, b\}$. In order to show that $\mathcal{S}$ is automatic we find the presentation for $\mathcal{S}$ and construct automata for $L_=$ and L_a.

First, the language z is a star, so $z^k = z^p$ for all $k, p \geqslant 1$. Second, since x is the set of all words over Σ that are ended by the letter a and both y and z contain the empty word we have equations $(y+z)^k x = x$. Finally, the language $xz = (a+b)^*ab^*$ is the set of all words over Σ that contain at least one letter a, so we have relations $x(y+z)^k z(y+z)^p = xz$ for all $k, p \geqslant 0$. The semigroup $\mathcal{S}$ satisfies no other relations. Thus, the semigroup $\mathcal{S}$ is given by the presentation

$$\langle x, y, z \mid yx = x, zx = x, z^2 = z, xzy = xz, xy^k z = xz \ (k \geqslant 1) \rangle.$$

We show now that $\mathcal{S}$ is automatic. First, let us construct the set L of normal forms, such that each element of the semigroup is represented by only one word in L. If a word $w \in L$ contains x then x is the first letter of the word. If w contains both x and z then w has the from $x^k z$ by the last three relations. If $w \in L$ does not contain x, then w is the set of all words over $\{y, z\}$ that does not contain zz as a subword. So we have

$$L = xx^*(y^* + z) + yy^*(zyy^*)^*(\varepsilon + z) + z(yy^*z)^*y^*.$$

If every element of a semigroup is represented by exactly one word in L then the set $L_= = \{(u, u) \mid u \in L\}$ is regular.

Now consider the multiplication by generators. The product of a word $u \in L$ by x equals to x if the word u does not contain x, and equals to x^{k+1} if $u = x^k u'$ (the word u' does not contain x). We have

$$\begin{aligned} L_x = {} & (x,x)(x,x)^* \, [(\$,x) + (y,x)(y,\$)^* + (z,x)] + \\ & (y,x)(y,\$)^*[(z,\$)(y,\$)(y,\$)^*]^*[\varepsilon + (z,\$)] + \\ & (z,x) \, [(y,\$)(y,\$)^*(z,\$)]^* \, (y,\$)^*. \end{aligned}$$

The first line above corresponds to the first part of $L_=$, i.e. words that start with x, the second and the third lines – to words that start with y and z, respectively.

Similarly we can construct the languages L_y and L_z. We have

$$\begin{aligned} L_y = {} & (x,x)(x,x)^*[(y,y)^*(\$,y) + (z,z)] + \\ & (y,y)(y,y)^*[(z,z)(y,y)(y,y)^*][\varepsilon + (z,z)](\$,y) + \\ & (z,z)[(y,y)(y,y)^*(z,z)]^*(y,y)^*(\$,y) \end{aligned}$$

and

$$\begin{aligned} L_z = {} & (x,x)(x,x)^*[(\$,z) + (y,z)(y,\$)^* + (z,z)] + \\ & (y,y)(y,y)^*[(z,z)(y,y)(y,y)^*][(\$,z) + (z,z)] + \\ & (z,z)[(y,y)(y,y)^*(z,z)]^*[(y,y)(y,y)^*(\$,z) + (z,\$)]. \end{aligned}$$

The language $L_=$ and all languages L_a for $a \in \Delta$ are regular so the semigroup $\mathcal{S}$ is automatic.

Theorem 3.2 states that in general (A, Δ^+) is not an automatic structure for a semigroup of regular languages. Nevertheless, we expect that

Conjecture 4.1. *Every semigroup of regular languages is automatic.*

Although automatic structure may be easily constructed for a particular semigroup of regular languages, just like in the above example, this conjecture seems not to be trivial. For example, let $\mathcal{S}$ be a semigroup of finite languages. Every finite language may be factorized into prime languages, but this factorization is not unique (e.g. $\{\varepsilon, a, a^2, a^3\} = \{\varepsilon, a\}^3 = \{\varepsilon, a\}\{\varepsilon, a^2\}$), thus prime factors may satisfy nontrivial relations. In the general case the situation become more complicated.

References

1. S. Afonin and E. Hazova, Membership and finiteness problems for rational sets of regular languages, Proceedings of the Developments in Language Theory 2005, C. De Felice and A. Restivo, Eds., *Lecture Notes in Computer Science* **3572**, Springer, 88–99 (2005).
2. J. Brzozowski, K. Culik, and A. Gabrielian, Classification of noncounting events, *Journal of Computer and System Sciences* **5**, 41–53 (1971).
3. D. Calvanese, G. De Giacomo, M. Lenzerini and M. Vardi, Rewriting of regular expressions and regular path queries, *Journal of Computer and System Sciences* **64**, 443–465 (2002).
4. C. M. Campbell, E. F. Robertson, N. Ruškuc and R. M. Thomas, Automatic semigroups, *Theoretical Computer Science* **250 (1–2)**, 365–391 (2001).
5. K. Hashiguchi, Representation theorems on regular languages, *Journal of Computer and System Sciences* **27**, 101–115 (1983).
6. J. Karhumäki and L. P. Lisovik, The equivalence problem of finite substitutions on ab*c, with applications, *ICALP'02: Proceedings of the 29th International Colloquium on Automata, Languages and Programming*, Springer-Verlag, 812–820 (2002).

COMPLETE REDUCIBILITY OF PSEUDOVARIETIES

J. ALMEIDA

Departamento de Matemática Pura, Fac. de Ciências, Universidade do Porto, Rua do Campo Alegre, 687, 4169-007 Porto, Portugal
E-mail: jalmeida@fc.up.pt

J. C. COSTA

Centro de Matemática, Universidade do Minho, Campus de Gualtar, 4700-320 Braga, Portugal
E-mail: jcosta@math.uminho.pt

M. ZEITOUN

LaBRI, Université Bordeaux 1 – CNRS
351 cours de la Libération, 33405 Talence Cedex, France
E-mail: mz@labri.fr

The notion of reducibility for a pseudovariety has been introduced as an abstract property which may be used to prove decidability results for various pseudovariety constructions. This paper is a survey of recent results establishing this and the stronger property of complete reducibility for specific pseudovarieties.

1. Introduction

One of the most fruitful settings for the applications of the theory of finite semigroups in computer science has been formalized by Eilenberg in [25]. The classification of rational languages according to several natural combinatorial properties is translated in terms of the pseudovarieties of finite semigroups to which their syntactic semigroups belong. Several combinatorial constructions on rational languages correspond to algebraic operations on semigroups which have counterparts as operations on pseudovarieties. See [25, 28, 33, 1] for background and examples.

To establish decidability results for certain pseudovariety constructions, one is often led to a decision problem which consists in determining whether a system of equations of some suitable type with rational constraints admits a solution modulo every semigroup of a given pseudovariety V.

A standard compactness argument allows us to transfer this problem to deciding whether the system has a solution in a fixed free pro-V semigroup $\overline{\Omega}_A\mathsf{V}$. Since such semigroups are usually uncountable, these decision problems are hard to handle directly but a successful approach has been devised by Almeida and Steinberg [12, 11]. Under mild hypotheses on V (recursive enumerability) and on the type of equations (recursive enumerability of the corresponding signature, as well as computability of its operations), it is easy to exhibit a semi-algorithm to enumerate the systems which do not have solutions. So, the question amounts to determining whether there is also a semi-algorithm to enumerate those systems that do have solutions. Since there are too many candidates for solutions, the next idea is to reduce the universe where solutions need to be sought. This leads to the reducibility property: if the system admits a solution then it admits a solution of a special type. The universe of candidates for solutions that is most often encountered consists of the smallest subsemigroup of the free profinite semigroup $\overline{\Omega}_A\mathsf{S}$ containing the free generators which is closed under unary pseudo-inversion $s \mapsto s^{\omega-1}$. If the reducibility property holds for every finite system of equations, then we say that V is completely reducible. For the method to be successful, besides this reducibility property, one needs the decidability of a word problem so as to be able to determine whether a candidate for a solution is actually a solution.

This paper is a survey of reducibility results for pseudovarieties. We also present a sketch of a proof that the pseudovariety R, of all finite $\mathcal{R}$-trivial semigroups, is completely reducible. The proof is inspired by Makanin's algorithm to decide whether a finite system of word equations with rational constraints has a solution in the free semigroup [30, 31, 29]. It suggests new connections between Finite Semigroup Theory and Combinatorics on Words which deserve further investigation. The full details of the proof will appear elsewhere [6].

2. How we are led to systems of equations

We start by illustrating with two examples how decision problems for systems of equations come up when trying to prove decidability of pseudovariety constructions through bases of pseudoidentities for such pseudovarieties.

Let Sl denote the pseudovariety $[\![x^2 = x, xy = yx]\!]$ of all finite semilattices. Given any pseudovariety V, the *Basis Theorem* for semidirect products [16][a] gives the following basis of pseudoidentities for the semidirect product of Sl with V:

$$\mathsf{Sl} * \mathsf{V} = [\![wu^2 = wu,\ wuv = wvu :\ \mathsf{V} \models wu = wv = w]\!].$$

Thus, to check whether a given finite semigroup S belongs to $\mathsf{Sl} * \mathsf{V}$, it suffices to verify the following condition:

> let $\bar{w}, \bar{u}, \bar{v} \in S$ be such that at least one of the inequalities $\bar{w}\bar{u}^2 \neq \bar{w}\bar{u}$ and $\bar{w}\bar{u}\bar{v} \neq \bar{w}\bar{v}\bar{u}$ holds; then there are no elements $w, u, v \in \overline{\Omega}_A\mathsf{S}$ and evaluation of the generators A in S such that:
>
> (1) w, u, v are evaluated to $\bar{w}, \bar{u}, \bar{v}$, respectively;
> (2) $\mathsf{V} \models wu = wv = w$.

Thus, we are led to consider the system of equations $zx = zy = z$ upon whose variables x, y, z we impose constraints in the semigroup S. We would like to be able to decide whether there is some *solution of the system modulo* V in the sense that the above conditions (1) and (2) hold.

A similar example is provided by Mal'cev products. Bases of pseudoidentities for Mal'cev products have been described by Pin and Weil [34]:

$$\mathsf{Sl} \textcircled{m} \mathsf{V} = [\![u^2 = u,\ uv = vu :\ \mathsf{V} \models u^2 = u = v]\!].$$

Here, the system consists of the equations $x^2 = x = y$. But, otherwise, the nature of the decision problem is the same: to be able to decide whether, imposing constraints for the variables in a given finite semigroup, the system admits a solution modulo every semigroup from V.

The type of equations that appear depends on the operation on pseudovarieties that one is interested in computing and on a certain parameter from the "other" pseudovariety. In the above cases, the parameter is respectively a graph[b]

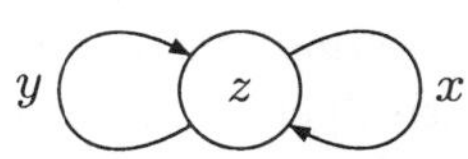

upon which a basis of pseudoidentities for the global[c] gSl may be written,

---

[a]The proof of the Basis Theorem is known to have a gap in its full generality, although its validity remains open. See [3, 42, 36] for further information.

[b]We associate a system of equations to a finite directed graph by viewing each edge and each vertex as a variable and writing the equation $xy = z$ for each edge $x \xrightarrow{y} z$.

[c]The global of a pseudovariety of semigroups is the pseudovariety of semigroupoids which it generates and the restriction under which the Basis Theorem for semidirect products is known to be valid is that the global of the first factor admit a basis of pseudoidentities over graphs with a bounded number of vertices.

and the "rank" of the pseudovariety SI, that is the minimum number of variables in a basis of pseudoidentities defining it.

In general, we are given a finite system of equations $u_i = v_i$ $(i \in I)$ over a finite set X of variables for which constraints are chosen in a given finite semigroup S: s_x $(x \in X)$. By a *solution of the system modulo* an A-generated profinite semigroup T we mean a mapping $\varphi : X \to \overline{\Omega}_A\mathsf{S}$ into the free profinite semigroup $\overline{\Omega}_A\mathsf{S}$ over the set A, together with a continuous homomorphism $\psi : \overline{\Omega}_A\mathsf{S} \to S$ such that the following conditions hold:

(1) $\forall x \in X,\ \psi(\varphi(x)) = s_x$;
(2) $\forall i \in I,\ \theta\hat{\varphi}(u_i) = \theta\hat{\varphi}(v_i)$,

where $\hat{\varphi}$ is the unique extension of φ to a continuous homomorphism $\overline{\Omega}_X\mathsf{S} \to \overline{\Omega}_A\mathsf{S}$ and $\theta : \overline{\Omega}_A\mathsf{S} \to T$ is the unique continuous homomorphism determined by the choice of generators. In case $T = \overline{\Omega}_A\mathsf{V}$, we speak of a *solution modulo* V. The problem is to decide whether such a solution exists.

There are a number of reformulations and generalizations which we proceed to present. See [3] for further details. First, it suffices to consider onto continuous homomorphisms $\psi : \overline{\Omega}_A\mathsf{S} \to S$, in which case the existence of a solution modulo V is independent of the finite set A. Second, for a fixed onto continuous homomorphism $\psi : \overline{\Omega}_A\mathsf{S} \to S$, the constraints may be lifted to constraint sets in $\overline{\Omega}_A\mathsf{S}$ which are therefore clopen subsets of $\overline{\Omega}_A\mathsf{S}$, that is closures of rational languages of the free semigroup A^+. In this form, the problem is formulated entirely as a problem in the free profinite semigroup $\overline{\Omega}_A\mathsf{S}$: the analogous problem with constraints given by clopen subsets of a fixed free profinite semigroup $\overline{\Omega}_A\mathsf{S}$, where solutions modulo V are sought, is equivalent to the original problem. It may be useful to have variables for which there is no room for choice for their values, that is they play the role of *parameters*. The equations $u_i = v_i$ may be given by pseudowords,[d] that is we may consider *pseudo-equations* instead of *word equations*.

As mentioned in the Introduction, it is not hard to obtain a semi-algorithm for non-solvability. If the system has a solution in $\overline{\Omega}_A\mathsf{S}$ modulo V then it also has a solution modulo any A-generated semigroup from V: every solution modulo V has that property. By a compactness theorem, the converse is also true. For a specific A-generated semigroup T from V, the problem of existence of solutions modulo T can be solved by checking

[d]Elements of $\overline{\Omega}_A\mathsf{S}$ may be called *pseudowords* when they are viewed as combinatorial entities generalizing finite words, or *implicit operations* if they are identified with such operations via their natural interpretation as operations on finite semigroups.

a finite number of candidates. Thus, the existence of solutions modulo V for finite systems of word equations is (theoretically) decidable if we can also exhibit a semi-algorithm that enumerates the solvable systems. The difficulty is that, $\overline{\Omega}_A\mathsf{S}$ being uncountable for every non-empty set A, there are too many candidates for solutions. Moreover, we need to be able to determine whether a candidate for a solution modulo V actually has this property, namely whether it satisfies the constraints and the equations, modulo V. The first difficulty is overcome if we can reduce the existence of solutions modulo V in $\overline{\Omega}_A\mathsf{S}$ to the existence of solutions modulo V in some recursively enumerable subset of $\overline{\Omega}_A\mathsf{S}$. A setting for performing such a reduction was proposed in [11]: a subalgebra $\Omega_A^\sigma\mathsf{S}$ of $\overline{\Omega}_A\mathsf{S}$ for an *implicit signature* σ, that is a signature consisting of binary multiplication together with some implicit operations, which have a natural interpretation in every finite semigroup. The computational requirements for such a signature are: (1) it should be recursively enumerable (so that we may enumerate the members of $\Omega_A^\sigma\mathsf{S}$); (2) its operations should be computable in finite semigroups (so that we may check the constraints); (3) the word problem for $\Omega_A^\sigma\mathsf{V}$ should be solvable (so that we may verify whether the equations hold modulo V).

We say that V is *σ-reducible* with respect to a class of equation systems if the existence of a solution modulo V of any system in the class entails the existence of a solution in σ-terms. In case the class consists of all finite systems of equations of σ-terms (with parameters also given by σ-terms), we say that V is *completely σ-reducible.* If the class consists of all systems of equations associated with finite graphs, then we say that V is *σ-reducible.*

An example of a common candidate for such a signature consists of multiplication together with the unary pseudo-inversion $x \mapsto x^{\omega-1}$. It is called the *canonical signature* and denoted κ; whether it is suitable or not depends on the pseudovariety V, as we need the word problem for $\Omega_A^\kappa\mathsf{V}$ and the appropriate κ-reducibility property. Here are some examples: the pseudovariety G of all finite groups is κ-reducible [17][e] but not completely

[e]For groups, κ-reducibility admits a different type of formulation which was originally established by Ash; the equivalence between the two formulations can be found in [11]. Ash obtained his results as a means to prove the *Rhodes Type II Conjecture*, whose history and relevance is explained in [26]. Independently and roughly at the same time, the conjecture was also proved by Ribes and Zalesskiĭ [37] through the theory of profinite groups. In turn, their result was translated into a result in Model Theory which was extended by Herwig and Lascar [27] into a deep result about the existence of extensions to automorphisms (of perhaps larger finite structures) of partial automorphisms of finite relational structures, together with a technical formulation of the same result as a

κ-reducible [24]; for a prime p, the pseudovariety G_p of all finite p-groups is not κ-reducible but it is σ-reducible for a certain infinite signature σ [2]; the pseudovariety Ab of all finite Abelian groups is completely κ-reducible [9]; the pseudovariety OCR of all finite orthodox completely regular semigroups is κ-reducible [13]; the pseudovariety CR of all finite completely regular semigroups is κ-reducible [14][f]; the pseudovariety LSl of all finite semigroups S whose local subsemigroups eSe are semilattices is κ-reducible [23]; the pseudovariety R is κ-reducible [5]; the pseudovariety J of all finite $\mathcal{J}$-trivial semigroups is completely κ-reducible [3]. The κ-reducibility of the pseudovariety A of all finite aperiodic semigroups was announced by J. Rhodes in 1997 but no proof has yet been published. The word problem for $\Omega_A^\kappa \mathsf{A}$ was solved by McCammond [32] and, independently, by Zhil'tsov [43].

Although the join operation is not as amenable to decidability proofs through reducibility arguments as the semidirect and Mal'cev products, there have been investigations in this direction. Both proofs of decidability of $\mathsf{J} \vee \mathsf{G}$ [4, 38], obtained independently, use some form of reducibility of G and J. The same approach has also been used to study other joins [40, 5].

3. Simplifications

There are a number of simplifications of the problem which we proceed to examine. See [6] for details.

A first simplification consists in observing that parameters may be captured by adding extra variables and constraining them suitably: σ-reducibility for systems without parameters implies σ-reducibility for systems with parameters given by σ-terms.

Say that a pseudovariety is *weakly cancellable* if, whenever it satisfies the pseudoidentity $u_1\#u_2 = v_1\#v_2$, where the letter $\#$ does not occur in u_1, u_2, v_1, v_2, it also satisfies the pseudoidentities $u_1 = v_1$ and $u_2 = v_2$. Many familiar pseudovarieties are weakly cancellable: A, R, J, CR, DA (finite semigroups in which regular elements are idempotent), DO (finite semigroups in which regular $\mathcal{D}$-classes are orthodox subsemigroups), DS (finite semigroups in which regular $\mathcal{D}$-classes are subsemigroups), and lo-

property about free groups, which explains the connection with the Ribes and Zalesskiĭ Theorem. The formal equivalence of the latter with Ash's Theorem was recognized in [7, 8]. The connections between the two approaches to the Type II Conjecture have been extensively investigated by Steinberg, later joined by Auinger [41, 39, 20, 18, 21].

[f]As has been observed by K. Auinger in a private communication, the stronger version of κ-reducibility for G which is needed in [14] can be established using the methods of [7, 8].

cally extensible pseudovarieties of groups in the sense of [22].[g] If V is weakly cancellable and σ-reducible for systems consisting of just one equation of σ-terms, without any parameters, then V is completely σ-reducible.

Another simplification stems from the relationship between the canonical signature κ and the alternative signature in which the unary pseudo-inversion is replaced by the ω-power operation $x \mapsto x^\omega = x^{\omega-1}x$. Since, for finite aperiodic semigroups, the two operations coincide, the following result is not surprising, although it does require a proof: if V is an aperiodic pseudovariety, then V is κ-reducible for an arbitrary system if and only if it is reducible for the same system with respect to the signature consisting of multiplication and the operation $x \mapsto x^\omega$.

4. Further simplifications for the case of R

In this paper, we pay special attention to the case of the pseudovariety R, for which there are also some specific simplifications of the reducibility problem which apply.

From the general simplifications of the preceding section, we know that, if R is κ-reducible for systems consisting of a single equation of κ-terms without parameters, then R is completely κ-reducible. In fact, it suffices to consider word equations. The idea is to express that an initial subterm t is an ω-power of u by the word equation $ut = t$. This leads to a finite system of word equations which may then be transformed into a single word equation taking into account that R is weakly cancellable.

For a pseudoword $w \in \overline{\Omega}_A\mathsf{S}$, let $c(w)$ be the set of all letters $a \in A$ which are factors of w and let $\vec{c}(w) = \{a \in A : \mathsf{R} \models wa = w\}$.

A solution δ modulo R of the equation $u = v$ is said to be R-*reduced with respect to* $u = v$ if it has the following property: for every factor xy of uv, where x and y are variables, if z is the first letter of $\delta(y)$, then $\mathsf{R} \not\models \delta(x)z = \delta(x)$. Suppose that R is κ-reducible for systems of word equations without parameters which involve one general equation $u = v$ and all other equations of the form $xy = x$, where x and y are variables, and which admit solutions modulo R which are R-reduced with respect to the equation $u = v$. Then R is completely κ-reducible. The idea here is to factorize each $\delta(x)$ as $a_1u_1a_2u_2\cdots a_{n_x}u_{n_x}$ where the a_i are letters and indicate their leftmost occurrences in $\delta(x)$. One may introduce n_x new variables $y_{x,i}$ to represent the intermediate factors u_i (depending on x) as well as variables z_a to

[g]See the Appendix for a characterization of weak cancellability in pseudovarieties of groups.

represent the individual letters a from the alphabet. Upon the variable $y_{x,i}$ is imposed a constraint which requires that $c(y_{x,i}) \subseteq \{a_1, \ldots, a_i\}$. In turn, the variables z_a are constrained to be equal to a. In the original equation, for each two-letter factor x_1x_2, we expand the variable x_2 according to the factorization of its value in a solution δ modulo R, replacing x_2 by the associated product $z_{a_m} y_{x_2,m} \cdots z_{a_{n_{x_2}}} y_{x_2,n_{x_2}}$, where a_m is the first letter in $\delta(x_2)$ which does not belong to $\vec{c}(\delta(x_1))$, dropping x_2 altogether at that position in the equation if $c(\delta(x_2)) \subseteq \vec{c}(\delta(x_1))$. The resulting finite system of word equations may be compressed into a single word equation by the tricks of the preceding section. To retain the information about the value of each $\vec{c}(\delta(x))$, we add the equations $y_{x,n_x} z_a = y_{x,n_x}$ whenever $a \in \vec{c}(\delta(x))$.

5. Complete reducibility of R

The aim of the remainder of the paper is to sketch a proof of the following result from [6]. Weaker forms were previously established in [10] and [5].

Theorem 5.1. *The pseudovariety* R *is completely* κ*-reducible.*

In the sequel, we try as much as possible to formulate the arguments in a more general setting, thus referring to a general pseudovariety V.

Let $u = x_1 \cdots x_r$, $v = x_{r+1} \cdots x_s$, where the x_i are not necessarily distinct variables from a set X. Suppose that $\varphi : X \to \overline{\Omega}_A\mathsf{S}$ is a solution of the equation $u = v$ modulo a given pseudovariety V, satisfying prescribed constraints in a finite semigroup S. Suppose that V determines some kind of unique factorization in the free profinite semigroup $\overline{\Omega}_A\mathsf{S}$ and that we may assume that the solution is such that the resulting factorizations of u and v under the solution are of that kind. Then the two factorizations must match. For example, if we have a solution of the equation $xyzx = yzxy$, then the two factorizations of the common value of the words $xyzx$ and $yzxy$ must match, say as indicated in the following diagram:

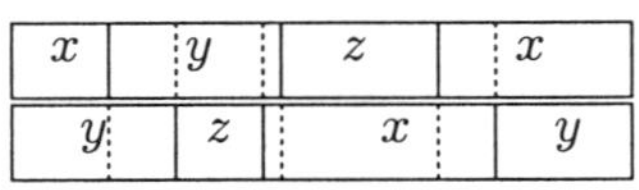

The factorizations of the value of a variable corresponding to its different occurrences in the equation must also be matched and this leads to the successive refinement of factorizations. How to manage the propagation of these factorizations which, for pseudowords, may perhaps have to be carried *ad infinitum*?

For R the propagation of factorizations has been successfully handled in [5] in the case of systems of equations associated with finite graphs. The case of arbitrary word equations is much more delicate. The management of the propagation of factorizations is done by adapting ideas from Makanin's algorithm to decide whether a finite system of word equations with rational constraints admits a solution in the free monoid [30, 31]. There is a recent more efficient (PSPACE) algorithm, due to Plandowski [35]. Since we are concerned at present with an abstract property rather than the construction of an algorithm, there is no complexity issue for us, and so we preferred to use Makanin's ideas, with which we are more familiar, and which, perhaps therefore, seem more adjusted to the current problem.

One of the simple ideas in Makanin's algorithm is to organize the matching of factorizations by only matching a couple of factorizations of the same word at a time. For instance, for the equation $xyzx = yzxy$, the matching might be done as indicated in the following diagram:

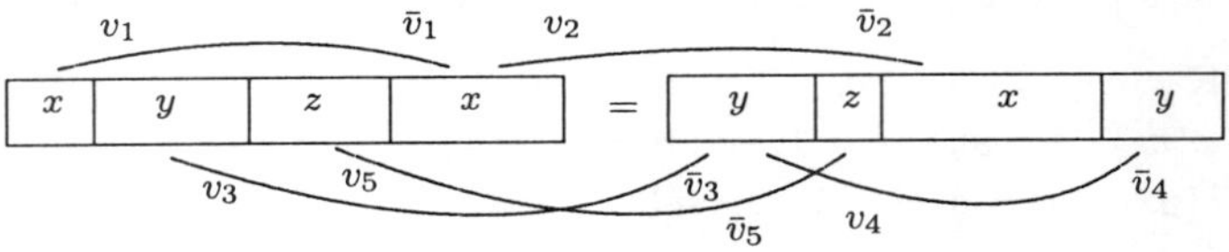

The variables v_0 and $\bar{v}_0$ are used to match the common value of both sides of the equation. Each *box* is identified by the *position* i of its beginning (its *left*) together with the new *variable* v_k or $\bar{v}_k$ that determines it:

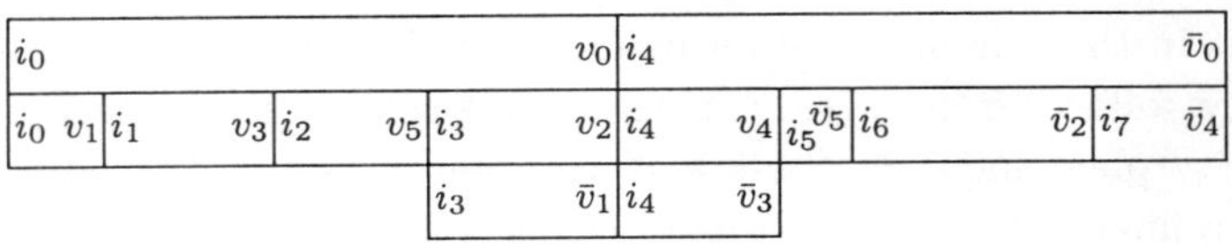

The *right* of a box is where it ends. A quadruple of the form $(i, v, j, \bar{v})$ is called a *boundary equation*. Each of the pairs (i, v) and $(j, \bar{v})$ that constitute it corresponds to a box in the diagram and thus to a pseudoword under the given solution of the original equation. The two pseudowords thus obtained define a pseudoidentity which is valid in V.

If we are working with finite words, as in Makanin's algorithm, when we use a boundary equation $(i, v, j, \bar{v})$ to match two segments of a solution, the words are actually equal and therefore we do not have to worry about carrying along the constraint value. For pseudowords and solutions modulo V, the situation is more complicated: under the solution, the two sides are not really equal but only equal over V. One might formulate the constraints in terms of conditions in $\overline{\Omega}_A\mathsf{V}$ but then what we get are in general only closed sets, rather than clopen sets, and thus a finiteness condition is lost which turns out to be essential in our reduction arguments.

Suppose the constraints are given by values in a finite A-generated semigroup. Although initially we only have one constraint for each pair of consecutive positions, corresponding to the value assigned to a variable under a solution of the equation modulo V, as we start refining factorizations the constraint values must be factorized accordingly, and in S the factorization will not be unique. In other words, the pseudowords coming from the solution of the original equation show that the constraining subsets must be V-*pointlike*. This leads to the following special case of κ-reducibility for R which can be found in [5] in a slightly different form.

Proposition 5.1. *Let* $\varphi : \overline{\Omega}_A\mathsf{S} \to S$ *be a continuous homomorphism and let* $u_1, \ldots, u_n \in \overline{\Omega}_A\mathsf{S}$ *be pseudowords such that* $\mathsf{R} \models u_1 = \cdots = u_n$*. Then there exist* $w_1, \ldots, w_n \in \Omega_A^\kappa\mathsf{S}$ *such that the following conditions hold:*

(1) $\mathsf{R} \models w_1 = \cdots = w_n$*;*
(2) $\varphi(u_i) = \varphi(w_i)$ $(i = 1, \ldots, n)$*;*
(3) $c(u_i) = c(w_i)$ $(i = 1, \ldots, n)$*;*
(4) $\vec{c}(u_i) = \vec{c}(w_i)$ $(i = 1, \ldots, n)$*.*

Unlike the case of finite words, factorizations of pseudowords may continue forever. However, due to periodicity phenomena in the constraints, one may hope to control infinite refinements through the replacement of segments in the original solution by ω-terms. In Makanin's algorithm, decidability follows from a very delicate and complicated analysis of how periodicity phenomena in S allow to compute a bound for the number of times a refinement needs to be performed. Plandowski [35] describes it as *one of the most complicated termination proofs existing in the literature.*

6. General strategy of the proof

The basic reason why appropriate factorizations exist for the pseudovariety R are the following. We say that a pseudoword is *end-marked* if it is of the form wa with $\mathsf{R} \not\models wa = w$, where a is a letter. End-marked pseudowords enjoy some important properties which we quote from [5], where further references to related literature may also be found. If ua and vb are end-marked pseudowords such that $ua\,\mathcal{R}\,vb$, then $a = b$ and $u = v$ ($\mathcal{R}$-triviality). There are no infinite ascending $\leq_{\mathcal{R}}$-chains of end-marked pseudowords over a finite alphabet (well-foundedness). Suppose that u and v are two prefixes of the same element of $\overline{\Omega}_A \mathsf{S}$. Then one of them is a prefix of the other (unambiguous $\mathcal{R}$-order). This provides another proof of the characterization of A-pseudowords over R as "reduced A-labeled ordinals" found in [15].

The positions in the factorizations will thus be determined by certain ordinals smaller than the ordinal of the given solution. Now, the basic strategy of the proof should be clear: to use the boundary equations to reduce the maximum of the positions which appear in boxes or the number of boxes which end at that maximum. In an ordinal, such a procedure can only be carried out a finite number of times. The difficulty is that, unlike what happens for finite words, we may very well have $\mathsf{R} \models u = v$ with u a proper suffix of v, but not a proper prefix, assuming that in all factorizations that we consider factors stop just short of the last letter of an end-marked prefix. Yet such cases lead to periodicity phenomena which we have managed to handle. A boundary equation $(i, v, j, \bar{v})$ is said to be *elastic* if it has the following form:

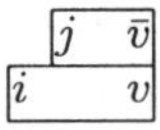

To proceed, we distinguish three cases which require different strategies. The description of the strategy will be essentially pictorial, which makes it somewhat imprecise. Also, we will make no further reference to the crucial detail of how the constraints need to be factorized as the factorizations for the values of each variable are merged. Full details are provided in [6].

Case A. Suppose that there is a "rightmost" boundary equation $(i, v, j, \bar{v})$ which is elastic and such that, under the given solution, not all letters which occur in the box (i, v) occur in the factor between the positions i and j. Then one may introduce a new position k which corresponds to the first letter in the box (i, v) which does not occur in the factor between the positions i and j and replace the boundary equation $(i, v, j, \bar{v})$ by $(i, v', j, \bar{v}')$:

Case B. Suppose that Case A does not hold and that there is at least one boundary equation $(i, v, j, \bar{v})$ whose box (i, v) ends at a maximum position for all boxes and such that the box $(j, \bar{v})$ ends earlier. Among all such boundary equations, we may choose one such that i is minimum and, by an argument of pushing forward periods in elastic equations which is sketched in Case C, we may also assume that there are no elastic boundary equations for which one of the boxes includes the position i and ends at the same position as (i, v). Then we may proceed as in Makanin's algorithm: let c be the *critical boundary* defined by $c = \max\{c', i\}$ where $c' = \max\{\text{right}(w) : \text{left}(w) < i\}$. We have to transport the constraints of the segment $(c, \text{right}(v))$ to the corresponding segment $(c^\circ, \text{right}(\bar{v}))$, where, as ordinals, $p^\circ - j = p - i$. These segments are then handled by Proposition 5.1 and can be dropped from the boundary equation $(i, v, j, \bar{v})$. Additionally, we transport all boxes (k, w) crossing c to their corresponding segment of $(j, \text{right}(\bar{v}))$. The diagram of boxes might include those on the left, in which case we transform it to the one on the right:

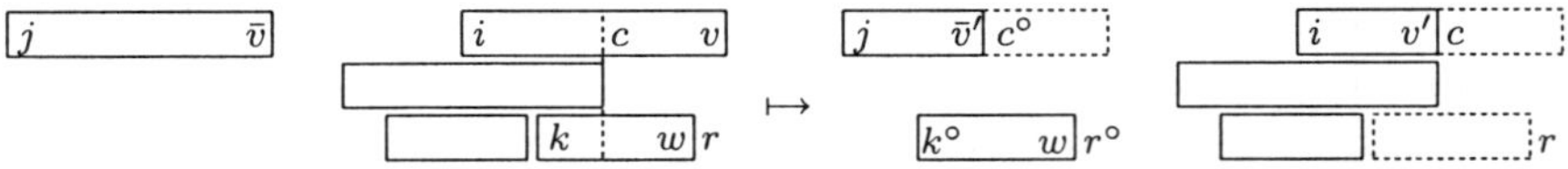

Case C. Suppose that all boundary equations which have a box which ends at the maximum position where boxes end are elastic and that none of the previous cases hold. Under the given solution, each such boundary equation $(k_i, v_i, \ell_i, \bar{v}_i)$ $(i = 1, \ldots, m)$ determines a pseudoidentity of the form $u_i w_i = w_i$ such that $\mathsf{R} \models u_i w_i = w_i$, where, assuming that $k_i < \ell_i$, u_i corresponds to the box which starts at position k_i and ends just short of position ℓ_i, while w_i corresponds to the box $(\ell_i, \bar{v}_i)$. Since Case A does not hold, we must have $c(u_i) = c(w_i)$ and so the pseudoidentity $u_i w_i = w_i$ is equivalent, for R, to $w_i = u_i^\omega$, which forces a periodicity phenomenon. This periodicity has to be carefully combined with periodicity in the constraints.

The first step consists in synchronizing the periods of the various elastic boundary equations involved so that a similar situation is produced with all k_i equal. This can be achieved by breaking up the boxes by a process which we call *pushing forward the period* and which is depicted in the following diagram which, for simplicity, considers the case of two boundary equations:

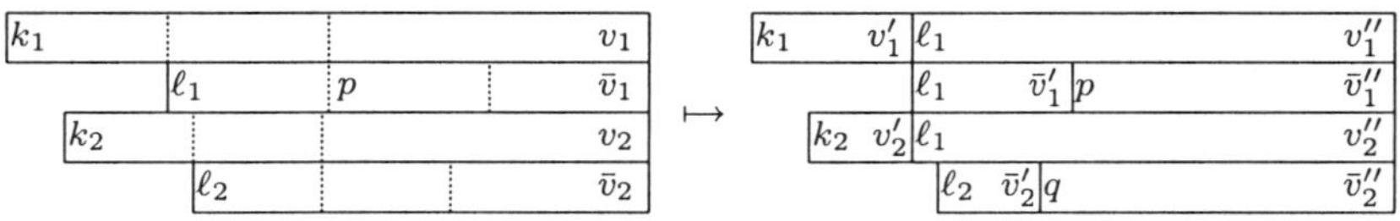

The positions p and q are such that the factors corresponding to the pairs of boxes $(k_1, v_1'), (\ell_1, \bar{v}_1')$ and $(k_2, v_2'), (\ell_2, \bar{v}_2')$ determine pseudoidentities which are valid in R. The boundary equations $(k_1, v_1, \ell_1, \bar{v}_1)$ and $(k_2, v_2, \ell_2, \bar{v}_2)$ are replaced by new boundary equations $(k_1, v_1', \ell_1, \bar{v}_1')$, $(\ell_1, v_1'', p, \bar{v}_1'')$, $(k_2, v_2', \ell_2, \bar{v}_2')$, and $(\ell_1, v_2'', q, \bar{v}_2'')$.

The same strategy works in general and hence we may assume that all k_i are equal to the same k, which implies that R satisfies all pseudoidentities of the form $u_i w_i = w_i = w_j$, so that the w_i have a value w over R which is independent of i and $\mathsf{R} \models w = u_i^\omega$ for $i = 1, \ldots, m$. To handle this situation, we have the following multi-periodicity result.

Proposition 6.1. *Let $u_1, \ldots, u_m$ be pseudowords over A such that R satisfies $u_1^\omega = \cdots = u_m^\omega$. Assume that, for all i, the product $u_i u_i$ is reduced. Then, there exist $z \in \overline{\Omega}_A\mathsf{S}$, $r_i \in \overline{\Omega}_A\mathsf{S}^1$, and integers $k_i > 0$ such that R satisfies the pseudoidentities $u_i = z^{k_i} r_i$ and $z = r_i z$, for all $i = 1, \ldots, n$, where all the products and zz are reduced.*

We introduce new boundary equations to capture the refined period z given by Proposition 6.1, along with its periods r_i $(i = 1, 2, \ldots)$:

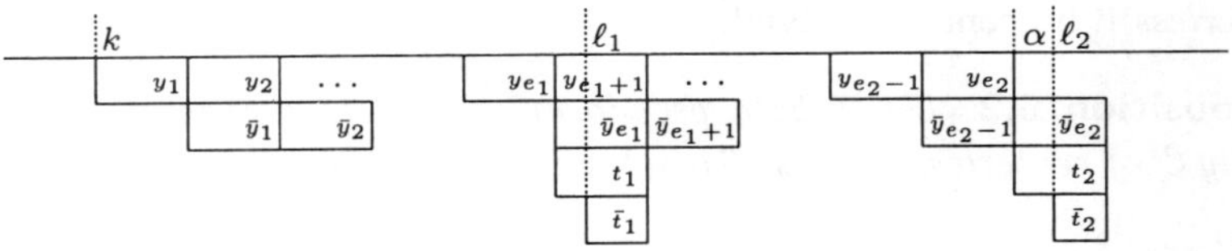

Finally, we indicate how the constraints are used to show that a solution modulo R in κ-terms must exist if there is some solution modulo R. It is well known that, for a finite semigroup S, there are integers h and p such that $1 < h < p$ and, for all $s_1, \ldots, s_p \in S$, $s_1 \cdots s_p = s_1 \cdots s_h (s_{h+1} \cdots s_p)^\omega$. We drop the boundary elastic equations whose boxes end at the maximum position where any boxes end and we introduce new boundary equations to capture the repetition p times of the longest period encountered so far:

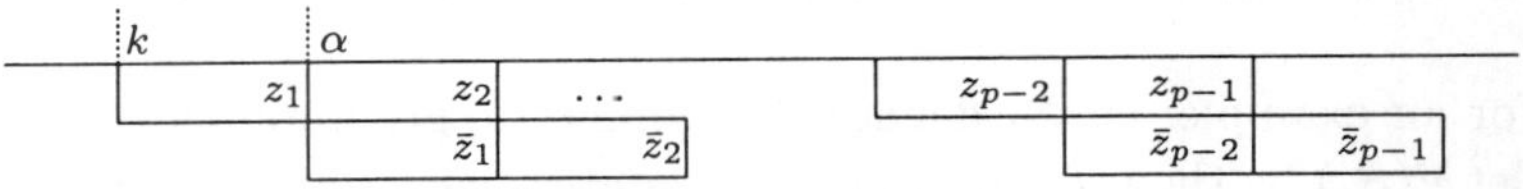

The proof of Theorem 5.1 is achieved by showing that each time we change our system of boundary equations we obtain a system which still admits a solution modulo V and, conversely, such that if the new system admits a solution in κ-terms then so does the old one.

Appendix

As has been pointed out by the anonymous referee, for pseudovarieties of groups, local extensibility is sufficient but not necessary for weak cancellability. Indeed, for a pseudovariety V of groups, the weak cancellation property may be reformulated as follows. Over V, the non-trivial pseudoidentity $u_1 \# u_2 = v_1 \# v_2$ is equivalent to one of the form $u\# = \#v$ for pseudowords u and v, with the letter $\#$ not occurring in them, such that V

does not satisfy both pseudoidentities $u = 1$ and $v = 1$. Substituting u for the letter #, we deduce the pseudoidentity $u^2 = uv$, so that $u = v$ holds in V. Hence, the original pseudoidentity $u_1\#u_2 = v_1\#v_2$ is equivalent to one of the form $u\# = \#u$ over V, where # is a letter not occurring in the non-trivial pseudoword u. This shows that V is not weakly cancellable if and only if there is a finitely generated free pro-V group with a non-trivial central element which does not use all free generators. The referee further asked whether it is sufficient for non-weak cancellability of V for $\overline{\Omega}_A\mathsf{V}$ to have a non-trivial center whenever A is a non-empty finite set.

Before giving a negative answer to the preceding question, we proceed to consider a special kind of pseudovarieties of groups. Say that a group is *centerless* if its center is trivial.

Proposition 6.2. *Let* V *be a pseudovariety which is generated by some family* $\mathcal{C}$ *of centerless groups. Then* V *is weakly cancellable.*

Proof. Let u be a non-trivial element of $\overline{\Omega}_A\mathsf{V}$ which belongs to the closed subgroup generated by $A \setminus \{a\}$ for some $a \in A$. Then there is some group G in $\mathcal{C}$ and some continuous homomorphism $\varphi : \overline{\Omega}_{A\setminus\{a\}}\mathsf{V} \to G$ such that $\varphi(u) \neq 1$. Since G has trivial center, there is some $g \in G$ which does not commute with $\varphi(u)$. Now, we may extend φ to a continuous homomorphism $\psi : \overline{\Omega}_A\mathsf{V} \to G$ by letting $\psi(a) = g$ and $\psi(b) = \varphi(b)$ for $b \in A \setminus \{a\}$. Since $\psi(a)$ and $\psi(u)$ do not commute, we conclude that u is not central in $\overline{\Omega}_A\mathsf{V}$. Hence the pseudovariety V is weakly cancellable by the referee's remark. □

For an example, let S_3 denote the symmetric group on three symbols and let $\mathsf{V}(S_3)$ be the pseudovariety it generates. By Proposition 6.2, $\mathsf{V}(S_3)$ is weakly cancellable, while, as any locally finite pseudovariety of groups, it is not locally extensible. We claim that $\overline{\Omega}_A\mathsf{V}(S_3)$ has a non-trivial center for every non-empty finite set A, which provides a negative answer to the question raised by the referee. The claim is proved by recursively exhibiting central elements.

Lemma 6.1. *Let* $A_n = \{x_1, \ldots, x_n\}$ *and define recursively a sequence* u_n *by taking* $u_1 = x_1$ *and* $u_{n+1} = (u_n x_{n+1}^3 u_n)^2$. *Then* u_n *is a non-trivial central element in the group* $G_n = \overline{\Omega}_{A_n}\mathsf{V}(S_3)$.

Proof. Let $w \mapsto \overline{w}$ denote an arbitrary homomorphism $G_n \to S_3$. If we take $\overline{x}_2 = \cdots = \overline{x}_n = 1$, then $\overline{u}_n = \overline{x}_1^{4^{n-1}}$ so that $\overline{u}_n \neq 1$ if we choose for $\overline{x}_1$ a 3-cycle. Hence $u_n \neq 1$.

To prove that u_n is in the center of G_n, we proceed by induction on n, the case $n = 1$ being trivial. Given elements $\overline{x}_1, \dots, \overline{x}_{n+1} \in S_3$, denote by Z_k the center of the subgroup H_k generated by $\overline{x}_1, \dots, \overline{x}_k$. We assume that, given $\overline{x}_1, \dots, \overline{x}_{n+1} \in S_3$, $\overline{u}_n \in Z_n$ and we claim that $\overline{u}_{n+1} \in Z_{n+1}$. If $\overline{u}_n$ commutes with $\overline{x}_{n+1}$ then $\overline{u}_{n+1} = (\overline{u}_n \overline{x}_{n+1}^3 \overline{u}_n)^2 = \overline{u}_n^4 \overline{x}_{n+1}^6 = \overline{u}_n^4$, which shows that $\overline{u}_{n+1} \in Z_{n+1}$. Hence, we may assume that $\overline{x}_{n+1}$ does not commute with $\overline{u}_n$, which implies that $\overline{u}_n$ does not belong to the subgroup generated by $\overline{x}_{n+1}$ and, by induction hypothesis, that $\overline{x}_{n+1} \notin H_n$. We claim that, under these circumstances, $\overline{u}_{n+1} = 1$.

Indeed, if $\overline{u}_n$ is a 3-cycle, then $\overline{x}_{n+1}$ is a 2-cycle and so $\overline{u}_n \overline{x}_{n+1}^3 \overline{u}_n = \overline{x}_{n+1}$ and $\overline{u}_{n+1} = \overline{x}_{n+1}^2 = 1$. Assume next that $\overline{u}_n$ is a 2-cycle. If $\overline{x}_{n+1}$ is a 3-cycle, then $\overline{u}_{n+1} = \overline{u}_n^4 = 1$. If $\overline{x}_{n+1}$ is also a 2-cycle, then $\overline{u}_n \overline{x}_{n+1}^3 \overline{u}_n$ is again a 2-cycle and so $\overline{u}_{n+1} = 1$. □

The above lemma serves only to handle a very special example. We do not know how far it can be generalized, that is which non-trivial finite groups G have the property that the center of $\overline{\Omega}_{A_n}\mathsf{V}(G)$ is non-trivial for every $n \geq 1$. But, of course, this is a remotely marginal question for the theme of this paper.

Acknowledgments

The work of J. Almeida was (partially) supported by the *Centro de Matemática da Universidade do Porto* (CMUP), financed by FCT (Portugal) through the programmes POCTI and POSI, with national and European Community structural funds. The work of J. C. Costa was supported, in part, by FCT through the *Centro de Matemática da Universidade do Minho*. The work of M. Zeitoun was partly supported by the European research project HPRN-CT-2002-00283 GAMES.

References

1. J. Almeida, *Finite Semigroups and Universal Algebra*, World Scientific, Singapore (1995). English translation.
2. J. Almeida, Dynamics of implicit operations and tameness of pseudovarieties of groups, *Trans. Amer. Math. Soc.* **354**, 387–411 (2002).
3. J. Almeida, Finite semigroups: an introduction to a unified theory of pseudovarieties, in *Semigroups, Algorithms, Automata and Languages*, G. M. S. Gomes, J.-E. Pin, and P. V. Silva, eds., Singapore, World Scientific, 3–64 (2002).
4. J. Almeida, A. Azevedo and M. Zeitoun, Pseudovariety joins involving $\mathcal{J}$-trivial semigroups, *Int. J. Algebra Comput.* **9**, 99–112 (1999).

5. J. Almeida, J. C. Costa and M. Zeitoun, Tameness of pseudovariety joins involving R, *Monatsh. Math.* **146**, 89–11 (2005).
6. J. Almeida, J. C. Costa and M. Zeitoun, *Complete reducibility of systems of equations with respect to R.* In preparation.
7. J. Almeida and M. Delgado, Sur certains systèmes d'équations avec contraintes dans un groupe libre, *Portugal. Math.* **56**, 409–417 (1999).
8. J. Almeida and M. Delgado, Sur certains systèmes d'équations avec contraintes dans un groupe libre—addenda, *Portugal. Math.* **58**, 379–387 (2001).
9. J. Almeida and M. Delgado, Tameness of the pseudovariety of Abelian groups, *Int. J. Algebra Comput.* **15**, 327–338 (2005).
10. J. Almeida and P. V. Silva, SC-hyperdecidability of **R**, *Theor. Comp. Sci.* **255**, 569–591 (2001).
11. J. Almeida and B. Steinberg, On the decidability of iterated semidirect products and applications to complexity, *Proc. London Math. Soc.* **80**, 50–74 (2000).
12. J. Almeida and B. Steinberg, Syntactic and Global Semigroup Theory, a Synthesis Approach, in *Algorithmic Problems in Groups and Semigroups*, J. C. Birget, S. W. Margolis, J. Meakin, and M. V. Sapir, eds., Birkhäuser, 1–23 (2000).
13. J. Almeida and P. G. Trotter, Hyperdecidability of pseudovarieties of orthogroups, *Glasgow Math. J.* **43**, 67–83 (2001).
14. J. Almeida and P. G. Trotter, The pseudoidentity problem and reducibility for completely regular semigroups, *Bull. Austral. Math. Soc.* **63**, 407–433 (2001).
15. J. Almeida and P. Weil, Free profinite $\mathcal{R}$-trivial monoids, *Int. J. Algebra Comput.* **7**, 625–671 (1997).
16. J. Almeida and P. Weil, Profinite categories and semidirect products, *J. Pure Appl. Algebra* **123**, 1–50 (1998).
17. C. J. Ash, Inevitable graphs: a proof of the type II conjecture and some related decision procedures, *Int. J. Algebra Comput.* **1**, 127–146 (1991).
18. K. Auinger, A new proof of the Rhodes type II conjecture, *Int. J. Algebra Comput.* **14**, 551–568 (2004).
19. K. Auinger and B. Steinberg, On the extension problem for partial permutations, *Proc. Amer. Math. Soc.* **131**, 2693–2703 (2003).
20. K. Auinger and B. Steinberg, The geometry of profinite graphs with applications to free groups and finite monoids, *Trans. Amer. Math. Soc.* **356**, 805–851 (2004).
21. K. Auinger and B. Steinberg, A constructive version of the Ribes-Zalesskii product theorem, *Math. Z.* **250**, 287–297 (2005).
22. K. Auinger and B. Steinberg, Hall varieties of finite supersolvable groups. To appear in *Math. Ann.*.
23. J. C. Costa and M. L. Teixeira, Tameness of the pseudovariety LSl, *Int. J. Algebra Comput.* **14**, 627–654 (2004).
24. T. Coulbois and A. Khélif, Equations in free groups are not finitely approximable, *Proc. Amer. Math. Soc.* **127**, 963–965 (1999).
25. S. Eilenberg, *Automata, Languages and Machines*, vol. B, Academic Press,

New York (1976).
26. K. Henckell, S. Margolis, J.-E. Pin and J. Rhodes, Ash's type II theorem, profinite topology and Malcev products. Part I, *Int. J. Algebra Comput.* **1**, 411–436 (1991).
27. B. Herwig and D. Lascar, Extending partial automorphisms and the profinite topology on free groups, *Trans. Amer. Math. Soc.* **352**, 1985–2021 (2000).
28. G. Lallement, *Semigroups and Combinatorial Applications*, Wiley, New York (1979).
29. M. Lothaire, *Algebraic Combinatorics on Words*, Cambridge University Press, Cambridge, UK (2002).
30. G. S. Makanin, The problem of solvability of equations in a free semigroup, *Mat. Sb. (N.S.)* **103 (2)**, 147–236 (1977). In Russian. English translation in: *Math. USSR-Sb.* 32, 128-198 (1977).
31. G. S. Makanin, Equations in a free semigroup, *Amer. Math. Soc. Transl.* (II Ser.) **117**, 1–6 (1981).
32. J. McCammond, Normal forms for free aperiodic semigroups, *Int. J. Algebra Comput.* **11**, 565–580 (2001).
33. J.-E. Pin, *Varieties of Formal Languages*, Plenum, London (1986). English translation.
34. J.-E. Pin and P. Weil, Profinite semigroups, Mal'cev products and identities, *J. Algebra* **182**, 604–626 (1996).
35. W. Plandowski, Satisfiability of word equations with constants is in PSPACE, *J. ACM* **51**, 483–496 (2004).
36. J. Rhodes and B. Steinberg, *The q-theory of finite semigroups*, 2001–2004. Book under preparation. Preliminary versions available through `http://mathstat.math.carleton.ca/~bsteinbg/qtheor.html`.
37. L. Ribes and P. A. Zalesskiĭ, On the profinite topology on a free group, *Bull. London Math. Soc.* **25**, 37–43 (1993).
38. B. Steinberg, On pointlike sets and joins of pseudovarieties, *Int. J. Algebra Comput.* **8**, 203–231 (1998).
39. B. Steinberg, Inevitable graphs and profinite topologies: some solutions to algorithmic problems in monoid and automata theory, stemming from group theory, *Int. J. Algebra Comput.* **11**, 25–71 (2001).
40. B. Steinberg, On algorithmic problems for joins of pseudovarieties, *Semigroup Forum* **62**, 1–40 (2001).
41. B. Steinberg, Inverse automata and profinite topologies on a free group, *J. Pure Appl. Algebra* **167**, 341–359 (2002).
42. P. Weil, Profinite methods in semigroup theory, *Int. J. Algebra Comput.* **12**, 137–178 (2002).
43. I. Y. Zhil'tsov, *On identities of finite aperiodic epigroups*. Ural State Univ. (1999).

FINITE GENERATION OF P-SEMIGROUPS WITH ALMOST G-INVARIANT IDEMPOTENTS

CATARINA A. CARVALHO

Mathematical Institute
University of St Andrews
St Andrews, Fife KY16 9SS, U.K.
E-mail: catarina@mcs.st-and.ac.uk

A P-semigroup is a semigroup construction that characterizes E-unitary inverse semigroups. We say that a P-semigroup $P(G, X, Y)$ has almost G-invariant idempotents if $X \backslash Y$ is finite. In this paper we give necessary and sufficient conditions for a P-semigroup with almost G-invariant idempotents to be finitely generated. We also show that the same conditions, although necessary, are not sufficient for an arbitrary P-semigroup to be finitely generated.

1. Introduction

An inverse semigroup S is E-unitary if for all $e \in E(S)$ and all $s \in S$ we have

$$es \in E(S) \;\Rightarrow\; s \in E(S),$$

where $E(S)$ is the set of idempotents of S. The E-unitary semigroups form one of the most important classes of inverse semigroups. Indeed, McAlister [6] proved that every inverse semigroup is an idempotent-separating homomorphic image of an E-unitary semigroup. Also many inverse semigroups are E-unitary, for example the free inverse semigroup, the bicyclic monoid, and semidirect products of semilattices by groups.

In 1976 McAlister [7] showed that every E-unitary inverse semigroup can be constructed in terms of groups and partially ordered sets. Let X be a poset. Let $Y \subseteq X$ be a semilattice and ideal of X, i.e. for all $x \in X$, $y \in Y$ if $x \leqslant y$ then $x \in Y$. Let G be a group acting on X by order preserving automorphisms. Furthermore assume that the action of G on X is such that $G \cdot Y = X$ and $g \cdot Y \cap Y \neq \emptyset$ for all $g \in G$. We say that the triple (G, X, Y) is a *McAlister triple.* This triple induces the semigroup

$$P(G, X, Y) = \{(e, g) \in Y \times G \;:\; g^{-1} \cdot e \in Y\},$$

subject to multiplication

$$(e, g)(f, h) = (e \wedge (g \cdot f),\ gh),$$

where $e_1 \wedge e_2$ denotes the meet of the elements $e_1, e_2 \in X$. These semigroups are called *P-semigroups*. They were introduced by McAlister and McFadden [8], and are a generalization of Scheiblich's [11, 12] characterization of the free inverse semigroup. In [3] we can find a proof that the multiplication defined above is indeed well-defined.

Since then different proofs of McAlister's result that every E-unitary inverse semigroup is a P-semigroup have been given, see for example [9, 5, 13, 14]. Although E-unitary inverse semigroups have been widely studied one important question has not yet been answered:

Open Question When are P-semigroups finitely generated?

More generally we can ask when are P-semigroups finitely presented. Some work has been undertaken on the problem of determining when a given semigroup construction is finitely generated and finitely presented, see [1] for a survey. Conditions have been given for finite generation, and presentability, of several semigroup constructions but not for P-semigroups. Our aim is to initiate the study of combinatorial algebraic properties of P-semigroups, in particular we aim to find necessary and sufficient conditions in terms of X, Y and G for the P-semigroups to be finitely generated. In this paper we look at P-semigroups $P(G, X, Y)$ with $X \backslash Y$ finite. In [3] we looked at some particular cases of P-semigroups with $X \backslash Y$ infinite.

We define a new notion of generation for the semilattice Y, the notion of being generated with respect to the action of the group G. Given a subset E of Y, let $\langle E \rangle$ be the subsemilattice of Y generated by E, under the meet operation $\wedge$, and let $G \cdot \langle E \rangle = \{g \cdot e \ : \ g \in G,\ e \in \langle E \rangle\}$. Fixing a set $Y_0 \subseteq Y$, we define for each $i \geqslant 1$ a set $Y_i = G \cdot \langle Y_{i-1} \rangle \cap Y$. We say that Y is *generated by* Y_0 *with respect to the action of* G if $Y = \bigcup_{i \geqslant 0} Y_i$. Naturally, the semilattice Y is *finitely generated with respect to the action of* G if Y_0 can be chosen to be finite.

We say that a P-semigroup $P(G, X, Y)$ has *almost G-invariant idempotents* if the difference between the sets X and Y is finite. In this paper we give necessary and sufficient conditions, in terms of X, Y and G, for the P-semigroups with almost G-invariant idempotents to be finitely generated:

Main Theorem *Let* $P = P(G, X, Y)$ *be a P-semigroup with almost G-invariant idempotents. The semigroup P is finitely generated if and only* *if*

(i) *the group G is finitely generated;*
(ii) *the semilattice Y is finitely generated with respect to the action of G;*
(iii) *Y has finitely many maximal elements that cover all the semilattice.*

This result generalizes Dombi and Ruškuc's result, that appears in [4], for the semidirect product of semilattices by groups, and it does not generalize to an arbitrary P-semigroup. Indeed in Section 3 we give an example that shows that extra conditions are needed to obtain necessary and sufficient conditions for an arbitrary P-semigroup to be finitely generated.

2. Proof of the Main Theorem

Let X, Y and G be such that $P(G,X,Y)$ is a P-semigroup. For each $x \in X$ the *orbit* of x, under the action of G, is the set $orb(x) = \{g \cdot x \ : \ g \in G\}$. The subset of G that maps an element $x \in X$ into itself is called the *stabilizer* of x, we denote it by $G_x = \{g \in G \ : \ g \cdot x = x\}$. Note that the set G_x, for any $x \in X$ is a subgroup of G.

Let $P = P(G,X,Y)$ be a P-semigroup with almost G-invariant idempotents. Since G acts on X by order preserving automorphisms we know that given $x_1, x_2 \in X$ and $g \in G$, if $x_1 \wedge x_2$ exists then $g \cdot x_1 \wedge g \cdot x_2$ exists and

$$g \cdot x_1 \wedge g \cdot x_2 = g \cdot (x_1 \wedge x_2).$$

First assume that P is finitely generated and let

$$C = \{(e_{\xi_1}, g_{\xi_1}), (e_{\xi_2}, g_{\xi_2}), \ldots, (e_{\xi_k}, g_{\xi_k})\}$$

be a finite generating set for it. Let $y \in Y$ be arbitrary. Since $1^{-1} \cdot y = y \in Y$ the element $(y,1) \in Y \times G$ belongs to P, where 1 is the identity of G. So we have

$$(y,1) = (e_1, g_1)(e_2, g_2) \ldots (e_t, g_t),$$

for some $(e_1, g_1), \ldots, (e_t, g_t) \in C$. It follows that

$$(y,1) = (e_1 \wedge g_1 \cdot e_2 \wedge (g_1 g_2) \cdot e_3 \wedge \ldots \wedge (g_1 \ldots g_{t-1}) \cdot e_t, \ g_1 g_2 \ldots g_t),$$

so we have

$$y = g_1 \cdot (g_1^{-1} \cdot e_1 \wedge g_2 \cdot (g_2^{-1} \cdot e_2 \wedge g_3 \cdot (g_3^{-1} \cdot e_3 \wedge \ldots \wedge g_{t-1} \cdot (g_{t-1}^{-1} \cdot e_{t-1} \wedge e_t)))).$$

Since $(e_i, g_i) \in P$ we know that $g_i^{-1} \cdot e_i \in Y$ for all $i = 1, \ldots, t$. The elements $g_i^{-1} \cdot e_i$ and e_{i+1} belong to Y so their meet $g_i^{-1} \cdot e_{i+1} \wedge e_i$ is defined. It follows that $g_i \cdot (g_i^{-1} \cdot e_i \wedge e_{i+1})$ also exists. We have

$$g_i \cdot (g_i^{-1} \cdot e_i \wedge e_{i+1}) = e_i \wedge g_i \cdot e_{i+1} \leqslant e_i$$

so $g_i \cdot (g_i^{-1} \cdot e_i \wedge e_{i+1})$ belongs to Y for all $i = 1, \ldots, t$. Hence Y is finitely generated with respect to the action of G by the set $Y_0 = \{e_{\xi_1}, \ldots, e_{\xi_k}\}$. From the decomposition of y above we can also see that $y \leqslant e_1$, so each element of Y is less than or equal to an element in the set $\{e_{\xi_1}, e_{\xi_2}, \ldots, e_{\xi_k}\}$. Thus the maximal elements elements of Y are the maximal elements of the set $\{e_{\xi_1}, e_{\xi_2}, \ldots, e_{\xi_k}\}$, and they cover all the semilattice. So the semigroup P satisfies conditions (ii) and (iii).

Let $g \in G$ be arbitrary. We know that $g^{-1} \cdot Y \cap Y \neq \emptyset$ so there exists $e \in Y$ such that $g^{-1} \cdot e \in Y$. This implies that the $(e, g) \in P$, and we have

$$(e, g) = (e_1, g_1)(e_2, g_2) \ldots (e_r, g_r),$$

for some $(e_1, g_1), \ldots, (e_r, g_r) \in A$. It follows that $g = g_1 g_2 \ldots g_r$. Hence G is generated by the finite set $\{g_{\xi_1}, g_{\xi_2}, \ldots, g_{\xi_k}\}$, and P satisfies condition (i).

Remark 1. Note that in this proof we did not use the fact that the semigroup P has almost G-invariant idempotents. So conditions (i), (ii) and (iii) are necessary conditions for any P-semigroup to be finitely generated.

Conversely, suppose that G is finitely generated by a set A, the semilattice Y is finitely generated with respect to the action of G by a set F and Y has finitely many maximal elements $M = \{m_1, m_2, \ldots, m_s\}$ that cover the semilattice.

Before presenting a generating set for the semigroup P we need to prove some results regarding the orbits of the elements of X.

Lemma 2.1. *There are only finitely many elements in the semilattice Y that contain elements from $X \backslash Y$ in their orbits. Consequently the orbits of the elements in $X \backslash Y$ are finite.*

Proof. Since Y contains finitely many maximal elements and $X \backslash Y$ is finite it is clear that each maximal element of Y has a finite orbit, for the action of G on Y is order preserving.

For each maximal element m_i, $i = 1, \ldots, s$, let $orb(m_i) = \{m_i^1, m_i^2, \ldots, m_i^{j_i}\} \subseteq X$ be the orbit of m_i, with $m_i^1 = m_i$. For each $j = 1, 2, \ldots, j_i$ define

$$G_{i,j} = \{g \in G \;:\; g \cdot m_i = m_i^j\}.$$

We clearly have $G_{i,j} = r_{ij}G_{i,1}$, see for example [2]. Hence for all $i = 1, \ldots, s$ the subgroup $G_{i,1}$ has exactly j_i cosets, thus it is a subgroup with finite index. Let $K = \cap_{i=1}^{s} G_{i,1}$. We know that K is a subgroup of G with finite index, for it is the intersection of finitely many subgroups with finite index, see for example [10]. Let $r_1, r_2, \ldots, r_\eta$ be coset representatives of K in G, we have $G = \cup_{j=1}^{\eta} r_j K$. Now let $e \in Y$ be such that $orb(e) \cap X\backslash Y \neq \emptyset$. We prove that e has finite orbit. Assume, for a contradiction, that $orb(e)$ is infinite. Let $x \in orb(e) \cap X\backslash Y$ be arbitrary. There must exist an $h \in K$ such that $h \cdot x \in Y$, since if for all $g \in K$ we had $g \cdot x \in X\backslash Y$, because $X\backslash Y$ is finite, this would imply that $K \cdot x$ is a finite set. It would then follow that $G \cdot x = \bigcup_{j=1}^{\eta}(r_j K) \cdot x$ is finite, but this contradicts the fact that x, and consequently e, has infinite orbit. Since $h \in K$, $h \cdot x \in Y$ and Y is covered by the maximal elements $m_1, \ldots, m_s$, we know that $h \cdot x < m_i$ for some $i = 1, \ldots, s$, but $h \in K \subseteq G_{i,1}$ so $h^{-1} \in G_{i,1}$, i.e. $h^{-1} \cdot m_i = m_i$, and we have

$$h \cdot x \leqslant m_i \;\Rightarrow\; h^{-1}h \cdot x \leqslant h^{-1} \cdot m_i \;\Leftrightarrow x \leqslant m_i,$$

implying that $x \in Y$ since Y is an ideal and $m_i \in Y$, a contradiction.

We conclude that e has finite orbit and as a consequence there are only finitely many elements in Y whose orbit intersects the set $X\backslash Y$. □

Let $D = \{\delta_1, \delta_2, \ldots, \delta_p\}$ be the set of elements from Y whose orbit intersects $X\backslash Y$. For each $i = 1, \ldots, p$ let $orb(\delta_i) = \{\delta_i^1, \delta_i^2, \ldots, \delta_i^{j_i}\}$ be the orbit of δ_i and $G_{i,1}$ its stabilizer. For each $i = 1, \ldots, p$ and each element δ_i^j in the orbit of δ_i define

$$G_{i,j} = \{g \in G \;:\; g \cdot \delta_i = \delta_i^j\},$$

where $\delta_i^1 = \delta_i$. The cosets of the subgroup $G_{i,1}$ are the sets $G_{i,j}$ with $j = 1, \ldots, j_i$, so this subgroup has finite index, and thus, by the Reidemeister-Schreier Theorem (see for example [10]), it is finitely generated.

For each $i = 1, \ldots, p$ let $R_i = \{r_{i1}, r_{i2}, \ldots, r_{ij_i}\}$ be a set of coset representatives of the subgroup $G_{i,1}$, where $r_{i1} = 1$ is the identity of G. Let A_i be a finite generating set for $G_{i,1}$, with $i = 1, \ldots, p$.

Define a set $H \subseteq G$ in the following way

$$H = A \cup (\bigcup_{i=1}^{p} A_i) \cup (\bigcup_{i=1}^{p} R_i) \cup (\bigcup_{i=1}^{p} R_i{}^{-1}),$$

where $R_i^{-1} = \{r^{-1} \;:\; r \in R_i\}$.

Note that the subsemilattice $\langle D \rangle$ of Y is finite since D is finite. Let

$\bigwedge X\backslash Y$ be the set of formal products of elements from $X\backslash Y$, i.e.

$$\bigwedge X\backslash Y = \{x_1 \wedge x_2 \wedge \ldots \wedge x_n \ : \ x_1, \ldots, x_n \in X\backslash Y\}.$$

The reason we define this set is because by computing the meet of one of these formal products with an element from Y we might obtain an element of the semilattice Y. Consider now the set of all elements of Y that belong to one of these sets or can be obtained by meeting elements of $\bigwedge D$ with elements of $\bigwedge X\backslash Y$:

$$D_{X\backslash Y} = \{\delta' \wedge x' \ : \ \exists\delta' \wedge x',\ \delta' \in (\bigwedge D) \cup \{1\},\ x' \in (\bigwedge X\backslash Y) \cup \{1\}\} \cap Y.$$

We use the notation $\exists\delta' \wedge x'$ to mean that the element $\delta' \wedge x'$ belongs to Y. In the set $D_{X\backslash Y}$ we allow both δ' and x' to be equal to 1 so that this set contains both sets $\langle D\rangle$ and $(\bigwedge X\backslash Y) \cap Y$. So $D_{X\backslash Y}$ contains all elements from $\langle D\rangle$, all formal products from $\bigwedge X\backslash Y$ that belong to Y, and all products of elements from $\langle D\rangle$ and $\bigwedge X\backslash Y$ that are defined. Note that this set is finite since D is finite and by multiplying elements from D and $X\backslash Y$, that is a finite set, we obtain only finitely many defined products.

Claim 1. Let $e \in Y\backslash D$ and $x \in X\backslash Y$ be arbitrary. The element $e \wedge x$ belongs to Y.

Proof. Since $e \in Y$ and $e \notin D$ we know that $g \cdot e \in Y$ for all $g \in G$. By the definition of the action of G on X we know that x contains at least one element from Y in its orbit. So there exists $h \in G$ such that $h \cdot x \in Y$. We have

$$h \cdot (e \wedge x) = h \cdot e \wedge h \cdot x,$$

and $h \cdot e, h \cdot x \in Y$ so their meet is defined, and consequently $h \cdot (e \wedge x)$ is defined. It follows that $e \wedge x = h^{-1} \cdot (h \cdot (e \wedge x))$ is defined. Since $e \wedge x \leqslant e$ and Y is an ideal we have $e \wedge x \in Y$. □

Define a set $Z \subseteq X$ as follows

$$Z = D \cup M \cup (\bigcup_{i=1}^{p} (R_i^{-1} \cdot (F \cup D')),$$

where

$$D' = D_{X\backslash Y} \cup \{f \wedge \delta_i \ : \ f \in F,\ i = 1, \ldots, p\} \cup \{f \wedge x \ : \ f \in F\backslash D,\ x \in X\backslash Y\}.$$

Note that all elements in D' belong to Y, by Claim 1. Note also that we assumed that the set R_i contains the identity of G, and so the sets D' and

F belong to Z. We now define a set B to be the set of all pairs in $Z \times H$ that belong to the semigroup P, i.e. $B = (Z \times H) \cap P$. Recall that Z and H are finite sets, so B is a finite set. We show that the semigroup P is generated by the set B.

First we show that for all $i = 1, \dots, p$ and $h \in G$ such that $h^{-1} \cdot \delta_i \in Y$, the element $(\delta_i, h) \in P$ can be written as a product of elements from B. We know that $G = \cup_{j=1}^{j_i} r_{ij} G_{i,1}$ and $h^{-1} \in G$. So we can write $h = (r_{ij}h')^{-1}$, for some $j = 1, \dots, j_i$ and some $h' \in G_{i,1}$. It follows that

$$(\delta_i,\ h) = (\delta_i,\ (r_{ij}h')^{-1}) = (\delta_i,\ h'^{-1} r_{ij}^{-1}) = (\delta_i,\ h'^{-1})(\delta_i,\ r_{ij}^{-1}).$$

Note that $h^{-1} \cdot \delta_i \in Y$ so $(r_{ij}h') \cdot \delta_i \in Y$, but $h' \in G_{i,1}$ so $h' \cdot \delta_i = \delta_i$, thus $r_{ij} \cdot \delta_i,\ h' \cdot \delta_i \in Y$. It follows that $(\delta_i, r_{ij}^{-1}) \in B$ and $(\delta_i, h'^{-1}) \in P$. The group $G_{i,1}$ is generated by A_i and $h'^{-1} \in G_{i,1}$ so $h'^{-1} = h_1 h_2 \dots h_n$ for some $h_1, h_2, \dots, h_n \in A_i$. Thus we have

$$(\delta_i,\ h'^{-1}) = (\delta_i,\ h_1 h_2 \dots h_n) = (\delta_i,\ h_1)(\delta_i,\ h_2) \dots (\delta_i,\ h_n),$$

note that $h_j \in A_i$ for all $j = 1, \dots, n$ so $h_j \cdot \delta_i = \delta_i$. It follows that

$$(\delta_i, h) = (\delta_i,\ h_1 h_2 \dots h_n) = (\delta_i,\ h_1)(\delta_i,\ h_2) \dots (\delta_i,\ h_n)(\delta_i,\ r_{ij}^{-1}),$$

thus $(\delta_i, h) \in \langle B \rangle$.

Let $m \in M$ be arbitrary and assume that $m \notin D$, i.e. it does not contain elements from $X \backslash Y$ in its orbit. Let $g \in G$ be arbitrary, we show that $(m, g) \in \langle B \rangle$. The group G is generated by the set A so we have $g = g_1 g_2 \dots g_n$ for some $g_1, \dots, g_n \in A$. Note that since m does not contain elements from $X \backslash Y$ in its orbit and G acts on Y by automorphisms we have $g \cdot m \in M$ for all $g \in G$. It follows that (m, g) can be decomposed as a product of elements from P as follows

$$\begin{aligned}(m,\ g) = (m,\ g_1)(g_1^{-1} \cdot m,\ g_2)((g_2^{-1} g_1^{-1}) \cdot m,\ g_3) \dots \\ \dots ((g_{n-1}^{-1} g_{n-2}^{-1} \cdots g_1^{-1}) \cdot m,\ g_n).\end{aligned}$$

We know that $g_1^{-1} \cdot m$, $(g_2^{-1} g_1^{-1}) \cdot m$, $\dots$, $(g_{n-1}^{-1} g_{n-2}^{-1} \cdots g_1^{-1}) \cdot m$ are maximal elements of Y, i.e. belong to M, and $g_1, \dots, g_n \in A$, so all the components of this product belong to B. Hence $(m,\ g) \in \langle B \rangle$.

Let $f \in F \cup D'$ be such that $y \notin D$ and f is not a maximal element of Y. Let $g \in G$ be arbitrary. Since $f \notin D$ we know that $g^{-1} \cdot f \in Y$, so $(f,\ g) \in P$. We show that $(f,\ g)$ can be written as a product of elements from B. Since the maximal elements of Y cover it, we know that there exists $m \in M$ such that $g^{-1} \cdot f < m$. If $m \notin D$ then $g \cdot m = m'$ for some $m' \in M$. The maximal element m' is above f, since $g^{-1} \cdot f < m$ implies

$f < g \cdot m$. Also the orbit of m' does not contain elements from $X\backslash Y$, for m' and m belong to the same orbit. Hence we can decompose the element (f, g) as follows:

$$(f,\ g) = (f,\ 1)(m',\ g)$$

and we have seen that $(m',\ g) \in \langle B \rangle$. Since $(f, 1) \in B$ it follows that $(f,\ g) \in \langle B \rangle$. Suppose now that $m = \delta_i$ for some $i = 1, \ldots, p$. Recall that $G = \cup_{j=1}^{j_i} r_{ij} G_{i,1}$, where $G_{i,1}$ is the stabilizer of δ_i. So we have $g = r_{ij} h$ for some $j = 1, \ldots, j_i$ and some $h \in G_{i,1}$. It follows that $f < r_{ij} \cdot \delta_i$. Since $r_{ij}^{-1} \cdot f < \delta_i$ we know that $r_{ij}^{-1} \cdot f \in Y$. It follows that $(f,\ r_{ij})$, $(r_{ij}^{-1} \cdot f,\ 1) \in B$ and we have

$$(f,\ g) = (f,\ r_{ij})(r_{ij}^{-1} \cdot f,\ 1)(\delta_i,\ h).$$

We know that $(\delta_i, h) \in \langle B \rangle$, so $(f,\ g) \in \langle B \rangle$.

Like above, let $f \in F \cup D'$ be such that $f \notin D \cup M$. Let $h, g \in G$ be arbitrary. The element $(h^{-1} \cdot f,\ g)$ belongs to P. We show that it can be written as a product of elements from B. Let $m \in M$ be such that $h^{-1} \cdot f < m$. We consider two different cases regarding the orbit of m.

Case 1 *The orbit of m does not intersect $X\backslash Y$, i.e. $m \notin D$.* In this case $(m, h^{-1}) \in P$ and we have

$$(h^{-1} \cdot f,\ g) = (m',\ h^{-1})(f,\ hg).$$

We have seen that all the components of these products can be written as a product of elements from B. Thus $(h^{-1} \cdot f,\ g) \in \langle B \rangle$.

Case 2 *The orbit of m intersects $X\backslash Y$, i.e. $m = \delta_i$ for some $i = 1, \ldots, p$.* We write h as a product of an element from the stabilizer, $G_{i,1}$, of δ_i and a coset representative of this subgroup. Let $h = r_{ij} w$ for some $j = 1, \ldots, j_i$ and some $w \in G_{i,1}$. We have

$$(h^{-1} \cdot f,\ 1) = (w^{-1} r_{ij}^{-1} \cdot f,\ 1) = (\delta_i,\ w^{-1})(r_{ij}^{-1} \cdot f,\ 1)(\delta_i,\ w).$$

Note that $w \in G_{i,1}$ so $(\delta_i,\ w^{-1})$, $(\delta_i,\ w) \in P$, and we have seen that these elements can be written as a product of elements from B. Since r_{ij} is a coset representative of $G_{i,1}$ we know that $(r_{ij}^{-1} \cdot f,\ 1) \in B$. Hence $(h^{-1} \cdot f,\ 1) \in \langle B \rangle$. Now we write $(h^{-1} \cdot y,\ g)$ as a product of elements from B. We have

$$(h^{-1} \cdot f,\ g) = (\delta_i,\ w^{-1})(r_{ij}^{-1} \cdot f,\ wg).$$

Let $m' \in M$ be such that $(g^{-1}h^{-1})\cdot f < m'$. If $m' \notin D$ then $(wg)\cdot m' = m''$ for some $m'' \in M$, and we have

$$(r_{ij}^{-1}\cdot f,\ wg) = (r_{ij}^{-1}\cdot f,\ 1)(m'',\ g'),$$

since $r_{ij}^{-1}\cdot f < m''$. We have seen that $(\delta_i,\ w^{-1})$ and $(m'',\ wg)$ can be written as products of elements from B. The element $(r_{ij}^{-1}\cdot f,\ 1)$ is in B, so $(h^{-1}\cdot f,\ g) \in \langle B\rangle$. If $m' = \delta_b$ for some $b = 1,\ldots,p$ then let $g'^{-1} = u^{-1}r_{bl}^{-1}$ with $u \in G_{b,1}$ and $l = 1,\ldots,k_b$. We have $u^{-1}r_{bl}^{-1}r_{ij}^{-1}\cdot f < \delta_b$ and u belongs to the stabilizer of δ_b, so $r_{bl}^{-1}r_{ij}^{-1}\cdot f < \delta_b$. It follows that $(r_{bl}^{-1}r_{ij}^{-1})\cdot f \in Y$, so $(r_{ij}^{-1}\cdot f,\ r_{bl})$, $((r_{bl}^{-1}r_{ij}^{-1})\cdot f,\ 1) \in P$ and we have

$$(r_{ij}^{-1}\cdot f,\ g') = (r_{ij}^{-1}\cdot f,\ r_{bl}u) = (r_{ij}^{-1}\cdot f,\ r_{bl})((r_{bl}^{-1}r_{ij}^{-1})\cdot f,\ 1)(\delta_b,\ u).$$

The element $(r_{ij}^{-1}\cdot f,\ r_{bl})$ belongs to B, and we have seen that $(r_{bl}^{-1}r_{ij}^{-1}\cdot f,\ 1)$, $(\delta_b, u) \in \langle B\rangle$. Thus $(h^{-1}\cdot f,\ g)$ can be written as a product of elements from B.

Finally, we consider an arbitrary element $(y,\ g) \in P$. Since Y is generated, with respect to the action of G, by the set F we know that

$$y = g_1\cdot(g_2\cdot(f_{1,1}\wedge\cdots f_{n,1})\wedge\cdots\wedge g_m\cdot(f_{1,k}\wedge\cdots\wedge f_{l,k})),$$

for some $g_1,\ldots,g_m \in G$, and $f_{1,1},\ldots,f_{l,k} \in F_i$ for some $i \geqslant 0$, where $F_i = G\cdot\langle F_{i-1}\rangle \cap Y$ (for $i \geqslant 1$) and $F_0 = F$. By simplifying this product we obtain

$$y = h_1\cdot e_1 \wedge h_2\cdot e_2\wedge\ldots\wedge h_l\cdot e_l$$

for some $e_1,\ldots,e_l \in F$ and $h_1,\ldots,h_l \in G$. Note that the elements $h_i\cdot e_i$, $i = 1,\ldots,l$, do not necessarily belong to Y. We consider two cases:

Case 1 *For all $i = 1,\ldots,l$ we have $h_i\cdot e_i \in Y$.* If there exists an $i = 1,\ldots,l$ such that $e_i \notin D$ then we have

$$\begin{aligned}(y,g) &= (h_1\cdot e_1\wedge\cdots\wedge h_{i-1}\cdot e_{i-1}\wedge h_{i+1}\cdot e_{i+1}\wedge\cdots\wedge h_l\cdot e_l,\ 1)(h_i\cdot e_i,\ g)\\ &= (h_1\cdot e_1,\ 1)\cdots(h_{i-1}\cdot e_{i-1},\ 1)(h_{i+1}\cdot e_{i+1},\ 1)\cdots(h_l\cdot e_l,\ 1)(h_i\cdot e_i,\ g).\end{aligned}$$

We have seen that $(h_i\cdot e_i,\ g)$, $(h_j\cdot e_j,\ 1) \in \langle B\rangle$ for all $e_i \in F\backslash D$ and all $j = 1,\ldots,i-1,i+1,\ldots,l$. Note that if for some $j = 1,\ldots,l$ we have $e_j \in D$ then $h_j\cdot e_j \in D$, and if we have $e_j \in M\backslash D$ then $h_j\cdot e_j \in M$ and $h_j\cdot e_j \notin D$. Thus $(y,\ g)$ can be written as a product of elements from B.

Now suppose that for all $i = 1,\ldots,l$ we have $e_i \in D$. Consequently we have $h_1\cdot e_i, h_2\cdot e_2,\ldots,h_l\cdot e_l \in D$, so $y \in D'$. It follows, by above, that (y,g) can be written as a product of elements from B.

Case 2 *There exists* $i = 1, \ldots, l$ *such that* $h_i \cdot e_i \in X \backslash Y$. Assume, without loss of generality that $h_i \cdot e_i, \ldots, h_l \cdot e_l \in X \backslash Y$ and all the other terms belong to Y. If all elements $h_1 \cdot e_1, \ldots, h_{i-1} \cdot e_{i-1}$ belong to D then y belongs to D', and it follows, by above, that $(y, g) \in \langle B \rangle$. Now assume, without loss of generality, that $h_1 \cdot e_1 \notin D$, i.e. $h_1 \cdot e_1$ does not contain elements from $X \backslash Y$ in its orbit. We can rewrite y as follows:

$$y = (h_1 \cdot e_1 \wedge x_i) \wedge \ldots \wedge (h_1 \cdot e_1 \wedge x_l) \wedge h_1 \cdot e_1 \wedge h_2 \cdot e_2 \wedge \ldots \wedge h_{i-1} \cdot e_{i-1},$$

where $x_i = h_i \cdot e_i, \ldots, x_l = h_l \cdot e_l \in X \backslash Y$. Since $h' \cdot (h_1 \cdot e_1) \in Y$ for all $h' \in G$ we know, by Claim 1, that $h_1 \cdot e_1 \wedge x \in Y$ for all $x \in X \backslash Y$. So $(h_1 \cdot e_1 \wedge x_j,\ 1) \in P$ for all $j = i, \ldots, l$. Hence we can decompose (y, g) into the following product of elements from P:

$$\begin{aligned}(y, g) = (h_1 \cdot e_1 \wedge x_i,\ 1)(h_1 \cdot e_1 \wedge x_{i+1},\ 1) \ldots \\ \ldots (h_1 \cdot e_1 \wedge x_l,\ 1)(h_1 \cdot e_1 \wedge \ldots \wedge h_{i-1} \cdot e_{i-1},\ g).\end{aligned}$$

By Case 1 we know that $(h_1 \cdot e_1 \wedge \ldots \wedge h_{i-1} \cdot e_{i-1},\ g) \in \langle B \rangle$, since all elements $h_1 \cdot e_1, \ldots, h_{i-1} \cdot e_{i-1}$ belong to Y. For all $j = i, \ldots, l$ we have

$$(h_1 \cdot e_1 \wedge x_j,\ 1) = (h_1 \cdot (e_1 \wedge h_1^{-1} \cdot x_j),\ 1).$$

Since $x_j \in X \backslash Y$, for all $j = i, \ldots, l$, we know that $h_1^{-1} \cdot x_j$ belongs to either $X \backslash Y$ or D. In any case $e_1 \wedge h_1^{-1} \cdot x_j$ belongs to D' since $e_1 \in F$. It follows that $(h_1 \cdot (e_1 \wedge h_1^{-1} \cdot x_j),\ 1) \in \langle B \rangle$, and consequently $(y, g) \in \langle B \rangle$.

We conclude that the set B generates the semigroup P, and so P is finitely generated.

3. Example

Next we give an example of a P-semigroup with infinite complement, that satisfies conditions (i), (ii) and (iii) of the Main Theorem, but is not finitely generated. This shows that these conditions are not always sufficient for a P-semigroup to be finitely generated.

Let X be the poset $Y \cup \{x_i : i \in \mathbb{Z}\}$, where $Y = \{m, 0\} \cup \{e_i : i \in \mathbb{Z}\}$ is a semilattice with maximal element m, zero 0 and the elements e_i, $i \in \mathbb{Z}$, form an anti-chain. Also for each $x \in \{x_i : i \in \mathbb{Z}\}$ and $e \in \{e_i : i \in \mathbb{Z}\}$ we have $x > e$. For simplicity assume that $m = x_0$. Let G be the free group on one generator a. Define an action of G on X as follows: $a \cdot x_i = x_{i+1}$, $a \cdot e_i = e_{i+1}$ for all $i \in \mathbb{Z}$, and $a \cdot 0 = 0$. This action is described in Figure 1, where the elements of $X \backslash Y$ are represented by the non-filled dots, the elements of Y by the filled dots, and the action of the generator a is represented by the arrows.

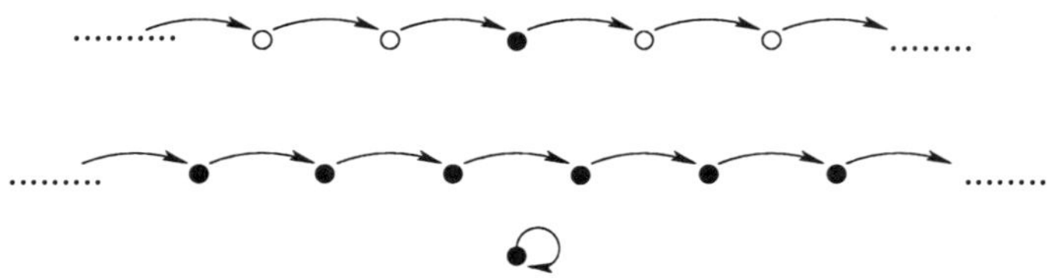

Figure 1. Action of G on X

The triple (G, X, Y) is clearly a McAlister triple, and it yields the P-semigroup

$$P = \{(m, 1)\} \cup \{(e_i, g) \ : \ g \in G, \ i \in \mathbb{Z}\} \cup \{(0, g) \ : \ g \in G\}.$$

We now show that this semigroup is not finitely generated. First we show that P is generated by the infinite set

$$A' = \{(m, 1)\} \cup \{(a^i \cdot e_0, \ a), (a^i \cdot e_0, a^{-1}) \ : \ i \in \mathbb{Z}\}.$$

Let $(e_i, g) \in P$ be arbitrary. Since G is the free cyclic group on the generator a, we have $g = a^k$, and by the action of G on Y we know that $e_i = a^i \cdot e_0$ for some $k, i \in \mathbb{Z}$. Thus

$$(e_i, g) = (a^i \cdot e_0, \ a)(b^{i-1} \cdot e_0, \ a)(b^{i-2} \cdot e_0, \ a) \dots (b^{i-k+1} \cdot e_0, \ a),$$

if $k \geqslant 0$, if $k < 0$ then we replace a by a^{-1} in the above product, so $(e_i, g) \in \langle A' \rangle$. Let $g \in G$ be arbitrary, we have

$$(0, g) = (e_0 \wedge e_1, \ g) = (e_0, 1)(e_1, g),$$

so $(0, g) \in \langle A' \rangle$. Hence the set A' generates the semigroup P.

Suppose, for a contradiction, that P is finitely generated. Then there exists an integer $s \geqslant 1$ such that the set

$$A = \{(m, 1)\} \cup \{(a^i \cdot e_0, \ a), (a^i \cdot e_0, \ a^{-1}) \ : \ -s < i < s\}$$

generates P. It follows that given $i \in \mathbb{Z}$ and $g \in G$ we have $(a^i \cdot e_0, g) = u_1 u_2 \dots u_n$, for some $u_1, \dots, u_n \in A$. We prove by induction of n that $-s < i < s$. If $n = 1$ then $(a^i \cdot e_0, g) \in A$, so $g = a^{\pm 1}$ and $-s < i < s$. Assume that for all $(a^j \cdot e_0, g')$, with $j \in \mathbb{Z}$ and $g' \in G$, such that $(b^j \cdot e_0, g') = u_1 u_2 \dots u_n$ for some $u_1, \dots, u_n \in A$ and $n < l$ we have $-s < j < s$. Let $i \in \mathbb{Z}$ and $g \in G$ be such that $(a^i \cdot e_0, g) = u_1 u_2 \dots u_l$, with $u_1, \dots, u_l \in A$. If $u_l = (m, 1)$ then

$$(a^i \cdot e_0, g) = (y, g')(m, 1) = (y, \ g').$$

It follows that $a^i \cdot e_0 = y$, and by the inductive hypothesis we have $-s < i < s$, since $(y, g') = u_1 u_2 \ldots u_{l-1}$. If $u_l = (a^j \cdot e_0, a^{\pm 1})$ for some $-s < j < s$ then

$$(a^i \cdot e_0, g) = (y, g')(a^j \cdot e_0, a^{\pm 1}) = (y \wedge (g'a^j) \cdot e_0,\ g'a^{\pm 1}).$$

This implies that $a^i \cdot e_0 \leqslant y$. So either $y = m$ in which case $g' = 1$, and $a^i \cdot e_0 = a^j \cdot e_0$, or else $y = a^i \cdot e_0$ which implies, by the inductive hypothesis, that $-s < i < s$. It follows by induction that any element $(a^i \cdot e_0, g) \in \langle A \rangle$ satisfies $-s < i < s$, for all $g \in G$. Thus the element $(a^s \cdot e_0, 1)$ cannot be written as a product of elements from A. Since this element belongs to the semigroup P we can conclude that A does not generate P, and so P is not finitely generated.

Acknowledgments

The author would like to thank Prof. Nik Ruškuc for his suggestions regarding this work, and Fundação para a Ciência e a Tecnologia for the financial support.

References

1. Isabel M. Araújo, Finite presentability of semigroup constructions, *Internat. J. Algebra Comput.* **12 (1-2)**, 19–31 (2002).
2. Peter J. Cameron, *Combinatorics: topics, techniques, algorithms*, Cambridge University Press, Cambridge (1994).
3. C. A. Carvalho, *Generation and Presentations of Semigroup Constructions: Bruck–Reilly Extensions and P-semigroups*, Ph.D. Thesis, University of St. Andrews (2005).
4. E. Dombi, *Automatic S-acts and inverse semigroup presentations*, Ph.D. Thesis, University of St. Andrews (2004).
5. Stuart W. Margolis and John C. Meakin, E-unitary inverse monoids and the Cayley graph of a group presentation, *J. Pure Appl. Algebra* **58 (1)**, 45–76 (1989).
6. D. B. McAlister, Groups, semilattices and inverse semigroups I, *Trans. Amer. Math. Soc.* **192**, 227–244 (1974).
7. D. B. McAlister, Groups, semilattices and inverse semigroups II, *Trans. Amer. Math. Soc.* **192**, 351–370 (1976).
8. D. B. McAlister and R. McFadden, Zig-zag representations and inverse semigroups, *J. Algebra* **32**, 178–206 (1974).
9. W. D. Munn, A note on E-unitary inverse semigroups, *Bull. London Math. Soc.* **8 (1)**, 71–76 (1976).
10. Derek J. S. Robinson, A course in the theory of groups, *Graduate Texts in Mathematics* **80**, 2nd edition, Springer-Verlag, New York (1996).

11. H. E. Scheiblich, Free inverse semigroups, *Semigroup Forum* **4**, 351–359 (1972).
12. H. E. Scheiblich, Free inverse semigroups, *Proc. Amer. Math. Soc.* **38**, 1–7 (1973).
13. Boris M. Schein, A new proof for the McAlister "P-theorem", *Semigroup Forum* **10 (2)**, 185–188 (1975).
14. Benjamin Steinberg, McAlister's P-theorem via Schützenberger graphs, *Comm. Algebra* **31 (9)**, 4387–4392 (2003).

OPEN PROBLEMS ON REGULAR LANGUAGES: A HISTORICAL PERSPECTIVE

LAURA CHAUBARD AND JEAN-ÉRIC PIN

LIAFA, Université Paris VII and CNRS, Case 7014, 2 Place Jussieu, 75251 Paris Cedex 05, France

E-mail: {Laura.Chaubard | Jean-Eric.Pin}@liafa.jussieu.fr

Operations on regular languages have been studied for fifty years, but several major problems remain wide open. This paper surveys the semigroup approach to these problems. We consider successively the star-height problem, the Straubing-Thérien's concatenation hierarchy and the shuffle operation. On the algebraic side, we present Eilenberg's variety theory and its successive improvements, including the recent notion of $\mathcal{C}$-variety.

Recall that a *language* is a subset of a finitely generated free monoid. The aim of this paper is to discuss various instances of the following general problem.

Problem. *Given a "basis" of languages, a set of operations and some rules to use them, describe the languages expressible from the basis by using the operations according to the rules.*

In practice, a basis of languages will consist of a set of very simple languages, such as the languages of the form $\{a\}$, where a is a letter of the alphabet. There are many possible choices for the operations, but we shall restrict ourselves to nine of them, that we now introduce.

1. Operations on languages

Let A be a finite alphabet and let A^* be the free monoid on A. Let us describe the operations we have in mind.

(1) *Boolean operations*, which comprise

   (a) finite *union* and finite *intersection* (these operations are also called the *positive Boolean operations*),

(b) *complement* (denoted by $L \to L^c$).

(2) *Residual*: given a language L and a word u of A^*, $u^{-1}L = \{v \mid uv \in L\}$ and $Lu^{-1} = \{v \mid vu \in L\}$.

(3) *Star*: L^* is the submonoid of A^* generated by L. Thus $L^* = \{u_1u_2 \cdots u_n \mid n \geqslant 0, u_1, \ldots, u_n \in L\}$.

(4) *Product*: the product of two languages L_1 and L_2 is the languages $L_1L_2 = \{u_1u_2 \mid u_1 \in L_1, u_2 \in L_2\}$.

(5) *Marked product*: given letters $a_1, \ldots, a_k$ of A and languages $L_0, L_1, \ldots, L_k$ of A^*, the marked product $L_0a_1L_1 \cdots a_kL_k$ is the language $\{u_0a_1u_1 \cdots a_ku_k \mid u_0 \in L_0, \ldots, u_k \in L_k\}$.

(6) *Shuffle product.* The shuffle of two words u and v of A^* is the set u Ш v of words of A^* of the form $u_1v_1 \cdots u_nv_n$, with $n \geqslant 0$, $u_1, \ldots, u_n, v_1, \ldots, v_n \in A^*$, $u_1 \cdots u_n = u$, $v_1 \cdots v_n = v$. For instance,

$$ab \text{ Ш } ba = \{abba, baab, abab, baba\}$$

The shuffle of two languages L_1 and L_2 of A^* is the set

$$L_1 \text{ Ш } L_2 = \bigcup_{u_1 \in L_1,\ u_2 \in L_2} u_1 \text{ Ш } u_2$$

(7) *Morphisms.* Let A and B be two alphabets, and let φ be a function from A into B^*. Then φ extends in a unique way into a morphism from A^* into B^*. If L is a language of A^*, $\varphi(L) = \{\varphi(u) \mid u \in L\}$ is a language of B^*.

(8) *Inverse morphisms.* If $\varphi\colon A^* \to B^*$ is a morphism and L is a language of B^*, then $\varphi^{-1}(L) = \{u \in A^* \mid \varphi(u) \in L\}$ is a language of A^*.

In our context, a *positive Boolean algebra* will be a class of languages closed under finite union and finite intersection. Since the empty language $\emptyset$ (resp. the full language A^*) can be considered as the union (resp. intersection) of an empty family of languages, they belong to all positive Boolean algebras. A Boolean algebra is a positive Boolean algebra closed under complement.

2. Rational and recognisable languages

Our first example is the class of *rational languages.* It is obtained by taking the languages $\{a\}$, for each letter a, as the basis and by allowing the use of only three operations, union, product and star, with no particular rules. If $A = \{a, b\}$, languages like A^*abaA^* or $(aba)^*ba \cup (bb(aa)^*ba)^*$ are rational.

Rational languages were characterised by Kleene in a seminal paper published in 1956 [13]. Kleene's theorem states that the rational languages are exactly the recognisable languages, that can be defined in (at least) three equivalent ways. We recall here two of these definitions, one relying on deterministic automata and one using finite monoids. A third possibility would be to make use of nondeterministic automata, but we shall not consider this approach in this paper.

A *finite automaton* is a quintuple $\mathcal{A} = (Q, A, E, q_0, F)$ where Q is a finite set (the set of *states*), A is an alphabet, E is a subset of $Q \times A \times Q$ (the set of *transitions*), q_0 is an element of Q (the *initial state*) and F is a subset of Q (the set of *final* states). Two transitions (p, a, q) and (p', a', q') are *consecutive* if $q = p'$. A *successful path* in $\mathcal{A}$ is a finite sequence of consecutive transitions starting in the initial state and ending in some final state

$$q_0 \xrightarrow{a_1} q_1 \xrightarrow{a_2} q_2 \ \cdots \ q_{n-1} \xrightarrow{a_n} q_n \in F$$

The word $a_1a_2 \cdots a_n$ is its *label*. The language recognised by $\mathcal{A}$ is the set of labels of all the successful paths in $\mathcal{A}$. A language is *recognisable* if it is recognised by some finite automaton.

A finite automaton is *deterministic* if for each state $p \in Q$ and each letter $a \in A$, there is at most one state q such that $(p, a, q) \in E$. This unique state q is denoted by $p \cdot a$. Thus each letter a induces a partial function $p \to p \cdot a$ from Q into itself. One can show that every recognisable language can be recognised by a deterministic automaton.

The definition involving monoids is more abstract. A monoid morphism $\varphi \colon A^* \to M$ *recognises* a language L of A^* if there is a subset P of M such that $L = \varphi^{-1}(P)$. By a slight abuse of language, we also say in this case that M recognises L.

It is not too difficult to show that the two definitions, by automata and by monoids, are equivalent. We can now reformulate Kleene's theorem as follows.

Theorem 2.1. **[Kleene 1956]** *For a language L, the following conditions are equivalent:*

(1) L is rational,

(2) L is recognised by a finite monoid,

(3) L is recognised by a finite automaton.

The term *regular* is also frequently usually used in the literature as an equivalent to *recognisable* or *rational*. It is important, however, to distinguish the latter two notions. First, both of them can be extended to arbitrary monoids, but they do not coincide in general. Secondly, depending on the problem, it might be more appropriate to take one definition or the other. Precise examples are given in the next paragraph.

A consequence of Kleene's theorem is that the class of recognisable languages is closed under the nine operations considered in Section 1. The importance of Kleene's theorem stems from the fact that some closure properties are transparent for rational expressions while others are much easier to prove using automata or monoids. For instance, it is straighforward to see that the class of rational languages is closed under union, (marked) product, star and morphisms. On the other hand, it is easy to see that recognisable languages are closed under complement, residuals, shuffle and inverse morphims. It is also possible, although slightly more difficult, to prove directly that recognisable languages are closed under (marked) product and star, but proving that rational languages are closed under complement without invoking Kleene's theorem is a real challenge. The skeptical reader may try to find a rational expression for the complement of the language $(((ab)^*aba)^*ba)^*$ to apprehend the difficulty of the problem.

The proof of Kleene's theorem is interesting for itself, since it provides an algorithm to convert a rational expression into a finite automaton and back. In the sequel, we shall meet several decidability problems of the form *decide whether a given regular language satisfies a certain property.* By Kleene's theorem, the solution of such a decision problem is independent of the representation chosen for the regular language, since descriptions by a rational expression, a finite automaton or a finite monoid can be translated one into another. However, the chosen representation has a strong influence on the complexity of the decision algorithms, a problem that we shall not address in this paper.

3. Star-height

In this section, we focus our attention on the star operation.

3.1. *Star-free languages*

The class of *star-free* languages is obtained by taking the languages $\{1\}$ and $\{a\}$, for each letter a, as the basic class $\mathcal{B}$ and by allowing Boolean operations and product. According to our general definition of a Boolean algebra, the languages $\emptyset$ and A^* are star-free. If $A = \{a, b\}$, the following languages are also star-free:

$$
\begin{aligned}
a^* &= (A^*bA^*)^c \\
(ab)^* &= (bA^* \cup A^*aaA^* \cup A^*bbA^* \cup A^*a)^c
\end{aligned}
$$

One can also show that the languages $\{ab, ba\}^*$ and $(a(ab)^*b)^*$ are star-free but that the languages $(aa)^*$ and $\{aba, b\}^*$ are not. Deciding whether a given rational language is star-free is a difficult problem, which was solved by Schützenberger in 1965.

Before stating this result, we need to introduce a few definitions. Let $\mathcal{A}$ be a finite deterministic automaton recognising a language L of A^*. A state q is called *accessible* if there exists a path from the initial state to q, and *coaccessible* if there is a path from q to some final state. By removing the states of $\mathcal{A}$ which are not simultaneously accessible and coaccessible, one obtains a *trim* automaton $\mathcal{B}$ that also recognises L. A further reduction consists in identifying two states p and q whenever, for every $u \in A^*$, $p \cdot u$ is final if and only if $q \cdot u$ is final. Performing this equivalence on the set of states of $\mathcal{B}$, one obtains a new automaton, called the *minimal automaton* of L, which also recognises L. For instance, the minimal automaton of the language $\{a, b\}^*aA^*b\{b, c\}^*$ is pictured in Figure 3.1.

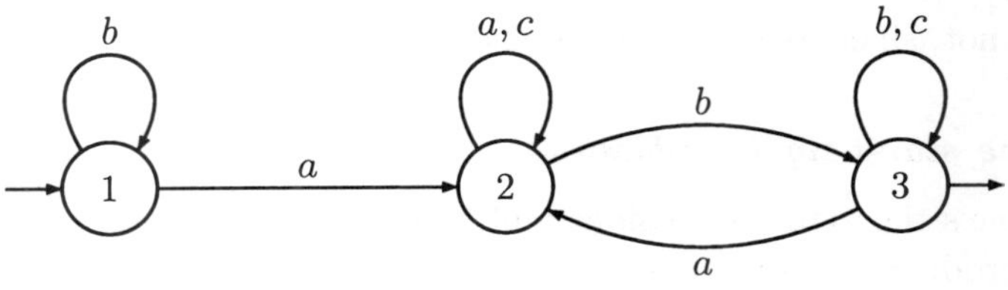

Figure 3.1. The minimal automaton of $\{a, b\}^*aA^*b\{b, c\}^*$.

The *syntactic monoid* of a language L can be defined in two equivalent ways. First, it is the transition monoid of the minimal automaton of L. Secondly, it is the quotient of A^* by the *syntactic congruence* of L, defined on A^* as follows: $u \sim_L v$ if and only if, for every $x, y \in A^*$,

$$xvy \in L \Leftrightarrow xuy \in L.$$

The syntactic monoid of the language $L = \{a, b\}^*aA^*b\{b, c\}^*$ and its $\mathcal{J}$-

class structure are given below

1	1	2	3
a	2	2	2
b	1	3	3
c	–	2	3
ab	3	3	3
bc	–	3	2
ca	–	2	2

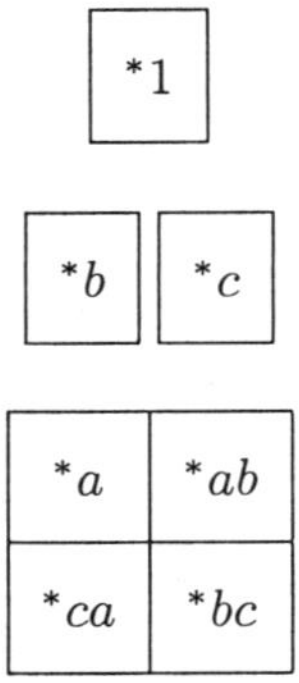

Schützenberger [29] used the syntactic monoid to characterise the star-free languages. Recall that a finite monoid M is *aperiodic* if for each $x \in M$, there exists $n \geqslant 0$ such that $x^{n+1} = x^n$. Equivalently, a monoid is aperiodic if all the groups it contains are trivial, or if the Green's relation $\mathcal{H}$ is the equality.

Theorem 3.1. **[Schützenberger 1965]** *A language is star-free if and only if its syntactic monoid is finite and aperiodic.*

A consequence of Schützenberger's theorem is that one can effectively decide whether a given regular language is star-free. For instance, $\{a,b\}^*aA^*b\{b,c\}^*$ is star-free, since its syntactic monoid is aperiodic, but $(A^2)^*$ is not, since its syntactic monoid is the cyclic group of order 2.

3.2. *The star-height problem*

By Kleene's theorem, expressions built from letters by using Boolean operation, product and star represent regular languages. Such expressions are called *extended rational expressions.* The *star-height* of such an expression is the maximum number of nested stars occurring in the expression. For instance, the expression

$$(\{a, ba, abb\}^* bba \cap (aa\{a, ab\}^*))^c bbA^*$$

is of star-height one, while the expression

$$\left(a(ba)^* abb\right)^* bba \cap (aa\{a, ab\}^*))^c bbA^*$$

is of star-height two. The *star-height of a language* is the minimal star-height of an expression representing the language. In particular a language of star-height 0 is a star-free language.

We have seen that the language $(A^2)^*$ is not star-free. Since $(A^2)^*$ is an expression of star-height 1, this language has star-height exactly one. However, it is an open problem to know whether there are languages of star-height 2. We shall come back on this problem in Section 5.

4. Concatenation hierarchies

In this section, we introduce a hierarchy among star-free languages of A^*, known as the *Straubing-Thérien's hierarchy*, or *concatenation hierarchy.*[a] For historical reasons, this hierarchy is indexed by half-integers. The level 0 consists of the languages $\emptyset$ and A^*. The other levels are defined inductively as follows:

(1) the level $n+1/2$ is the class of union of marked products of languages of level n;

(2) the level $n+1$ is the class of Boolean combination of languages of marked products of level n.

We call the levels n (for some nonegative integer n) the *full levels* and the levels $n+1/2$ the *half levels.*

It is not clear at first sight whether the Straubing-Thérien's hierarchy does not collapse, but this question was solved in 1978 by Brzozowski and Knast [6].

Theorem 4.1. [Brzozowski and Knast 1978] *The Straubing-Thérien's hierarchy is infinite.*

It is a major open problem on regular languages to know whether one can decide whether a given star-free language belongs to a given level.

Problem 2. *Given a half integer n and a star-free language L, decide whether L belongs to level n.*

One of the reasons why this problem is particularly appealing is its close connection with finite model theory, first explored by Büchi in the early sixties. Büchi's logic comprises a relation symbol $<$ and, for each letter $a \in A$ a predicate symbol $\mathbf{a}$. First order formulas are built in the usual way by using these symbols, the equality symbol, (first order) variables,

[a] A similar hierarchy, called the *dot-depth hierarchy* was previously introduced by Brzozowski, but the Straubing-Thérien's hierarchy is easier to define.

Boolean connectives and quantifiers. Formal definitions can be found for instance in [32], but here we shall just present on an example how sentences are interpreted on finite words. The sentence

$$\varphi_1 = \exists x \ \exists y \ \big((x < y) \wedge (\mathbf{a}x) \wedge (\mathbf{b}y)\big),$$

can intuitively be interpreted on a word u by the English sentence "there exist two integers $x < y$ such that, in u, the letter in position x is an a and the letter in position y is a b". Therefore, the set of words satisfying φ_1 is $A^*aA^*bA^*$. McNaughton and Papert [15] showed that a language is first-order definable if and only if it is star-free. Thomas [32] (see also [16]) refined this result by showing that the concatenation hierarchy of star-free languages corresponds, level by level, to a hierarchy of first order formulas, the Σ_n-hierarchy. This hierarchy can be defined inductively as follows:

(1) Σ_0 consists of the quantifier-free formulas
(2) Σ_{n+1} consist of the formulas of the form $\exists x_1 \ldots \exists x_p \forall y_1 \ldots \forall y_q \ \varphi$, where $p, q \geqslant 0$ and φ is a Σ_n-formula.
(3) $\mathcal{B}\Sigma_n$ denotes the class of formulas that are Boolean combinations (that is, conjunctions of disjunctions) of Σ_n-formulas.

For instance, $\exists x_1 \exists x_2 \forall x_3 \forall x_4 \forall x_5 \exists x_6 \, \varphi$, where φ is quantifier free, is in Σ_3. The next theorem summarizes the results of [15, 32, 16].

Theorem 4.2.

(1) A language is first-order definable if and only if it is star-free.
(2) A language is Σ_n-definable if and only if it is of level $n - 1/2$.
(3) A language is $\mathcal{B}\Sigma_n$-definable if and only if it is of level n.

Thus deciding whether a language has level n is equivalent to a very natural problem in finite model theory. The first decidabilty result was obtained by I. Simon [30].

Theorem 4.3. [Simon 1972] *A language has level* 1 *if and only if its syntactic monoid is finite and $\mathcal{J}$-trivial.*

As in the case of star-free languages, the characterisation is given by a property of the syntactic monoid. This raises the question whether other families of regular languages can be described by an algebraic property of their syntactic monoid. The solution to this question was given by Eilenberg [10] in his variety theorem. We shall see in particular that the full levels of the concatenation hierarchy are varieties in Eilenberg's sense and thus can be described by some properties of their syntactic monoid. However,

Eilenberg's theory does not apply to half levels, because they are not closed under complement. The solution, proposed in [20], consists of using the syntactic ordered monoid in place of the syntactic monoid. We briefly describe this extension before stating the variety theorem and its extended version.

First recall that an *ordered monoid* is a monoid equipped with an order $\leqslant$ compatible with the multiplication: $x \leqslant y$ implies $zx \leqslant zy$ and $xz \leqslant yz$.

We now give two equivalent definitions of the syntactic ordered monoid. We start with the algorithmic definition, which is probably easier to understand. Consider a minimal deterministic automaton $\mathcal{A} = (Q, A, \cdot, i, F)$. One defines a partial order $\leqslant$ on Q by $p \leqslant q$ if and only if, for each $u \in A^*$, $q \cdot u \in F \Rightarrow p \cdot u \in F$. For instance, for the automaton pictured in Figure 4.2, the partial order is $2 \leqslant 4$ and $1, 2, 3, 4 \leqslant 0$.

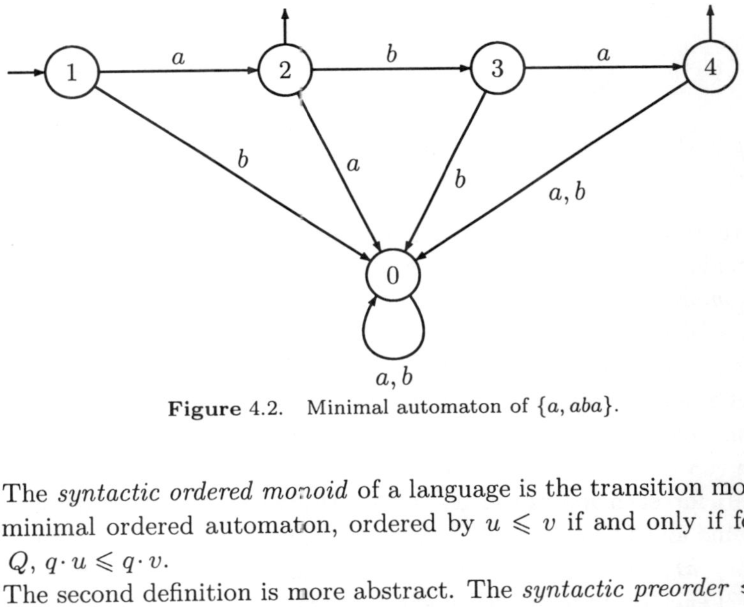

Figure 4.2. Minimal automaton of $\{a, aba\}$.

The *syntactic ordered monoid* of a language is the transition monoid of its minimal ordered automaton, ordered by $u \leqslant v$ if and only if for each $q \in Q$, $q \cdot u \leqslant q \cdot v$.

The second definition is more abstract. The *syntactic preorder* $\leqslant_L$ of a language L is defined as follows: $u \leqslant_L v$ iff, for every $x, y \in A^*$,

$$xvy \in L \Rightarrow xuy \in L$$

This preorder induces a partial order on the syntactic monoid of L, called the *syntactic order* of L. Thus the syntactic ordered monoid of L is equal to $(A^*/{\sim_L}, {\leqslant_L}/{\sim_L})$.

Example 4.1. The syntactic monoid and the syntactic order of the language $\{a,b\}^*aA^*b\{b,c\}^*$ are pictured below:

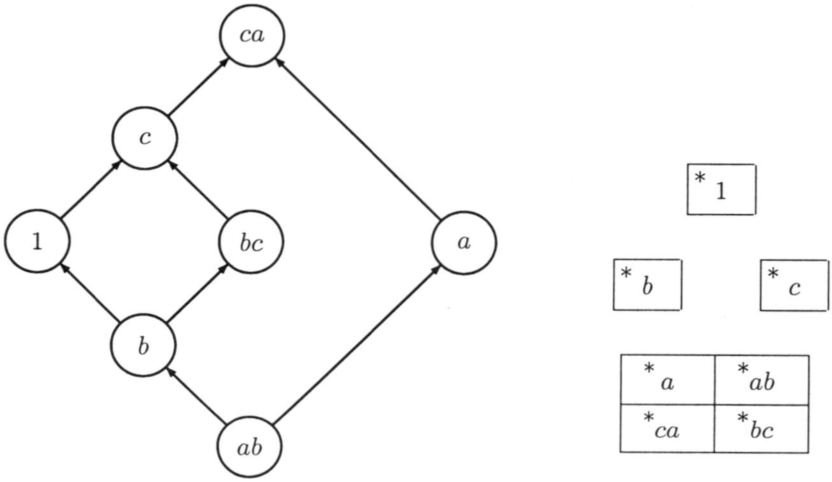

Thus ab is the smallest element in the syntactic order of L, and ca is the greatest.

A *variety of finite monoids* is a class of finite monoids closed under taking submonoids, quotients and finite products. Similarly, a *variety of finite ordered monoids* is a class of finite ordered monoids closed under taking ordered submonoids, quotients and finite products. Varieties of finite (ordered) *semigroups* are defined analogously. There is an abundant literature on varieties and we refer the reader to the books [1, 10, 19] for more details.

A convenient way to define varieties of finite monoids is to use identities. Let u, v be words of the free monoid A^*. A monoid M satisfies the *identity* $u = v$ if, for each morphism $\varphi\colon A^* \to M$, $\varphi(u) = \varphi(v)$. Similarly, an ordered monoid $(M, \leqslant)$ *satisfies the identity* $u \leqslant v$ if, for each morphism $\varphi : A^* \to M$, $\varphi(u) \leqslant \varphi(v)$. A variety of (ordered) monoids satisfies an identity if each of its monoids satisfies it.

The definition of an identity can be extended to *profinite identities*, which are formal equalities of the form $u = v$ (or $u \leqslant v$) where u and v are *profinite words*. We shall not attempt to define here profinite words nor profinite topology and the reader is referred to [2, 3, 33] for more details. We shall however define ω-terms, a special case of profinite words. An *ω-term* on an alphabet A is built from the letters of A using the usual concatenation product and the unary operator $x \to x^\omega$. Thus, if $A = \{a, b, c\}$, abc, a^ω and $((ab^\omega c)^\omega ab)^\omega$ are examples of ω-terms.

Let $\varphi : A^* \to M$ be a morphism from A^* into a finite monoid. The image $\varphi(t)$ of an ω-term t is defined recursively as follows. If t is a letter, then $\varphi(t)$ is already defined. If t and t' are ω-terms, then $\varphi(tt') = \varphi(t)\varphi(t')$. If $t = u^\omega$, then $\varphi(t)$ is the unique idempotent power of $\varphi(u)$.

Reiterman's theorem [28] ensures that a class of finite monoids is a variety if and only if it can be defined by a set of profinite identities. A similar result holds for varieties of ordered monoids. We refer to [3] for a detailed survey of this theory.

It is easy to prove directly that the class of finite (ordered) monoids (semigroups) satisfying a given set E of profinite identities is a variety of finite (ordered) monoids (semigroups), denoted by $[\![E]\!]$. Usually the context suffices to decide whether we are dealing with varieties of monoids or varieties of semigroups. For instance $[\![x^2 = x, xy = yx]\!]$ is the variety of finite idempotent and commutative monoids and $[\![x^\omega y x^\omega \leqslant x^\omega]\!]$ is the variety of all finite ordered semigroups S such that, for all $s \in S$ and $e \in E(S)$, $ese \leqslant e$.

A *positive variety of languages* is a class of recognisable languages $\mathcal{V}$ such that for any alphabets A and B,

(1) $\mathcal{V}(A^*)$ is a positive Boolean algebra,

(2) if $L \in \mathcal{V}(A^*)$ and $a \in A$ then $a^{-1}L, La^{-1} \in \mathcal{V}(A^*)$,

(3) if $\varphi \colon A^* \to B^*$ is a morphism, $L \in \mathcal{V}(B^*)$ implies $\varphi^{-1}(L) \in \mathcal{V}(A^*)$.

A *variety of languages* is a positive variety $\mathcal{V}$ such that, for each alphabet A, $\mathcal{V}(A^*)$ is closed under complement. We can now state the two variety theorems.

Theorem 4.4. [**Eilenberg 1976**] *Let* $\mathbf{V}$ *be a variety of finite monoids. For each alphabet* A, *let* $\mathcal{V}(A^*)$ *be the set of all languages of* A^* *whose syntactic monoid is in* $\mathbf{V}$. *Then* $\mathcal{V}$ *is a variety of languages. Further, the correspondence* $\mathbf{V} \to \mathcal{V}$ *is a bijection between varieties of finite monoids and varieties of languages.*

Theorem 4.5. [**Pin 1995**] *Let* $\mathbf{V}$ *be a variety of finite ordered monoids. For each alphabet* A, *let* $\mathcal{V}(A^*)$ *be the set of all languages of* A^* *whose syntactic ordered monoid is in* $\mathbf{V}$. *Then* $\mathcal{V}$ *is a positive variety of languages. Further, the correspondence* $\mathbf{V} \to \mathcal{V}$ *is a bijection between varieties of finite ordered monoids and positive varieties of languages.*

The next proposition shows that the variety approach is relevant for studying the concatenation hierarchy.

Proposition 4.1.

(1) The star-free languages form a variety of languages.

(2) Each full level of the concatenation hierarchy is a variety of languages.

(3) Each half level of the concatenation hierarchy is a positive variety of languages.

We shall denote by $\mathbf{V}_n$ the variety of finite monoids corresponding to the languages of level n and by $\mathbf{V}_{n+1/2}$ the variety of ordered monoids corresponding to the languages of level $n+1/2$.

Unfortunately, very few decidability results are known. It is obvious that a language has level 0 if and only if its syntactic monoid is trivial. The level 1/2 is also easy to study.

Theorem 4.6. **[Pin-Weil 1995]** *A language has level* $1/2$ *if and only if its ordered syntactic monoid* M *satisfies the identity* $x \leqslant 1$.

We already mentioned Simon's characterisation of languages of level 1. The decidability of level 3/2 was first proved by Arfi [4, 5] and the algebraic characterisation was found by Pin-Weil [24]. We need to introduce the Mal'cev product to state this result precisely.

Let $\mathbf{V}$ be variety of finite ordered semigroups and let M and N be two ordered monoids. A relational morphism $\tau : M \to N$ is a *$\mathbf{V}$-relational morphism* if, for every ordered subsemigroup T of N in $\mathbf{V}$, the ordered semigroup $\tau^{-1}(T)$ belongs to $\mathbf{V}$. Given a variety of finite monoids $\mathbf{W}$, the class of all ordered monoids M such that there exists a $\mathbf{V}$-relational morphism from M into an ordered monoid of $\mathbf{W}$ is a variety of ordered monoids, denoted by $\mathbf{V} \circledM \mathbf{W}$ and called the *Mal'cev product* of $\mathbf{V}$ and $\mathbf{W}$.

Theorem 4.7. **[Pin-Weil 2001]** *A language is of level* $3/2$ *if and only if its ordered syntactic monoid belongs to the Mal'cev product* $[\![x^\omega y x^\omega \leqslant x^\omega]\!] \circledM [\![x^2 = x, xy = yx]\!]$. *This condition is decidable.*

The decidability of level 2 is a major open problem in automata theory. An algebraic characterisation of $\mathbf{V}_2$ was given in [21], but it is not effective. Recall that a monoid M *divides* a monoid N if M is a quotient of a submonoid of N.

Theorem 4.8. **[Pin-Straubing 1981]** *A monoid belongs to* $\mathbf{V}_2$ *if and only if it divides a monoid of upper triangular Boolean matrices.*

Several partial results are known and a conjecture was proposed for the identities of $\mathbf{V}_2$, but its decidability is still open. See [27] for recent progress on this problem.

For the other levels, the decidability problem is also wide open. Pin and Weil [24, 26] established an algebraic connection between the varieties $\mathbf{V}_n$ and $\mathbf{V}_{n+1/2}$.

Theorem 4.9. [Pin-Weil 1995] *The variety* $\boldsymbol{V}_{n+1/2}$ *is equal to the Mal'cev product* $[\![x^\omega y x^\omega \leqslant x^\omega]\!] \circledM \boldsymbol{V}_n$.

Another result [25] describes, given the identities of a variety of finite monoids $\mathbf{V}$, a set of identities defining the variety $[\![x^\omega y x^\omega \leqslant x^\omega]\!] \circledM \mathbf{V}$.

Theorem 4.10. [Pin-Weil 1996] *The variety* $[\![x^\omega y x^\omega \leqslant x^\omega]\!] \circledM \boldsymbol{V}$ *is defined by the profinite identities* $u^\omega v u^\omega \leqslant u^\omega$, *where* u *and* v *are profinite words such that* $u = u^2$ *and* $u = v$ *are profinite identities of* $\boldsymbol{V}$.

These results illustrate the power of the algebraic approach, but do not suffice yet to show that if $\mathbf{V}_n$ is decidable, then $\mathbf{V}_{n+1/2}$ is decidable, except for $n = 1$.

5. Back to the star-height problem

Schützenberger's theorem gives a characterisation of the languages of star-height 0 and shows that they form a variety of languages. One may wonder whether this latter result also holds for the languages of star-height $\leqslant 1$. The answer to this question reduces to the existence of a language of star-height 2, as shown in [18]. Unfortunately, this problem is still open.

Theorem 5.1. [Pin 1978] *If the languages of star-height* $\leqslant 1$ *form a variety of languages, then there is no language of star-height* 2.

The closure properties of the languages of star-height $\leqslant n$ were analysed in [23]. Recall that a morphism between two free monoids is *length-preserving* if it maps each letter to a letter.

Theorem 5.2. [Pin, Straubing, Thérien 1989] *For each nonnegative integer* n, *the class of all languages of star-height* $\leqslant n$ *is closed under Boolean operations, residuals and inverse of length-preserving morphisms.*

Thus the languages of star-height $\leqslant n$ "almost" form a variety of languages. In fact, many other interesting classes of languages satisfy the two first conditions defining a variety of languages, but only a weak form of the third condition. Such examples include languages defined by fragments of first order logic or by temporal logic. Straubing [31] recently proposed a new extension of the notion of variety which covers these examples. A similar notion was introduced independently by Ésik and Ito [11].

Let $\mathcal{C}$ be a class of morphisms between free monoids, closed under composition and containing all length-preserving morphisms. Examples include the classes of all *length-preserving* morphisms, of all *length-multiplying* morphisms (morphisms such that, for some integer k, the image of any letter is a word of length k), all *non-erasing* morphisms (morphisms for which the image of each letter is a nonempty word), all *length-decreasing* morphisms (morphisms for which the image of each letter is either a letter or the empty word) and all morphisms.

A *positive $\mathcal{C}$-variety of languages* is a class $\mathcal{V}$ of recognisable languages satisfying the two first conditions defining a positive variety of languages and a third condition

(3′) if $\varphi\colon A^* \to B^*$ is a morphism in $\mathcal{C}$, $L \in \mathcal{V}(B^*)$ implies $\varphi^{-1}(L) \in \mathcal{V}(A^*)$.

A *$\mathcal{C}$-variety of languages* is a positive $\mathcal{C}$-variety of languages closed under complement. When $\mathcal{C}$ is the class of all (resp. length-preserving, length-multiplying, non-erasing, length-decreasing) morphisms, we use the term *all*-variety (resp. *lp*-variety, *lm*-variety, *ne*-variety, *de*-variety).

Theorem 5.2 gives an interesting example of *lp*-variety of languages.

Corollary 5.1. *For each $n \geqslant 0$, the languages of star-height $\leqslant n$ form an lp-variety of languages.*

The algebraic counterpart relies on a new syntactic invariant, the syntactic stamp. A *stamp* is a surjective morphism from A^* onto a finite monoid. The *syntactic stamp* of a regular language of A^* is the canonical morphism from A^* onto its syntactic monoid.

A stamp $\varphi : A^* \to M$ *$\mathcal{C}$-divides* a stamp $\psi : B^* \to N$ if there is a pair (f, η) (called a *$\mathcal{C}$-division*), where $f : A^* \to B^*$ is in $\mathcal{C}$, $\eta : N \to M$ is a partial surjective monoid morphism, and $\varphi = \eta \circ \psi \circ f$. If f is the identity on A^*, the pair (f, η) is simply called a division.

The *product* of two stamps φ_1 and φ_2 is the stamp φ defined by $\varphi(a) = (\varphi_1(a), \varphi_2(a))$. A *$\mathcal{C}$-variety of stamps* is a class of stamps closed under $\mathcal{C}$-division and finite products.

Straubing's $\mathcal{C}$-variety theorem [31] can now be stated as follows.

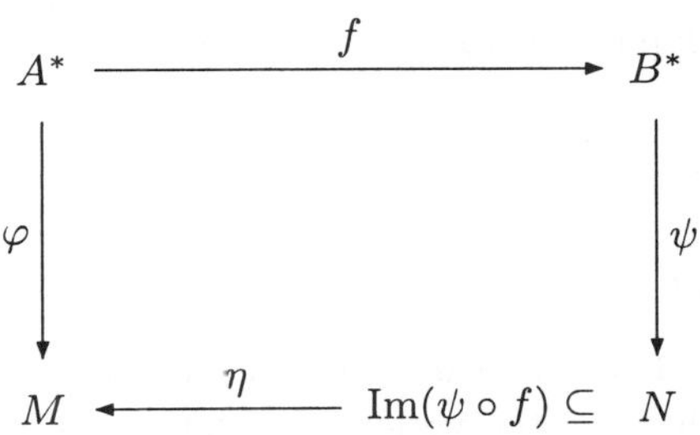

Figure 5.3. A division diagram.

Theorem 5.3. *Let* $\mathbf{V}$ *be a* $\mathcal{C}$*-variety of stamps. For each alphabet* A*, denote by* $\mathcal{V}(A^*)$ *the set of all languages of* A^* *whose syntactic stamp is in* $\mathbf{V}$*. Then* $\mathcal{V}$ *is a* $\mathcal{C}$*-variety of languages. Further, the correspondence* $\mathbf{V} \to \mathcal{V}$ *is a bijection between* $\mathcal{C}$*-varieties of stamps and* $\mathcal{C}$*-varieties of languages.*

The identity approach can be extended to $\mathcal{C}$-varieties of stamps as follows. Let u, v be two words of B^*. A stamp $\varphi : A^* \to M$ is said to *satisfy the* $\mathcal{C}$*-identity* $u = v$ if, for every $\mathcal{C}$-morphism $f\colon B^* \to A^*$, $\varphi \circ f(u) = \varphi \circ f(v)$. If M is ordered, we say that φ satisfies the $\mathcal{C}$-identity $u \leqslant v$ if, for every $\mathcal{C}$-morphism $f\colon B^* \to A^*$, $\varphi \circ f(u) \leqslant \varphi \circ f(v)$. By extension, we say that a language satisfies an identity if its syntactic stamp satisfies this identity.

Example 5.1. Let $\varphi : A^* \to M$ be a stamp. Consider the identity

$$xyx = x \tag{1}$$

If $\mathcal{C}$ is the class of all morphisms, φ satisfies (1) if and only if, for all $x, y \in A^*$, $\varphi(xyx) = \varphi(x)$. Now, if $\mathcal{C}$ is the class of length-preserving morphisms, φ satisfies (1) if and only if, for all $x, y \in A$, $\varphi(xyx) = \varphi(x)$. If $\mathcal{C}$ is the class of length-multiplying morphisms, φ satisfies (1) if and only if, for each $k \geqslant 0$ and for all $x, y \in A^k$, $\varphi(xyx) = \varphi(x)$.

The definition of identities can be extended to profinite identities to obtain a generalisation of Reiterman's theorem to $\mathcal{C}$-varieties [14, 22].

It follows from the previous results that the star-height problem amounts to showing that the *lp*-varieties of stamps corresponding to the languages of star-height $\leqslant n$ are decidable. But even if these varieties of stamps cannot be characterized precisely, one can still hope to find some identity satisfied by all languages of star-height $\leqslant 1$. It would then suffice to find a regular language not satisfying this identity to have an example of a language of star-height > 1.

For recent developments about $\mathcal{C}$-varieties, we refer the reader to the papers [8, 9, 22].

6. Shuffle product

Introducing the shuffle product into the picture leads to several interesting questions. First, what are the varieties of languages closed under shuffle? The commutative varieties of languages closed under shuffle were characterised by Perrot [17]: they correspond to the varieties of commutative monoids whose groups belong to a given variety of commutative groups. The variety of all rational languages is also closed under shuffle. Are there other examples? Esik and Simon [12] answered this question negatively. Let us say that a variety of languages is *proper* if it is not equal to the variety of all rational languages.

Theorem 6.1. **[Esik-Simon 1998]** *The variety of commutative languages is the largest proper variety of languages closed under shuffle.*

Is there a similar result for positive varieties of languages? That is, is there a largest proper positive variety of languages closed under shuffle? The answer was given in [7].

Theorem 6.2. **[Cano Gómez, Pin 2004]** *There is a largest positive variety not containing* $(ab)^*$. *It is also the largest proper positive variety closed under length preserving morphisms and the largest proper positive variety closed under shuffle.*

A characterisation of the corresponding variety of ordered monoids **W** was given in the same paper.

Theorem 6.3. **[Cano Gómez, Pin 2004]** *An ordered monoid belongs to* $\boldsymbol{W}$ *if and only if, for every pair* (a, b) *of mutually inverse elements, and for every element* z *of the minimal ideal of the submonoid generated by* a *and* b, $(abzab)^\omega \leqslant ab$. *In particular* $\boldsymbol{W}$ *is decidable.*

It would be interesting to know whether a similar result holds for lp-varieties of languages: is there a largest proper lp-variety of languages closed under shuffle? Is there a largest proper positive lp-variety of languages closed under shuffle?

7. Conclusion

The successive improvements over Eilenberg's variety theory have considerably enlarged the scope of the algebraic approach to the study of regular languages. It has been applied succesfully to a large range of applications, including logic and finite model theory, circuit complexity, abstract complexity, communication complexity, infinite words and other structures. However, several exciting problems remain unsolved and we would like to encourage the semigroup community to work on these questions.

References

1. J. Almeida, Finite semigroups and universal algebra, *Series in Algebra* **3**, World Scientific, Singapore (1994).
2. J. Almeida, Profinite semigroups and applications, in *Structural theory of automata, semigroups, and universal algebra, Proceedings of the NATO Advanced Study Institute, Montreal, Quebec, Canada, July 7-18, 2003.*, V. B. e. a. Kudryavtsev (ed.), *NATO Science Series II: Mathematics, Physics and Chemistry* **207**, Kluwer Academic Publishers, 1–45 (2005).
3. J. Almeida and P. Weil, Relatively free profinite monoids: an introduction and examples, in *NATO Advanced Study Institute Semigroups, Formal Languages and Groups*, J. Fountain (ed.), **466**, Kluwer Academic Publishers, 73–117 (1995).
4. M. Arfi, Polynomial operations on rational languages, in *STACS 87 (Passau, 1987), Lecture Notes in Comput. Sci.* **247**, Springer, Berlin, 198–206 (1987).
5. M. Arfi, Opérations polynomiales et hiérarchies de concaténation, *Theoret. Comput. Sci.* **91 (1)**, 71–84 (1991).
6. J. A. Brzozowski and R. Knast, The dot-depth hierarchy of star-free languages is infinite, *J. Comput. System Sci.* **16 (1)**, 37–55 (1978).
7. A. Cano Gómez and J.-É. Pin, Shuffle on positive varieties of languages, *Theoret. Comput. Sci.* **312 (2-3)**, 433–461 (2004).
8. L. Chaubard, Actions and wreath products of $\mathcal{C}$-varieties, in *LATIN'06 (Valdivia, 2006), Lecture Notes in Comput. Sci.* **3887**, Springer, Berlin, 274–285 (2006).
9. L. Chaubard, J.-É. Pin and H. Straubing, Actions, Wreath Products of $\mathcal{C}$-varieties and Concatenation Product, *Theoret. Comput. Sci.*, (2006), to appear.
10. S. Eilenberg, *Automata, languages, and machines. Vol. B*, Academic Press [Harcourt Brace Jovanovich Publishers], New York (1976). With two chapters ("Depth decomposition theorem" and "Complexity of semigroups and morphisms") by Bret Tilson, Pure and Applied Mathematics, Vol. 59.
11. Z. Ésik and M. Ito, Temporal Logic with Cyclic Counting and the Degree of Aperiodicity of Finite Automata, *Acta Cybernetica* **16**, 1–28 (2003).
12. Z. Ésik and I. Simon, Modeling Literal Morphisms by Shuffle, *Semigroup Forum* **56**, 225–227 (1998).

13. S. C. Kleene, Representation of Events in nerve nets and finite automata, in *Automata Studies*, C. Shannon and J. McCarthy (ed.), Princeton University Press, Princeton, New Jersey, 3–42 (1956).
14. M. Kunc, Equational description of pseudovarieties of homomorphisms, *Theoretical Informatics and Applications* **37**, 243–254 (2003).
15. R. McNaughton and S. Papert, *Counter-free automata* , The M.I.T. Press, Cambridge, Mass.-London (1971). With an appendix by William Henneman, M.I.T. Research Monograph, No. 65.
16. D. Perrin and J.-E. Pin, First-order logic and star-free sets, *J. Comput. System Sci.* **32 (3)**, 393–406 (1986).
17. J.-F. Perrot, Variétés de langages et operations, *Theoret. Comput. Sci.* **7**), 197–210 (1978).
18. J.-É. Pin, Sur le monoïde de L^* lorsque L est un langage fini, *Theoret. Comput. Sci.* **7**, 211–215 (1978).
19. J.-É. Pin, *Varieties of formal languages*, North Oxford, London and Plenum, New-York, 1986. (Traduction de Variétés de langages formels).
20. J.-É. Pin, A variety theorem without complementation, *Russian Mathematics (Izvestija vuzov.Matematika)* **39**, 80–90 (1995).
21. J.-E. Pin and H. Straubing, Monoids of upper triangular matrices, in *Semigroups (Szeged, 1981), Colloq. Math. Soc. János Bolyai* **39**, North-Holland, Amsterdam, 259–272 (1985).
22. J.-E. Pin and H. Straubing, Some results on $\mathcal{C}$-varieties, *Theoret. Informatics Appl.* **39**, 239–262 (2005).
23. J.-É. Pin, H. Straubing and D. Thérien, Some results on the generalized star-height problem, *Information and Computation* **101**, 219–250 (1992).
24. J.-É. Pin and P. Weil, Polynomial closure and unambiguous product, in *22th ICALP, Lect. Notes Comp. Sci.* **944**, Springer, Berlin, 348–359 (1995).
25. J.-É. Pin and P. Weil, Profinite semigroups, Mal'cev products and identities, *J. of Algebra* **182**, 604–626 (1996).
26. J.-É. Pin and P. Weil, Polynomial closure and unambiguous product, *Theory Comput. Systems* **30**, 1–39 (1997).
27. J.-É. Pin and P. Weil, A conjecture on the concatenation product, *ITA* **35**, 597–618 (2001).
28. J. Reiterman, The Birkhoff theorem for finite algebras, *Algebra Universalis* **14 (1)**, 1–10 (1982).
29. M.-P. Schützenberger, On finite monoids having only trivial subgroups, *Information and Control* **8**, 190–194 (1965).
30. I. Simon, Piecewise testable events, in *Proc. 2nd GI Conf.*, H. Brackage (ed.), *Lecture Notes in Comp. Sci.* **33**, Springer Verlag, Berlin/Heidelberg, 214–222 (1975).
31. H. Straubing, On logical descriptions of regular languages, in *LATIN 2002, Lect. Notes Comp. Sci.* **2286**, Springer Berlin/Heidelberg, 528–538 (2002).
32. W. Thomas, Classifying regular events in symbolic logic, *J. Comput. System Sci.* **25 (3)**, 360–376 (1982).
33. P. Weil, Profinite methods in semigroup theory, *Int. J. Alg. Comput.* **12**, 137–178 (2002).

SOLVING SYSTEMS OF EQUATIONS MODULO PSEUDOVARIETIES OF ABELIAN GROUPS AND HYPERDECIDABILITY

MANUEL DELGADO*

Centro de Matemática da Universidade do Porto
Rua do Campo Alegre, 687
4169-007 Porto, Portugal
E-mail: mdelgado@fc.up.pt

ARIANE MASUDA AND BENJAMIN STEINBERG†

School of Mathematics and Statistics, Carleton University
Ottawa, Ontario, K1S 5B6, Canada
E-mail: {ariane bsteinbg}@math.carleton.ca

Based on an algorithm to solve systems of equations modulo proper pseudovarieties of abelian groups given in this paper, we prove that decidable pseudovarieties of abelian groups are (completely) hyperdecidable. However these pseudovarieties are shown not to be reducible for the canonical signature.

1. Introduction

Recall that a pseudovariety of monoids is a class of finite monoids closed under formation of finite direct products, submonoids and homomorphic images. Generalizing the notions of pointlike sets [11] and Type I and Type II semigroups [14, 15] Almeida [1] introduced the notion of hyperdecidability for pseudovarieties. One of the nice properties of a hyperdecidable pseudovariety W is that under relatively mild hypothesis on a pseudovariety V (decidable with finite vertex rank) one can decide membership in the semidirect product pseudovariety $\mathsf{V} * \mathsf{W}$. Most proofs of hyperdecidability [4, 7, 16, 18] actually establish a somewhat stronger property called

*The first author gratefully acknowledges support of FCT through the CMUP. He also acknowledges a sabbatical grant of FCT used to visit Carleton University.

†The work of the third author was done under the auspices of an NSERC discovery grant.

tameness, introduced by Almeida and the third author [5, 6].

Let us recall the definitions. Using the symbol $\uplus$ to denote disjoint union, we define a *directed graph* to be a finite set $\Gamma = V \uplus E$ endowed with two adjacency functions $\alpha, \omega : E \to V$. The set V is called the vertex set, and the set E is called the edge set. The maps α, ω select respectively the initial and terminal vertices. A *labelling* of Γ by a monoid M is a function $\ell : \Gamma \to M$. The labelling is said to be *consistent* if

$$\ell(\alpha(e)) \cdot \ell(e) = \ell(\omega(e))$$

for every $e \in E$.

Let M and N be finite monoids. A relational morphism $\tau : M \multimap N$ is a map from M into the power set of N such that: $\tau(m) \neq \emptyset$, all $m \in M$; $1 \in \tau(1)$; and $\tau(m_1)\tau(m_2) \subseteq \tau(m_1 m_2)$, all $m_1, m_2 \in M$. Suppose that $\tau : M \multimap N$ is a relational morphism of monoids and let $\ell : \Gamma \to M$, $\delta : \Gamma \to N$ be labellings of Γ. We say that δ is *τ-related* to ℓ if $\delta(z) \in \tau(\ell(z))$ for every $z \in \Gamma$. Let V be a pseudovariety of monoids. The labelling $\ell : \Gamma \to M$ is said to be *V-inevitable* if, for every relational morphism $\tau : M \multimap N \in \mathsf{V}$, there is a consistent labelling $\delta : \Gamma \to N$ that is τ-related to ℓ. A pseudovariety V of semigroups is said to be *hyperdecidable* if there exists an algorithm to test whether a finite graph labelled by a finite monoid is V-inevitable.

This notion was generalized by Rhodes and the third author [13], see also Almeida [2]. Let E be a finite system of equations over an alphabet X. That is, E consists of equations $u_i = v_i$ with $u_i, v_i \in X^*$. (We remark that in $\mathfrak{q}$-theory [13] equations over free profinite monoids are also considered.) Let M be a finite monoid. If N is another monoid and $\tau : M \multimap N$ is a relational morphism, then the substitutions $\sigma : X \to M$ and $\sigma' : X \to N$ are said to be *τ-related* if $\sigma'(x) \in \tau(\sigma(x))$, for all $x \in X$. If V is a pseudovariety, then the substitution σ is said to be *(V, E)-inevitable* if, for all relational morphisms $\tau : M \multimap N \in \mathsf{V}$, there is a substitution $\sigma' : X \to N$ that is τ-related to σ and such that $\sigma' \models E$ (meaning the induced map $\sigma' : X^* \to N$ satisfies $\sigma'(u) = \sigma'(v)$ for all $u = v \in E$).

For instance, if $\Gamma = V \uplus E$ is a graph, then a labelling ℓ over M is a substitution. We take as a system of equations the set E_Γ of all equations of the form $\alpha(e) \cdot e = \omega(e)$; the system of equations obtained in this way is called the *consistency equations* of Γ. Then ℓ is V-inevitable if and only if it is (V, E_Γ)-inevitable.

We shall call V *completely hyperdecidable* if (V, E)-inevitability is decidable for all finite systems of equations E. Completely hyperdecidable pseudovarieties are certainly hyperdecidable and hence decidable; they also have decidable pointlikes and decidable idempotent-pointlikes [13]. However, not every decidable pseudovariety is hyperdecidable [12, 8].

Both Rhodes and Steinberg [13] and Almeida [2] consider a generalized notion of tameness in this context, but we shall not deal with it here. The motivation for these more general notions of inevitability comes from dealing with other operators than the semidirect product. In Rhodes and Steinberg [13] pseudovarieties of relational morphisms are defined and shown to give rise to operators. The operator associated to a pseudovariety of relational morphisms R is denoted by Rq. Such pseudovarieties can be defined by pseudoidentities of the appropriate sort. Complete hyperdecidability of V means that if R is a decidable pseudovariety of relational morphisms with a certain finiteness condition, then RqV is decidable.

Our main result is the following:

Theorem 1.1. *Let* H *be a pseudovariety of abelian groups. Then* H *is decidable if and only if it is completely hyperdecidable. In particular, decidable pseudovarieties of abelian groups are hyperdecidable.*

This result is in some sense sharp because Auinger and the third author [8] constructed an example of a decidable pseudovariety of metabelian groups that is not hyperdecidable.

The third author [18] showed that any pseudovariety V of J-trivial monoids with decidable word problem for free pro-V monoids with finite generating sets has a hyperdecidable join with any hyperdecidable pseudovariety of groups. The same proof works without change for joins with completely hyperdecidable pseudovarieties of groups (for instance equations of the form $u_1 = \cdots = u_n = u_1^2$ were also considered there).

Corollary 1.1. *Let* $\mathsf{V} \subseteq \mathsf{J}$ *be a pseudovariety of monoids with decidable word problem for each finitely generated free pro-*V *monoid. Let* H *be any decidable pseudovariety of abelian groups. Then* $\mathsf{V} \vee \mathsf{H}$ *is completely hyperdecidable. In particular* $\mathsf{J} \vee \mathsf{H}$ *is completely hyperdecidable.*

Recall that any aperiodic pseudovariety V of commutative monoids has decidable word problem for its free pro-V monoids on finite sets and that every decidable pseudovariety of commutative monoids is a join of an aperiodic pseudovariety of commutative monoids and a decidable pseudovariety of abelian groups. (See [18] for details.) Thus we obtain the following corollary.

Corollary 1.2. *A pseudovariety of commutative monoids is completely hyperdecidable if and only if it is decidable.*

Steinberg [19] proved that the pseudovariety of p-groups, p a prime, is hyperdecidable by showing that it is weakly κ-reducible. Almeida [3] then showed that p-group pseudovarieties are tame with respect to an infinite (but recursive) implicit signature. However, it turns out that every proper, non-locally finite pseudovariety of abelian groups is not tame with respect to the canonical signature [5, 6]; in fact no such pseudovariety is weakly κ-reducible (see Almeida and Steinberg [5, 6] for undefined terminology). This is the first example, as far as we know, of pseudovarieties of groups that are known to be hyperdecidable but are provably not weakly κ-reducible. Also, to the best of our knowledge there are currently no examples of hyperdecidable pseudovarieties that are not known in addition to be tame (for some implicit signature). Let us now formulate out second main result.

Theorem 1.2. *Let* H *be a proper pseudovariety of abelian groups that is not locally finite. Then* H *is not weakly* κ*-reducible.*

2. Solving systems modulo pseudovarieties of abelian groups

In this section we prove the previously announced theorems.

We begin with a lemma concerning solutions of systems of equations in a free abelian group modulo a sequence of integers. This lemma may be of interest in its own right. It builds on the same reparameterization trick used by Almeida and Delgado [4].

Recall [10] that a subset of $\mathbb{Z}^k$ is called *linear* if it can be expressed in the form $a + b_1\mathbb{N} + \cdots + b_p\mathbb{N}$ with $a, b_1, \ldots, b_p \in \mathbb{Z}^k$. The number p is called the *size* of this expression. A *semilinear* set [10] is a finite union of linear sets. If one is interested in a finite number of semilinear sets, then by taking, if necessary, some of the b_i's equal to $0 \in \mathbb{Z}^k$ we may suppose that all expressions of linear sets involved in these finitely many semilinear sets have the same size. If R is a ring, we use $M_{r,s}(R)$ to denote the set of $r \times s$ matrices with entries in R.

Lemma 2.1. *Let* $\mathcal{F}$ *be an infinite, recursive set of natural numbers closed under taking divisors and least common multiples. Let* $N = \{m_1, m_2, \ldots\} \subseteq \mathcal{F}$ *be such that* m_k *is a divisor of* m_{k+1}*, each* k*, and such that every element of* $\mathcal{F}$ *divides some* m_k*.*

Then given as input $B \in M_{r,st}(\mathbb{Z})$, $c \in \mathbb{Z}^r$ and, for each $i \in \{1, \ldots, s\}$, a semilinear subset $L_i \subseteq \mathbb{Z}^t$ it is decidable whether, for each k, there exists $X = \begin{pmatrix} x_1 \\ \vdots \\ x_s \end{pmatrix} \in (\mathbb{Z}^t)^s$ *such that $x_i \in L_i$ and $BX \equiv c \pmod{m_k}$.*

Proof. As observed, we may assume without loss of generality that the constraints L_i have the form $L_i = \cup_{j=1}^{r_i} L_j^{(i)}$ where

$$L_j^{(i)} = a_j^{(i)} + b_{1,j}^{(i)}\mathbb{N} + \cdots + b_{p,j}^{(i)}\mathbb{N}, \text{ and } a_j^{(i)}, b_{k,j}^{(i)} \in \mathbb{Z}^t.$$

We claim that it suffices to assume that $L_i = L_j^{(i)}$ for some j. Indeed, consider all possible instances of our algorithmic problem obtained by replacing each L_i by one of the $L_j^{(i)}$. Clearly if any of these finitely many new instances has a solution modulo m_k for all k (that is the algorithm outputs "yes" for such an instance), then the original problem has a solution modulo each m_k and hence a positive output. We show the converse by a standard compactness argument. Suppose our system with the original constraints has a solution modulo m_k for all k. Then for each k there is a vector $X_k \in (\mathbb{Z}^t)^s$ with the ith component in some $L_{j_k}^{(i)}$. Since there are infinitely many k but only finitely many indices i, and for each i there are only finitely many sets of the form $L_j^{(i)}$, there must be infinitely many k such that for each i the corresponding set $L_{j_k}^{(i)}$ is the same for these k. That is we can find, for each i, a j_i such that for infinitely many k there is a solution $X = \begin{pmatrix} x_1 \\ \vdots \\ x_s \end{pmatrix}$ to $BX \equiv c \pmod{m_k}$ with $x_i \in L_{j_i}^{(i)}$, all i. Since a solution modulo m_k is also a solution modulo m_ℓ for all $\ell \leq k$, as $m_\ell \mid m_k$, we see that, for the instance of our problem where L_i is replaced by $L_{j_i}^{(i)}$, we always have a solution modulo m_k, each k. Thus an algorithm for our original problem is simply to check all of the finitely many instances that we have created. Hence we may assume that each $L_i = L_j^{(i)}$ for some j, as claimed.

Say $L_i = a^{(i)} + b_1^{(i)}\mathbb{N} + \cdots + b_p^{(i)}\mathbb{N}$. We write $L_i = a^{(i)} + B_i\mathbb{N}^p$ where $B_i = (b_1^{(i)} | b_2^{(i)} | \cdots | b_p^{(i)}) \in M_{t,p}(\mathbb{Z})$. With this reparameterization, our system with constraints is equivalent to solving modulo each m_k the system

$$B \begin{pmatrix} a^{(1)} + B_1 Y_1 \\ \vdots \\ a^{(s)} + B_s Y_s \end{pmatrix} = c,$$

where each $Y_i \in \mathbb{N}^p$. But this is equivalent to solving modulo each m_k a system of the form $DY = e$ where $D \in M_{r,ps}(\mathbb{Z})$, $e \in \mathbb{Z}^r$ and $Y \in \mathbb{N}^{ps}$.

Equivalently we want to determine whether modulo each m_k it is possible to express e as a non-negative linear combination of the columns of D; that is whether modulo each m_k, e is in the submonoid of $\mathbb{Z}^r$ generated by the columns of D. As every submonoid of a finite group is a subgroup, our problem is equivalent to determining whether modulo each m_k, e is in the subgroup generated by the columns of D. But [17] if H is the pseudovariety of abelian groups with exponent in $\mathcal{F}$, then by the definition of N, this question is equivalent to asking whether e belongs to the pro-H closure in $\mathbb{Z}^r$ of the subgroup generated by the columns of D. Moreover, since $\mathcal{F}$ is recursive, the pseudovariety H is decidable [17]. Now, the results of the third author [17] show that it is decidable whether e belongs to the pro-H closure of the subgroup generated by the columns of D. □

Let us remark that the solvability of the system considered above depends only on $\mathcal{F}$ and not on the choice of N. We now proceed to reduce the complete hyperdecidability of a pseudovariety H of abelian groups to the previous lemma. Let M be a (perhaps infinite) monoid. *A system of equations* over M with variables in X is a set E of formal equalities $u = v$ between elements of $M \star X^*$ where $\star$ denotes the free product. The system is said to be *solvable modulo a pseudovariety* V with constraints $L_x \subseteq M$, $x \in X$, if: for each homomorphism $\psi : M \to N \in \mathsf{V}$, we can choose a substitution $\sigma : X \to M$ such that $\sigma(x) \in L_x$, all x, and the map $\overline{\psi} : M \star X^* \to N$ induced from ψ and $\psi \circ \sigma$ satisfies $\overline{\psi}(u) = \overline{\psi}(v)$ for all $u = v \in E$.

Let A be a finite alphabet. Recall that the set of rational subsets of A^* is the smallest collection of subsets of A^* containing the finite subsets that is closed under union, product and generation of submonoids. Solving systems of equations over A^* with rational constraints is very closely related to complete hyperdecidability. This motivates the next theorem.

Theorem 2.1. *Let* H *be a decidable pseudovariety of abelian groups. Then there is an algorithm which, given a finite alphabet* A*, a finite system of equations* E *over* A^* *with variables in* X *and given, for each variable* $x \in X$*, a rational subset* $L_x \subseteq A^*$*, determines whether* E *is solvable modulo* H *with constraints* L_x*,* $x \in X$.

Proof. If H is locally finite (that is, contains only finitely many A-generated groups for each finite set A), then the problem is trivial, so we suppose that this is not the case. Suppose $A = \{a_1, \ldots, a_t\}$ and

$X = \{x_1, \ldots, x_s\}$. Let $\eta : A^* \to \mathbb{N}^t$ be given by $\eta(w) = (|w|_{a_1}, \ldots, |w|_{a_t})$, where $|w|_{a_i}$ denotes the number of occurrences of the letter a_i in w. Denote by $\eta(E)$ the system $\{\eta(u) = \eta(v) \colon u = v \in E\}$. Then our problem is clearly equivalent to trying to solve $\eta(E)$ modulo H with the constraints $\eta(L_x)$, $x \in X$. But since each homomorphism of $\mathbb{N}^t$ to an abelian group extends uniquely to $\mathbb{Z}^t$, our problem is equivalent to trying to solve $\eta(E)$ in $\mathbb{Z}^t$ modulo H with the constraints $\eta(L_x)$, $x \in X$ (viewed as subsets of $\mathbb{Z}^t$). We can hence use subtraction to rewrite our system $\eta(E)$ as a linear system of the form $BX = c$ with $B \in M_{r,st}(\mathbb{Z})$ subject to the constraints $\eta(L_x)$, $x \in X$. A theorem of Eilenberg and Schützenberger [10] says that $\eta(L_x)$ is a semilinear set that can be algorithmically determined from L_x. A more efficient algorithm can be found in the work of the first author [9].

The next step is to obtain a set of positive integers as in Lemma 2.1. Let $\mathcal{F}_{\mathsf{H}} = \{n \mid \mathbb{Z}_n \in \mathsf{H}\}$; this is a recursive set closed under taking divisors and least common multiples [17]. Let us enumerate $\mathcal{F}_{\mathsf{H}}$ as a sequence in increasing order: $n_1 < n_2 < \ldots$. Since we are assuming that H is not locally finite, this sequence is infinite. We define another sequence of positive integers $(m_i)_{i \in \mathbb{N}}$ by setting $m_1 = n_1$ and $m_i = \mathrm{lcm}(m_{i-1}, n_i)$ for all $i > 1$. By construction $m_i \mid m_{i+1}$. Since $\mathcal{F}_{\mathsf{H}}$ is closed under taking least common multiples, $\{m_k\}_{k \in \mathbb{N}} \subseteq \mathcal{F}_{\mathsf{H}}$. By construction each element of $\mathcal{F}_{\mathsf{H}}$ divides some m_k.

Suppose that $\psi : \mathbb{Z}^t \to G \in \mathsf{H}$. Then if n denotes the exponent of G, we must have $n \in \mathcal{F}_{\mathsf{H}}$. Therefore $n \mid m_k$ for all sufficiently large m_k. Hence our system $\eta(E)$ has a solution modulo H if and only if $BX = c$ can be solved modulo each m_k subject to our constraints. But this can be decided according to Lemma 2.1. □

2.1. *Proof of Theorem 1.1*

Let H be a decidable pseudovariety of abelian groups. Let M be a finite monoid, E a finite system of equations in variables X. Let $\sigma : X \to M$ be a substitution. We must decide whether σ is (H, E)-inevitable. Choose a generating set A for M and consider the canonical projection $\varphi : A^* \to M$. A standard argument [13] shows that σ is (H, E)-inevitable if and only if, for all $\psi : A^* \to G \in \mathsf{H}$, there exists a substitution $\sigma' : X \to G$ such that $\sigma' \models E$ and σ' is $\psi\varphi^{-1}$-related to σ. Let $L_x = \varphi^{-1}(\sigma(x))$, for $x \in X$; by Kleene's theorem each L_x is a rational subset of A^*.

A substitution $\sigma' : X \to G$ is τ-related to σ if and only if we can choose for each $x \in X$, an element $w_x \in L_x$ such that $\psi(w_x) = \sigma'(x)$. With this in mind it is straightforward to see that σ is (H, E)-inevitable if and only if E is solvable modulo H with the constraints L_x, $x \in X$. But this can be algorithmically determined by Theorem 2.1. This establishes Theorem 1.1.

2.2. *Proof of Theorem 1.2*

In this proof we assume familiarity with the work of the third author [17] since our example is essentially from there. We will take the following definition of weak κ-reducibility for our non-locally finite pseudovariety H of abelian groups; its equivalence with the usual definition [5, 6] can be proved via a standard argument, which we leave to the reader. If A is a finite alphabet, we use $\eta_A : A^* \longrightarrow \mathbb{N}^{|A|}$ for the canonical projection.

Definition 2.1. A pseudovariety of groups H is *weakly κ-reducible* if given a finite alphabet A, a finite graph $\Gamma = V \uplus E$ and a rational subset $L_x \subseteq A^*$ for each $x \in \Gamma$, the following holds: the consistency equations E_Γ for Γ are solvable modulo H subject to the constraints L_x, $x \in \Gamma$, if and only if there exists for each $x \in X$ an element $w_x \in \overline{\eta_A(L_x)}$ such that $\{w_x\}_{x\in\Gamma}$ is a solution to E_Γ in $\mathbb{Z}^{|A|}$. Here, for $Y \in \mathbb{Z}^{|A|}$, $\overline{Y}$ denotes the closure of Y in the pro-H topology on $\mathbb{Z}^{|A|}$, which is the weakest topology making all homomorphisms from $\mathbb{Z}^{|A|}$ to groups in H continuous.

Let p be a prime such that $\mathbb{Z}_p \notin \mathsf{H}$. Let $A = \{a, b\}$. Set $\eta = \eta_{\{a,b\}}$. Consider the graph Γ given by

$$v_0 \xrightarrow{e} v_1 \xrightarrow{f} v_2$$

and consider the labelling

$$1 \xrightarrow{a^*} a^* \xrightarrow{(ab^p)^*} b.$$

The consistency equations are: $v_0 e = v_1$, $v_1 f = v_2$. After abelianization (that is taking the commutative image) the equations can be written as $v_0 + e = v_1$ and $v_1 + f = v_2$ and so they have the consequence $v_0 + e + f = v_2$. The algorithm [17] shows that $\overline{\eta(a^*)} = (1,0)\mathbb{Z}$ and $\overline{\eta((ab^p)^*)} = (1,p)\mathbb{Z}$. Since $\eta(b) = (0,1)$ is not in $\langle(1,0),(1,p)\rangle$ it follows that $v_0 + e + f = v_2$ cannot be solved in $\mathbb{Z}^{|A|}$ subject to the constrains $\overline{L_x}$, $x \in \Gamma$.

But if we follow the procedure of the proof of Lemma 2.1 we see that we just need that $(0,1)$ is in the pro-H closure of $\langle(1,0),(1,p)\rangle$ for the system to be solvable modulo H with our constraints. But the algorithm [17] shows that the closure of this subgroup is all of $\mathbb{Z}^2$. So the system is solvable modulo H. Thus H is not weakly κ-reducible.

Bibliography

1. J. Almeida, Hyperdecidable pseudovarieties and the calculation of semidirect products, *Internat. J. Algebra Comput.* **9**, 241–261 (1999).
2. J. Almeida, Finite semigroups: an introduction to a unified theory of pseudovarieties, *Semigroups, algorithms, automata and languages (Coimbra, 2001)*, World Sci. Publishing, River Edge, NJ, 3–64 (2002).
3. J. Almeida, Dynamics of implicit operations and tameness of pseudovarieties of groups, *Trans. Amer. Math. Soc.* **354**, 387–411 (2002).
4. J. Almeida and M. Delgado, Tameness of the pseudovariety of Abelian groups, *Internat. J. Algebra Comput.* **15**, 327–338 (2005).
5. J. Almeida and B. Steinberg, Iterated semidirect products with applications to complexity, *Proc. London Math. Soc.* **80**, 50–74 (2000).
6. J. Almeida and B. Steinberg, Syntactic and global semigroup theory, a synthesis approach, *Algorithmic Problems in Groups and Semigroups*, J.-C. Birget, S. Margolis, J. Meakin, M. Sapir eds., Birkhäuser, Basel, 1–23 (2000).
7. C. J. Ash, Inevitable graphs: A proof of the type II conjecture and some related decision procedures, *Internat. J. Algebra Comput.* **1**, 127–146 (1991).
8. K. Auinger and B. Steinberg, On the extension problem for partial permutations, *Proc. Amer. Math. Soc.* **131**, 2693–2703 (2003).
9. M. Delgado, Commutative images of rational languages and the abelian kernel of a monoid, *Theor. Inform. Appl.* **35**, 419–435 (2001).
10. S. Eilenberg and M. P. Schützenberger, Rational sets in commutative monoids, *J. Algebra* **13**, 173–191 (1969).
11. K. Henckell, Pointlike sets: The finest aperiodic cover of a finite semigroup, *J. Pure Appl. Algebra* **55**, 85–126 (1988).
12. J. Rhodes and B. Steinberg, Pointlike sets, hyperdecidability, and the identity problem for finite semigroups, *Internat. J. Algebra and Comput.* **9**, 475–481 (1999).
13. J. Rhodes and B. Steinberg, "The q-theory of finite semigroups", Springer-Verlag, to appear.
14. J. Rhodes and B. Tilson, Lower bounds for complexity of finite semigroups, *J. Pure Appl. Algebra* **1**, 79–95 (1971).
15. J. Rhodes and B. Tilson, Improved lower bounds for the complexity of finite semigroups, *J. Pure Appl. Algebra* **2**, 13–71 (1972).
16. B. Steinberg, On pointlike sets and joins of pseudovarieties, *Internat. J. Algebra Comput.* **8**, 203–231 (1998).
17. B. Steinberg, Monoid kernels and profinite topologies on the free Abelian group, *Bull. Austral. Math. Soc.* **60**, 391–402 (1999).
18. B. Steinberg, On algorithmic problems for joins of pseudovarieties, *Semigroup Forum* **62**, 1–40 (2001).
19. B. Steinberg, Inevitable graphs and profinite topologies: some solutions to algorithmic problems in monoid and automata theory, stemming from group theory, *Internat. J. Algebra Comput.* **11**, 25–71 (2001).

FINITE RESIDUE CLASS RINGS OF INTEGERS MODULO *N* FROM THE VIEWPOINT OF GLOBAL SEMIGROUP THEORY

ATTILA EGRI-NAGY* AND CHRYSTOPHER L. NEHANIV

School of Computer Science
University of Hertfordshire
College Lane, Hatfield, Herts AL10 9AB, U.K.
E-mail: {A.Nagy | C.L.Nehaniv}@herts.ac.uk

Krohn-Rhodes theory gives us a way of understanding structures described as semigroups or automata by hierarchical coordinate systems. Here we use our computational implementation of the holonomy method for the hierarchical algebraic decomposition of finite automata to study finite residue class rings R_n of integers modulo n. This computational exploration leads to allowing us to rigorously describe the detailed structure of the holonomy decomposition of R_n for all $n > 0$.

1. Introduction

The Krohn-Rhodes Theory [9] states that every automaton can be emulated by a hierarchical, cascaded product of simpler automata. The theory has a similar role in automata theory as the prime decomposition of integers in number theory. Despite the apparent importance of the theory, there had been no computational implementation before our creation of a computational tool to generate Krohn-Rhodes decompositions [5, 4]. With this tool we can decompose and coordinatize any structures that are amenable to a finite state automaton description. Here we choose to study residue class rings of the integers $\mathbb{Z}$, for three reasons. First, we would like to demonstrate how the computational tool helps in getting theoretical results by analysing actual decompositions. Second, since we decompose well-understood structures in a new way, we can easily get an intuition about the inner workings of the decomposition algorithm. The third reason for studying these rings

*This work was supported by a Hungarian National Foundation for Scientific Research grant (OTKA T049409) and the University of Hertfordshire Algorithms Research Group.

as automata is that they are especially interesting from the computer algebraic point of view, since the decompositions give us coordinatizations which are closely related to, but also unlike, our usual base 10 number system, and might be well-suited for applications in computer arithmetic. We shall regard the ring $R_n \cong \mathbb{Z}/n\mathbb{Z}$ of integers modulo n as a semigroup by considering all operations on $\{0,\ldots,n-1\}$ generated by the unary operations $+\beta$ and $\times\alpha$, where $\alpha, \beta \in R_n$.

The paper has the following structure. Section 2 introduces the mathematical notions needed for the Krohn-Rhodes Theory. Section 3 states the Krohn-Rhodes Prime Decomposition Theorem and describes one proof technique, namely the holonomy decomposition, in detail. Section 4 contains the theoretical results (with proofs) derived from insights resulting from study of the output of the computational tool for many examples. Section 5 briefly summarizes the computational experiments. Section 6 concludes the paper and mentions possible future directions.

2. Mathematical Preliminaries and Notations

For basic definitions of semigroup and group theory we refer the reader to standard texts [2, 8]. A semigroup S is *aperiodic* if for each element $s \in S$ there is a positive natural number n such that $s^n = s^{n+1}$; for finite S this means that S contains no nontrivial subgroups.

First we define one particular type of semigroups, namely the transformation semigroups, then we recall the close relationship of these semigroups to finite state automata. Next we describe how to build bigger automata from smaller pieces hierarchically, namely, via cascaded composition, or, generically, via the wreath product. Finally we define the notion of emulation (or, algebraically, of division). These are the most important notions for Krohn-Rhodes Theory.

Transformation Semigroups. For a nonvoid finite set A, a mapping $\varphi : A \to A$ is called a *transformation* of A. A is called the set of *states.* If the mapping is bijective, then it is a *permutation*, otherwise we say it *collapses* states (i.e. more than one state maps to some state). The *image* of φ is defined as $\{a\varphi : a \in A\}$ and denoted by $\mathrm{im}(\varphi)$. If the image of a mapping is a singleton then the mapping is called *constant.* The *rank* of a transformation is the cardinality of its image. The set T of all transformations of A forms a semigroup under the operation of function composition of transformations and is called the *full transformation semigroup* denoted by $\mathcal{T}_A = (A, T)$. If S is a subsemigroup of T then (A, S) is called a *trans-*

formation semigroup on A (or briefly a *ts*), and we say that S *acts* on A. (A, S) is a *permutation group* if each element $s \in S$ acts on A by a permutation. We write $a \cdot s$ for the image of state a under the transformation s, and we have $(a \cdot s_1) \cdot s_2 = a \cdot (s_1 s_2)$ for all $a \in A, s_1, s_2 \in S$. We use the notation $\mathbf{n}$ for the set of n points $\{0, \ldots, n-1\}$. If (A, S) is a transformation semigroup, we denote by $(A, \overline{S})$ the transformation semigroup with transformations $\overline{S} = \{t \mid t \in S \text{ or } t \text{ is constant}\}$. If $B \subseteq A$, then $B \cdot s$ denotes the set $\{b \cdot s \mid b \in B\}$.

Finite State Automata. By a finite state *automaton*, we mean a triple $\mathcal{A} = (A, X, \delta)$ where A is the (finite nonempty) *state set*, X is the *input alphabet* and $\delta : A \times X \to A$ is the *transition function*. We do not explicitly consider the output of the automaton as it can be recovered from the state and the input symbol. We tacitly use the state as the output.

We can naturally extend the transition function to words i.e. finite sequences of input symbols: for the empty word $\delta(a, \lambda) = a$, and for arbitrary words $u, v \in X^*$, $\delta(a, uv) = \delta(\delta(a, u), v)$. There is a natural equivalence relation on words defined by $u \equiv v$ if $\delta(a, u) = \delta(a, v)$ $\forall a \in A$, i.e. identifying words with the same action on A. It is called the *congruence induced by* $\mathcal{A}$. The *characteristic semigroup* $S(\mathcal{A})$, also called the *semigroup of the automaton*, is the set equivalence classes $X^+/\equiv$ of this congruence, with associative operation induced by concatenation. With the characteristic semigroup we can treat an automaton $\mathcal{A}$ as a transformation semigroup $(A, S(\mathcal{A}))$. Conversely if S is a semigroup then the corresponding automaton is $\mathcal{A}_S = (S^1, S, \delta_S)$, where the transition function δ_S is the natural action of S on S^1 by right multiplication. (Here S^1 denotes the smallest monoid (up to isomorphism) containing S — this either coincides with S or consists of $S \cup \{1\}$ where 1 is a new identity element adjoined to S.)

Using automata terminology, constant mappings in transformation semigroups are often called *resets*. A *permutation-reset* automaton is an automaton such that each of its inputs acts either as a permutation or a constant map on states.

The *state transition graph* $D(\mathcal{A})$ of an automaton $\mathcal{A} = (A, X, \delta)$ is the digraph with A as the set of vertices and (a, x, b) is a labelled edge if $a \cdot x = b$, where $a, b \in A$, $x \in X$.

We have two important types of automata here, one type is *identity-reset automata* whose inputs act either as the identity map on states or as constant maps ("resets"). This type is typified by the flip-flop automaton $\mathcal{F}$, which can be thought of as a device capable of storing one bit: we have two states $A = \{0, 1\}$ and three symbols in the input alphabet $X =$

{set0,set1,read}. The read is the identity operation, but since we consider the state as the output of the automaton we can think of it as retrieving whatever state was set before.

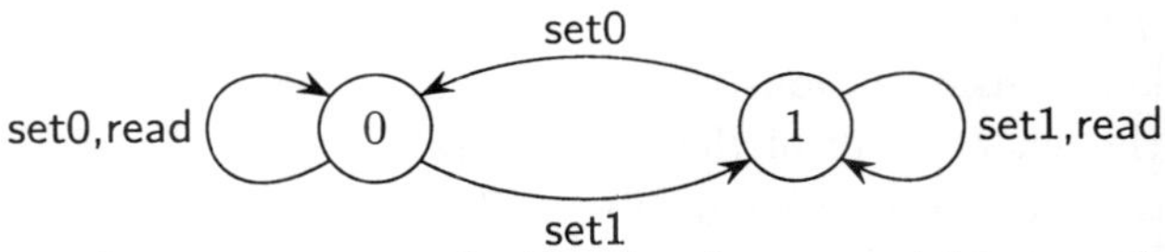

In semigroup-theoretic terms, the flip-flop has two right zeros (resets) and one identity. The other very important examples are *permutation automata*, whose inputs permute the automaton's states.

Wreath Product. In order to understand the Krohn-Rhodes decomposition, it is necessary to understand the wreath product. Although the concept of the wreath product is not so complicated, it is often difficult for newcomers to immediately grasp the intuitive idea of how this generic, loop-free cascaded product works. Together with the formal definition, Figure 1 may shed light on how state transitions happen in the product. It may also be helpful first to consider a simpler product with no dependence between the components.

Let $(A_n, S_n), \dots, (A_1, S_1)$ be transformation semigroups called *components*. The indices $1, \dots, n$ are called *coordinates*. The *direct product* $(A_n, S_n) \times \cdots \times (A_1, S_1)$ is the ts $(A_n \times \cdots \times A_1, S_n \times \cdots \times S_1)$ with the componentwise action $(a_n, \dots, a_1) \cdot (s_n, \dots, s_1) = (a_n \cdot s_n, \dots, a_1 \cdot s_1)$. Denote by $p_i : A \to A_i$ the ith projection map. Thus for each $i = 1, \dots, n$,

$$p_i((a_n, \dots, a_1) \cdot (s_n, \dots, s_1)) = a_i \cdot s_i.$$

Direct product is also called *parallel composition* as the components' state transitions do not depend on each other, and the order of the components does not really matter up to isomorphism.

Now we introduce an order-dependent connection between the components. Let $A = A_n \times \ldots \times A_1$ and $\mathcal{T}_A$ the full ts on A. Let S be the subsemigroup of $\mathcal{T}_A$ consisting of all transformations $s : A \to A$ satisfying the condition of *hierarchical dependence* of coordinates: For each $i = 1, \dots, n$, there exists $f_i : A_{i-1} \times \cdots \times A_1 \to S_i$ such that

$$p_i\big((a_n, \dots, a_1) \cdot s\big) = a_i \cdot f_i(a_{i-1}, \dots, a_1).$$

That is, the new ith coordinate resulting from the action of s depends only on the values of the old first i coordinates and on the transformation s. Moreover, it is given by acting on a_i with an element $f_i(a_{i-1}, \dots, a_1)$ of S_i which depends only on s and $(a_{i-1}, \dots, a_1)$. Any $s : A \to A$ of this form is called a *cascaded transformation.* We can write this transformation as the ordered list of these functions: $s = (f_n, \ldots, f_1)$, where f_i gives the component action in the ith position. We call these functions f_i the *dependency functions* of s.

Cascaded transformations are easily seen to be closed under composition, and the set S of all of them comprises a semigroup S acting on A. Then the transformation semigroup (A, S) denoted $(A_n, S_n) \wr \cdots \wr (A_1, S_1)$ is the *wreath product* of transformation semigroups $(A_n, S_n), \ldots, (A_1, S_1)$. Reading from left to right the last component is the top level of the hierarchy.

By a *cascaded state* we mean a tuple of component states as above, and by a *cascaded action* we mean an actual tuple $(s_n, \ldots, s_1)$ of component actions (this is not to be confused with the cascaded transformation, which is a tuple of dependency functions).

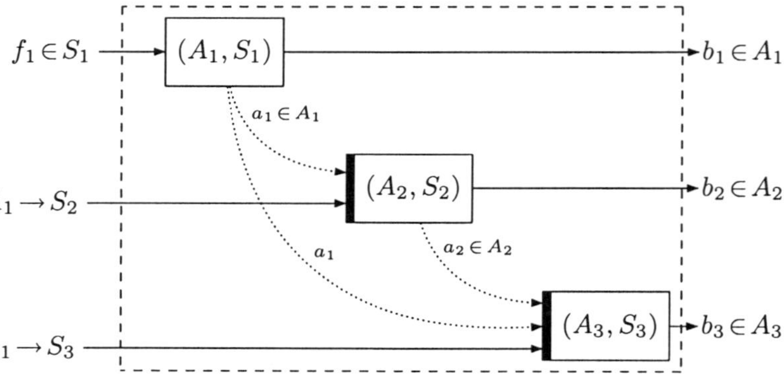

Figure 1. State transition in the wreath product $(A_3, S_3) \wr (A_2, S_2) \wr (A_1, S_1)$. The cascaded transformation (f_3, f_2, f_1) is applied to state (a_3, a_2, a_1) yielding $(b_3, b_2, b_1) = (a_3 \cdot f_3(a_2, a_1), a_2 \cdot f_2(a_1), a_1 \cdot f_1)$. The black bars denote the applications of functions f_2, f_3 according to hierarchical dependence. Note that the applications of these functions happen exactly at the same moment since their arguments are the previous states of other components, therefore there is no need to wait for the other components to calculate the new states. We use the state as the output of the automaton.

Division. We say that a transformation semigroup (A, S) *divides* (B, T), denoted by $(A, S) \mid (B, T)$, if we can choose for all $a \in A$ at least one $\tilde{a} \in B$ as a *lift* and for each $s \in S$ at least one $\tilde{s} \in T$ as a *lift*, such that the following hold:

(1) Each member of B (resp. T) is a lift of at most one element of A

(resp. S), i.e. the (non-empty) lift sets are non-intersecting,

(2) If $\tilde{a}$ is any lift of a and $\tilde{s}$ is any lift of s, then $\tilde{a} \cdot \tilde{s}$ is some lift of $a \cdot s$, i.e. the products are respected.

Denote the set of lifts of a state a by $\Lambda(a)$ and the lifts of a transformation s by $\Lambda(s)$. By (1), $\Lambda(a) \neq \emptyset$, and $\Lambda(a) \cap \Lambda(a') \neq \emptyset$ implies $a = a'$ for all $a, a' \in A$, and similarly for elements of S. By (2), $\Lambda(a) \cdot \Lambda(s) \subseteq \Lambda(a \cdot s)$, but equality need not hold.

An automaton $\mathcal{B}$ with states B *emulates* another one $\mathcal{A}$ with states A if every computation which can be done in $\mathcal{A}$ can be done in $\mathcal{B}$ as well, i.e. $(A, S(\mathcal{A}))$ divides $(B, S(\mathcal{B}))$. For semigroups S and T, S *divides* T if S is a homomorphic image of a subsemigroup of T.

3. Holonomy Decomposition Theorem

First we need to state the Krohn-Rhodes Prime Decomposition Theorem, as the holonomy decomposition is a special type of Krohn-Rhodes decomposition:

Theorem 3.1. *Given a finite automaton $\mathcal{A}$, then $\mathcal{A}$ can be emulated by a cascade product of components from $\{\mathcal{F}, \mathcal{A}_{G_1}, \ldots, \mathcal{A}_{G_n}\}$, where $\mathcal{F}$ is the flip-flop and G_i, $1 \leq i \leq n$, are groups. Moreover, the groups G_i can be chosen to be simple groups dividing the characteristic semigroup $S(\mathcal{A})$. Conversely, every simple group G dividing $S(\mathcal{A})$ divides some G_i of the decomposition.*

Informally, every finite automaton can be emulated by a cascaded product of simpler irreducible components.

The holonomy decomposition originates from improving[a] Zeiger's method of proving the Krohn-Rhodes Theorem [11, 7, 6, 3]. This algorithm works by the detailed study of how the semigroup S of an automaton (A, X, δ) acts on certain subsets of A. It looks for groups induced by S^1 permuting certain sets of these subsets of A. These groups are called the *holonomy groups.* These groups are the building blocks for the components of the decomposition. As we go deeper in the hierarchy of the cascade composition we have components that act on a set of subsets each having the same or smaller cardinality.

[a]The improvement is that the components in the decomposition occur in parallel when possible.

Sketch of the algorithm for holonomy decomposition: First we calculate the set of images of transformations in S. From now on, let $\mathcal{I}$ denote this set extended by A itself and its singletons. On $\mathcal{I}$ there is a preorder relation called *subduction* defined. A subset P is subduction related to a subset Q if P is contained in the set resulting from acting by some $s \in S^1$ on Q, i.e. $P \subseteq Q \cdot s$. The mutual relation of elements induces an associated equivalence relation $P \equiv Q \iff P \leq Q$ and $Q \leq P$. The set of equivalence classes are partially ordered by the subduction relation. The set of equivalence classes together with their partial order is called the *subduction picture.* For each $P \in \mathcal{I}$, we denote by $\overline{P}$ a fixed representative of its equivalence class. The *tiles* $\mathcal{B}_P$ of a subset P ($P \in \mathcal{I}, |P| > 1$) are its maximal proper subsets in $\mathcal{I}$. The tile-of relation is denoted by $\prec$. The union of its tiles equals P. The length of a longest strict path from a singleton to a subset P in the partial order of subduction equivalence classes defines the *height* of the subsets within the equivalence class of P. Consequently singletons have height 0. Equivalence classes with the same height are on the same hierarchical level. The height $h(\mathcal{A})$ of an automaton $\mathcal{A}$ is the height $h(A)$ of its state set A, and this will give the number of hierarchical levels in the holonomy decomposition. The inclusion relation of the sets of tiles for each element $Q \in \mathcal{I}$ forms the *tiling picture.* The holonomy group $\mathcal{H}_Q$ of Q is the group (arising from the action of the elements of S^1 on Q) permuting the tile set $\mathcal{B}_Q$ of Q. We have:

Fact 3.1. *If $\sigma \in \mathcal{H}_Q$ is a holonomy group element, then there is an $s \in S^1$ permuting Q that induces σ. Conversely, every $s \in S^1$ that permutes Q induces a permutation σ of the tiles of Q, i.e. $\sigma \in \mathcal{H}_Q$.*

Proof: By definition $\sigma : \mathcal{B}_Q \to \mathcal{B}_Q$ is a permutation of the tiles of Q and is given by $B \cdot \sigma = B \cdot s$ (for all $B \in \mathcal{B}_Q$), by some $s \in S^1$. Since σ is a permutation of a finite set, σ^n is the identity for some $n > 0$. Thus,

$$Q = \bigcup_{B \in \mathcal{B}_Q} B = \bigcup_{B \in \mathcal{B}_Q} (B \cdot \sigma^n) = \bigcup_{B \in \mathcal{B}_Q} (B \cdot s^n).$$

It follows s is surjective onto Q, and so permutes the (finite) set Q.

Conversely, if $s \in S^1$ permutes Q and has order n, then for any tile $B \prec Q$ if $T \prec Q$ is a tile containing $B \cdot s$, we have $B = (B \cdot s) \cdot s^{n-1} \subseteq T \cdot s^{n-1} \subseteq Q$. Since B is a tile and $T \neq Q$, the first inclusion is equality, and so applying s yields that $B \cdot s = T$, a tile. This proves s induces a map $\sigma : \mathcal{B}_Q \to \mathcal{B}_Q$, taking tiles to tiles, which is evidently a permutation since σ^n is the identity. $\square$

Due to this close relationship between elements s of S^1 permuting Q and holonomy group elements, we shall also write s for the element $\sigma \in \mathcal{H}_Q$ permuting $\mathcal{B}_Q$ which s induces if no confusion can result.

Finally the holonomy decomposition component $(\mathcal{B}_i, \overline{\mathcal{H}_i})$ of one hierarchical level i is the permutation-reset ts that is defined to be the direct product $(\mathcal{B}_i, \mathcal{H}_i) = \prod_{h(\overline{Q})=i}(\mathcal{B}_{\overline{Q}}, \mathcal{H}_{\overline{Q}})$ of the holonomy permutation groups $(\mathcal{B}_{\overline{Q}}, \mathcal{H}_{\overline{Q}})$ belonging to the representative elements $\overline{Q}$ of equivalence classes with height i, augmented with the constant mappings.

Theorem 3.2. *Let (A, S) be a finite transformation semigroup. Then (A, S) divides a wreath product of its holonomy permutation-reset transformation semigroups $(\mathcal{B}_1, \overline{\mathcal{H}_1}) \wr \cdots \wr (\mathcal{B}_h, \overline{\mathcal{H}_h})$.*

Note that the top level of the hierarchy is the component with highest index h, not with index 1. This strong formulation of part of the Krohn-Rhodes theorem is slightly different from the original since the components here are permutation groups extended with constants rather than the flip-flop and automata corresponding to simple groups. But these permutation-reset components can be easily decomposed into flip-flops and permutation group automata [9]. Moreover the permutation groups can be further decomposed using a cascaded series of simple group automata via the Lagrange Coordinate Decomposition Theorem and Jordan-Hölder Theorem [8, 9, 3]. See references [9, 6, 3] for more details on concepts of Krohn-Rhodes Theory.

4. Decomposition of the Rings of Integers Modulo n

For testing and evaluating the performance of our computational tool for holonomy decomposition [5, 4], we used several types of transformation semigroups (ranging from the straightforward example of the full ts to randomly generated ones). Studying the actual decompositions of the examples helps in understanding the inner workings of the holonomy method and can give insight into the automata being decomposed. Therefore it makes sense to decompose otherwise well-known examples. In this section we study the decompositions of rings of integers modulo n. The theoretical results presented here come from the analysis of the output of the tool (summarized in section 5).

4.1. *Representation*

Integer modulo n residue class rings R_n can be represented by automata. The state set is $\mathbf{n} = \{0, \ldots, n-1\}$, identified with the residue classes modulo n. The transformations correspond to the operations of the ring represented as one-argument functions such as $+1$, $\times 2$, $\times 3$ and so on. As generators, clearly we need only at most $+1$ and the multiplications by the prime elements smaller than n. The characteristic semigroup of this automaton is denoted by $S(R_n)$.

For example, the following automaton represents R_4, the residue class ring of integers modulo 4:

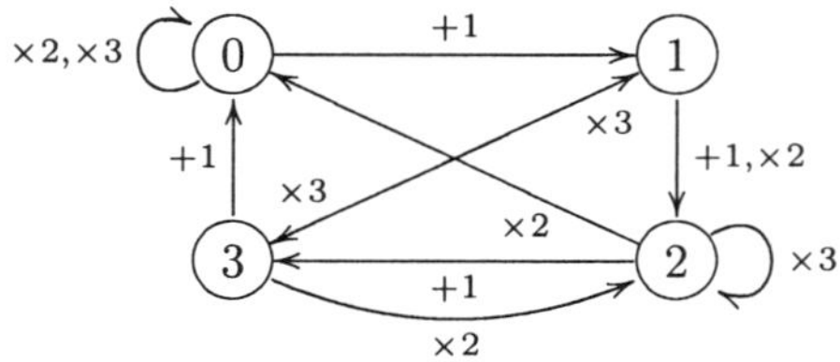

Proposition 4.1. $|S(R_n)| = n^2$.

Proof: $S(R_n)$ is a noncommutative semigroup for $n > 2$, and is a semidirect product of the additive group C_n acted on by the multiplicative monoid of R_n, since the elements are given by the affine linear transformations of the form $i \mapsto i \times \alpha + \beta$ which are closed under composition. It is easy see that all distinct pairs $\alpha, \beta \in \mathbf{n}$ give distinct transformations of $\mathbf{n}$, therefore we have n^2 transformations. □

Since it is easy to move between the ring and semigroup notation, for a general element of $S(R_n)$ we will use the notation $\alpha x + \beta$ instead of $x \mapsto (x \cdot \times\alpha) \cdot +\beta$, where x varies over $\mathbf{n}$ and α, β are elements of the ring R_n.

4.2. *The Extended Set of Images*

We get the images $\mathcal{I}$ by applying all transformations to the state set, therefore for $S(R_n)$ the images have the form $\alpha\mathbf{n} + \beta$. In this special case the image set equals $\mathcal{I}$, since $1\mathbf{n} = \mathbf{n}$, and $0\mathbf{n} + \beta$, $\beta \in \mathbf{n}$ gives the singletons.

Obviously, the same image can be the result of several transformations. The additive factor in an affine transformation always induces a permutation on $\mathcal{I}$ and the multiplicative factor will also induce a permutation if and only if (α, n) are relative primes. If α shares a divisor with n then collapsing of states occurs. Therefore the set $\mathcal{I}$ of proper images of $S(R_n)$ consists of the subsets $\alpha\mathbf{n} + \beta$ of $\mathbf{n}$, where $\gcd(\alpha, n) \neq 1$ and β is arbitrary (where all operations are done modulo n). We will use the fact that $\langle\alpha\rangle = \alpha\mathbf{n}$, where $\langle\alpha\rangle$ is the principal ideal of the ring generated by α. To make these intuitive statements more exact, we record the following facts:

Fact 4.1. *For all integers* α, β, k, $\alpha\boldsymbol{n} \equiv \alpha\boldsymbol{n}+\beta$, *and* $\alpha\boldsymbol{n}+\beta = \alpha\boldsymbol{n}+(\beta+k\alpha)$.

Fact 4.2. *Suppose* $\alpha \mid n$, $\alpha \in \boldsymbol{n}$. *Then multiplication by a prime* p *collapses some elements of* $\alpha\boldsymbol{n}$ *if* $p\alpha \mid n$, *but permutes the elements of* $\alpha\boldsymbol{n}$ *if* $p\alpha \nmid n$.

Proof: If $p\alpha \mid n$, then $|\alpha\mathbf{n}| = \frac{n}{\alpha}$ and $|p\alpha\mathbf{n}| = \frac{n}{p\alpha}$, thus $\langle p\alpha \rangle \subsetneq \langle \alpha \rangle$. Due to the strict inclusion we have collapsing under multiplication by p.

$p\alpha \nmid n$ case: Let $k\alpha, l\alpha \in \langle \alpha \rangle$. Since $\alpha \mid n$, we may assume that $0 \le k\alpha, l\alpha < n$. Now suppose $pk\alpha \equiv pl\alpha \mod n$. Then $p\alpha(k-l) = qn$ for some q, and $p \mid q$ since $\alpha \mid n$ and $p\alpha \nmid n$. So writing $q = q'p$, we have $k\alpha - l\alpha = q'n$, thus $k\alpha \equiv l\alpha \mod n$. Since $k\alpha$ and $l\alpha$ are smaller than n, therefore $k\alpha = l\alpha$, thus $\times p$ permutes $\langle \alpha \rangle = \alpha\mathbf{n}$. □

Next, we show that multiplication by integers that are not divisors of n does not produce any new images in $\mathcal{I}$.

Fact 4.3. *If* $\beta \nmid n, \beta \in \boldsymbol{n}$, *then* $\beta\boldsymbol{n} = \alpha\boldsymbol{n}$ *for some* $\alpha \mid n$. *In fact,* $\alpha = \gcd(\beta, n)$ *has this property.*

Proof: Let $\alpha = \gcd(\beta, n)$. If $\beta = 0$, then taking $\alpha = n$ the assertion holds. Otherwise we may write $\beta = \alpha k$ for some $k > 0$. Factoring k into primes, we write $k = p_1 \cdots p_m$ (possibly with repetitions). Since α is the greatest common divisor of β and n, then $p_i\alpha \nmid n$. It follows from Fact 4.2 that multiplication by each p_i permutes $\alpha\mathbf{n}$, hence so does multiplication by k. Thus, $\alpha\mathbf{n} = k\alpha\mathbf{n} = \beta\mathbf{n}$. □

Corollary 4.1. $\mathcal{I} = \{\alpha\boldsymbol{n} + \beta : \alpha \mid n$ *and* $0 \le \beta < \alpha \le n\}$.

4.3. *Subduction, Equivalence Relation, and the Tiling Picture*

Here we show how the subduction and equivalence relations are connected to the operations in R_n. The equivalence classes are determined by the multiplicative factors and the elements of a class are determined by the additive factors. Moreover, the partial order of the equivalence classes is the same as the inclusion relation of the principal ideals. This will allow to describe exactly how the operations of $S(R_n)$ are connected to the tiling picture.

Lemma 4.1. *Let* $P = \alpha_0\boldsymbol{n}+\beta_0$ *and* $Q = \alpha_1\boldsymbol{n}+\beta_1$ *be elements of* $\mathcal{I}$. *Then* $P \le Q \iff \alpha_1 \mid \alpha_0 \pmod{n}$.

Proof: ($\Rightarrow$) According to the definition of subduction relation $P \leq Q$ if $P \subseteq Q \cdot s$ for some $s \in S(R_n)$, i.e. $P \subseteq \alpha Q + \beta$. Thus we have $\alpha_0 \mathbf{n} + \beta_0 \subseteq \alpha(\alpha_1 \mathbf{n} + \beta_1) + \beta$. That is, $\alpha_0 \mathbf{n} \subseteq \alpha\alpha_1 \mathbf{n} + \beta_2$ for appropriate β_2. We know that 0 and $\alpha_0 \pmod{n}$ are in $\alpha_0 \mathbf{n}$, since $0, 1 \in \mathbf{n}$. Thus,

$$0 = \alpha\alpha_1 i + \beta_2 \pmod{n}$$
$$\alpha_0 = \alpha\alpha_1 j + \beta_2 \pmod{n}$$

for some $i, j \in \mathbf{n}$. Subtracting the first equation from the second we get $\alpha_0 = \alpha\alpha_1(j - i) + kn$ for some k. Therefore $\alpha_1 \mid \alpha_0 \pmod{n}$.

($\Leftarrow$) If $\alpha_1 \mid \alpha_0 \pmod{n}$ then $\alpha_0 = \alpha\alpha_1 + kn$ for some k, thus $\alpha_0 \mathbf{n} = (\alpha\alpha_1 + kn)\mathbf{n} = \alpha\alpha_1 \mathbf{n}$. Then $P = \alpha_0 \mathbf{n} + \beta_0 = \alpha\alpha_1 \mathbf{n} + \beta_0 = \alpha(\alpha_1 \mathbf{n} + \beta_1) + (\beta_0 - \alpha\beta_1) = \alpha Q + \beta'$, for appropriate β'. (Note that we have equality, not just inclusion!) □

We still have to show that there are no elements equivalent to $\alpha \mathbf{n}$ except those of the form $\alpha \mathbf{n} + \beta$, i.e. the equivalence classes correspond exactly to the principal ideals of the ring.

Lemma 4.2. *Let $\alpha \boldsymbol{n} + \beta \equiv \alpha' \boldsymbol{n} + \beta'$. Then $\alpha \boldsymbol{n} = \alpha' \boldsymbol{n}$.*

Proof: By Lemma 4.1 we have $\alpha' \mid \alpha \pmod{n}$ and $\alpha' \mid \alpha \pmod{n}$, thus $\alpha = \zeta\alpha' + kn$ and $\alpha' = \chi\alpha + k'n$ for some ζ, χ, k, k'. Therefore $\langle \alpha' \rangle \subseteq \langle \alpha \rangle$ and $\langle \alpha \rangle \subseteq \langle \alpha' \rangle$ in R_n yielding $\langle \alpha \rangle = \langle \alpha' \rangle$. □

We can summarize the previous results in the following theorem about the subduction relation of $S(R_n)$. As the choice of the representative is arbitrary, we may take it to have zero as the additive factor (Fact 4.1), thus to have the *canonical form* $\alpha \mathbf{n}$, where $\alpha \mid n$ by Fact 4.3.

Theorem 4.1. *Let n be written as $p_1^{\nu_1} \cdots p_m^{\nu_m}$, the product of powers of distinct primes, and P and Q be elements of $\mathcal{I}$. Then P and Q have representatives $\overline{P} = \alpha \boldsymbol{n}$ and $\overline{Q} = \beta \boldsymbol{n}$, respectively, where $\alpha = p_1^{\lambda_1} \cdots p_m^{\lambda_m}$ and $\beta = p_1^{\mu_1} \cdots p_m^{\mu_m}$, $0 \leq \lambda_i, \mu_i \leq \nu_i$. Moreover,*

(1) $P \leq Q$ if and only if $\lambda_i \geq \mu_i$ for all i with $1 \leq i \leq m$,
(2) $P < Q$ if and only if $\lambda_i \geq \mu_i$ for all i and $\lambda_i > \mu_i$ for some i,
(3) $P \equiv Q$ if and only if $\alpha = \beta$.

Proof: According to Fact 4.1 we can choose representatives with zero additive term. By Fact 4.3, these are equivalent to the representatives $\overline{P} = \alpha \mathbf{n}$ and $\overline{Q} = \beta \mathbf{n}$ such that α and β are divisors of n. By Lemma 4.1, it follows that $\overline{P} \leq \overline{Q}$ if and only if $\beta \mid \alpha$. i.e., $\mu_i \leq \lambda_i$ for all $1 \leq i \leq m$. This proves (1), from which (2) and (3) are immediate. □

It now follows that the set-theoretic description of elements of $\mathcal{I}$ given in Corollary 4.1 lists each element of the extended image set exactly once.

Corollary 4.2. *The sets $\alpha\mathbf{n} + \beta$ with $\alpha \mid n$ and $0 \leq \beta < \alpha \leq n$ are all the distinct elements of $\mathcal{I}$. Hence, $|\mathcal{I}| = \sum_{\alpha|n} \alpha$.*

In the notation of Theorem 4.1,

Corollary 4.3. *$\overline{P}$ is a tile of $\overline{Q}$ if and only if $\lambda_i = \mu_i + 1$ for some i and $\lambda_j = \mu_j$ for all $j \neq i$.*

In the vein of Facts 4.2 and 4.3, we can reformulate the description of the tile-of relation:

Corollary 4.4. *Suppose $\alpha \mid n$.*
(a) Let $\beta \mid n$. Then $\beta\mathbf{n} \prec \alpha\mathbf{n} \Longleftrightarrow \beta = p\alpha$ for some prime factor p of n.
(b) The tiles of $\alpha\mathbf{n}$ are the subsets of $\mathbf{n}$ of the form $\beta\mathbf{n} + \alpha i$ where p is prime, $\beta = p\alpha \mid n$ and $0 \leq i < p$.

Proof: Corollary 4.3 yields (a), and shows that any tile of $\alpha\mathbf{n}$ is represented by $p\alpha\mathbf{n}$ for some prime p with $p\alpha \mid n$. This tile is equivalent to $p\alpha\mathbf{n} + k$ for all $0 \leq k < p\alpha$ and the latter is a tile of $\alpha\mathbf{n}$ iff k is a multiple of α. □

4.4. *Number of Levels*

In studying hierarchical decompositions one of the key questions is the number of hierarchical levels in the cascaded product, or more precisely in the case of the holonomy decomposition the height of the automaton.

Theorem 4.2. *The height h of the holonomy decomposition of R_n is*

$$h = \sum_{i=1}^{k} \nu_i,$$

where $n = p_1^{\nu_1} \cdots p_m^{\nu_m}$ is the prime factorization of n.

Proof: We showed that the strict subduction and the inclusion relations are along the multiplications by prime factors, therefore the maximum length of strict chains in the tiling picture is the maximum length of the products of the prime factors with multiplicity. □

4.5. *Number of States*

Another important question is how many points the holonomy components act on. In the case of R_n, computational results suggested the following:

Theorem 4.3. *Let* $(\mathcal{B}_h, \overline{\mathcal{H}_h})$ *be the top level component of the holonomy decomposition of* R_n *with height* h*. Then*

$$|\mathcal{B}_h| = \sum_{i=1}^{k} p_i,$$

where $n = p_1^{\nu_1} \cdots p_k^{\nu_k}$ *is the prime factorization for* n*.*

Proof: By Corollary 4.4(b) for $\alpha = 1$, the tiles of $\mathbf{n}$ are of the form $p\mathbf{n} + k$, where p is a prime divisor of n and $k \in \mathbf{p}$. By Lemma 4.1, the equivalence classes of tiles $p\mathbf{n}$ and $q\mathbf{n}$ induced by two different prime divisors p and q of n cannot be subduction related. Therefore $\mathbf{n}$ has as many equivalence classes of tiles as there are prime divisors of n, and each equivalence class has as many elements as its corresponding prime, thus we have the sum as in the theorem. □

This can be generalized by considering the fact that the equivalence class representatives (the canonical ones of the form $\alpha\mathbf{n}$ with $\alpha \mid n$) correspond to principal ideals.

Theorem 4.4. *Let* $Q \in \mathcal{I}$ *be a canonical equivalence class representative* $\alpha\mathbf{n}$ *with* $\alpha \mid n$ *and its holonomy component be* $(\mathcal{B}_Q, \mathcal{H}_Q)$*. Then the number of states is given by*

$$|\mathcal{B}_Q| = \sum_{i=1}^{k} p_i$$

where $\frac{n}{\alpha} = p_1^{\alpha_1} \cdots p_k^{\alpha_k}$ *is the prime factorization for* $\frac{n}{\alpha}$*.*

Proof: The key point of the proof is to understand where $\frac{n}{\alpha}$ comes from. Let P be a tile of Q, $P \prec Q$. Then $P \equiv \overline{P} \prec Q = \alpha\mathbf{n}$, where $\overline{P}$ is the representative of the equivalence class of P. We may assume $\overline{P} = \beta\mathbf{n}$ with $\beta \mid n$. Then by Corollaries 4.3 and 4.4(a), $\beta = p_i\alpha$ for some prime divisor p_i of n, and $p_i\alpha\mathbf{n} \prec \alpha\mathbf{n}$ if and only if $p_i\alpha \mid n$, i.e. if and only if $p_i \mid \frac{n}{\alpha}$. We count that there are exactly p_i tiles of $\alpha\mathbf{n}$ equivalent to $p_i\alpha\mathbf{n}$ as shown by Corollary 4.4(b). Summing these as in Theorem 4.3 we have the result. □

4.6. *Holonomy Group Components*

Now we can characterize the invertible elements (permutations) of $S(R_n)$.

Proposition 4.2. *$\pi \in S(R_n)$ is invertible if and only if π can be represented as $\alpha x + \beta$, where $0 \leq \alpha, \beta < n$, with $\alpha = 0$ or $\gcd(\alpha, n) = 1$.*

These elements are the only ones which induce permutations on $\mathbf{n}$, so they also induce holonomy permutations, which can be fewer in number (since there might be two distinct permutations of $\mathbf{n}$ but acting exactly the same way on the set of tiles). We also can give a smaller generator set by considering only one generator of the additive cyclic subgroups. Therefore, in general, for each holonomy group component we have the following generator sets.

Proposition 4.3. *Let $Q = \alpha\mathbf{n} \in \mathcal{I}$ be a canonical equivalence class representative and its holonomy component be $(\mathcal{B}_Q, \mathcal{H}_Q)$. Then $\mathcal{H}_Q$ is generated by the set of transformations induced by the following elements of $S(R_n)$:*

$$\{\times\beta : \ \gcd(\beta, \frac{n}{\alpha}) = 1 \text{ where } 0 < \beta < n\} \cup \{+\alpha\}.$$

Now we present a theorem that basically says that the decomposition of $S(R_n)$ can be built from the unique top level components of the decompositions of smaller residue class rings R_m, where $m \mid n$.

Theorem 4.5. *Let $\alpha m = n$ and $\alpha\mathbf{n}$ be a canonical representative of its equivalence class in the tiling picture of $S(R_n)$. Then*

$$(\mathcal{B}_{\alpha\mathbf{n}}, \mathcal{H}_{\alpha\mathbf{n}}) \cong (\mathcal{B}_{\mathbf{m}}, \mathcal{H}_{\mathbf{m}}),$$

i.e. the holonomy group component of $\alpha\mathbf{n}$ in the decomposition of $S(R_n)$ is isomorphic to the holonomy group component of $\mathbf{m}$ in the decomposition of $S(R_m)$.

Proof: We explicitly construct a permutation group isomorphism, giving bijective mappings from the tiles and the holonomy group of $\mathbf{m}$ to those of $\alpha\mathbf{n}$. The holonomy group elements in question are induced by permutations on $\mathbf{m}$ and $\alpha\mathbf{n}$ given by elements of $S(R_m)$ and $S(R_n)$, respectively.

We have $\mathbf{m} = \{0, 1, \ldots, \frac{n}{\alpha} - 1\}$ and $\alpha\mathbf{n} = \{0, \alpha, \ldots, n - \alpha\}$. Let's consider the function $\phi : \mathbf{m} \to \alpha\mathbf{n}$ defined by $\phi(i) := \alpha i$ which is obviously bijective. ϕ induces a bijection of the powersets of $\alpha\mathbf{n}$ and $\mathbf{m}$ as well. Following Corollary 4.4(b) this gives a bijection of the corresponding tile sets, taking $p\mathbf{m} + i$ $(0 \leq i < p)$ to $p\alpha\mathbf{n} + \alpha i$ for every prime $p \mid m$.

Now we give the map for the holonomy groups. Define $\phi : \mathcal{H}_{\mathbf{m}} \rightarrow \mathcal{H}_{\alpha\mathbf{n}}$ as follows. We have for all tiles $T \prec \mathbf{m}$ a corresponding tile $T\phi \prec \alpha\mathbf{n}$. Let π be any element of $\mathcal{H}_{\mathbf{m}}$. Define $\phi(\pi)$ by $T\phi \mapsto (T \cdot \pi)\phi$. Now π is induced by some $\pi : x \mapsto ax + b \,(\bmod\, m)$ in $S(R_m)$ where $0 \leq a, b < m$ and $a = 0$ or $\gcd(a, m) = 1$. Writing $T = p\mathbf{m} + i$ $(0 \leq i < p)$, we have $(T\phi)\cdot\phi(\pi) = (p\alpha\mathbf{n}+\alpha i)\cdot\phi(\pi) = (T\cdot\pi)\phi = ((p\mathbf{m}+i)\cdot\pi)\phi = (ap\mathbf{m}+ai+b)\phi = ap\alpha\mathbf{n} + a\alpha i + \alpha b = a(p\alpha\mathbf{n} + \alpha i) + \alpha b = a(T\phi) + \alpha b$. Thus ϕ respects the action, and it is also clear that $\phi(\pi)$ is induced by the element π^* of $S(R_n)$ with $\pi^* : y \mapsto ay + \alpha b \pmod{n}$. Observe that π^* induces a permutation of the tiles of $\alpha\mathbf{n}$ by Prop. 4.3, since αb is a multiple of α and either $a = 0$ or $\gcd(a, \frac{n}{\alpha}) = 1$. Hence indeed $\phi(\pi) \in \mathcal{H}_{\alpha\mathbf{n}}$.

We show that ϕ is a homomorphism. Let $\pi = ax + b$ and $\rho = cx + d$ be elements of $S(R_m)$ permuting $\mathbf{m}$. We first show that $\pi^*\rho^* = (\pi\rho)^*$ as maps on $\alpha\mathbf{n}$. Let $i\phi$ $(0 \leq i < m)$ be an arbitrary element of $\alpha\mathbf{n}$. We have

$$\begin{aligned}((i\phi)\pi^*)\rho^* &= c(a(\alpha i) + \alpha b) + \alpha d \\ &= ca\alpha i + \alpha cb + \alpha d \\ &= (\alpha i)(cax + bc + d)^* \\ &= (\alpha i)(c(ax + b) + d)^* \\ &= (i\phi)(\pi\rho)^*.\end{aligned}$$

Hence $\pi^*\rho^*$ and $(\pi\rho)^*$ are equal, and, clearly, for the induced maps, $\phi(\pi)\phi(\rho)$ and $\phi(\pi\rho)$ are equal as permutations of the tiles of $\alpha\mathbf{n}$. This proves ϕ is a permutation group homomorphism.

Next we show that the homomorphism is one to one. Suppose $\phi(\pi) = \phi(\rho)$ in $\mathcal{H}_{\alpha\mathbf{n}}$. Then for all tiles $Y \prec \alpha\mathbf{n}$, we have $aY + \alpha b = cY + \alpha d$ (where a, b, c, d are as above). Thus, for all tiles $p\alpha\mathbf{n}+\alpha i$ of $\alpha\mathbf{n}$, $a(p\alpha\mathbf{n}+\alpha i)+\alpha b = c(p\alpha\mathbf{n}+\alpha i)+\alpha d$. Applying the state bijection ϕ^{-1}, we have $a(p\mathbf{m}+i)+b = c(p\mathbf{m} + i) + d$, for all tiles $p\mathbf{m} + i$ of $\mathbf{m}$, whence $\pi = \rho$ in $\mathcal{H}_{\mathbf{m}}$.

Finally we show that the homomorphism is onto. Let $\gamma \in \mathcal{H}_{\alpha\mathbf{n}}$. Then γ is induced by a permutation of $\alpha\mathbf{n}$ with the form $y \mapsto ay + b \pmod{n}$. We show this map is an image $\phi(\pi)$ of some π in $\mathcal{H}_{\mathbf{m}}$. It is immediate that b is a multiple of α since the image of 0 must lie in $\alpha\mathbf{n}$. Take $\pi : x \mapsto ax + \frac{b}{\alpha}$ $(\bmod\ m)$. It suffices to show that π permutes $\mathbf{m}$. Suppose π collapses states $i, j \in \mathbf{m}$. Then $ai = aj \pmod{m}$, whence $\alpha ai = \alpha aj \pmod{n}$, and so $a\alpha i + b = a\alpha j + b \pmod{n}$. Therefore $\alpha i = \alpha j$, since the map $ay + b$ permutes $\alpha\mathbf{n}$. It follows that $i = j$, so π is injective on $\mathbf{m}$ (and hence a permutation).

This establishes the isomorphism of permutation groups. □

5. Examples of Decomposition

We used our computational implementation [4] of the holonomy decomposition to systematically decompose the residue class rings of integers modulo n, starting from $n = 2$. Formulating and proving the theorems in the previous section was relatively easy after studying the output decompositions of the software. This is yet another proof of the concept that computers can be used in mathematical research.

Holonomy decompositions of these rings up to R_{20} are summarized in Table 1. The following notation is used: C_n is the cyclic group of order n, S_n is the symmetric group on n points. D_n is the dihedral group of order n. $G_{n:k}$ identifies any nontrivial semidirect product $C_n \rtimes C_k$. G_n denotes a group of order n with trivial Frattini subgroup[b], where the Frattini subgroup is the intersection of all maximal subgroups (or equivalently, the subgroup of non-generator elements [10]).

Table 1. The holonomy decompositions of first 20 nontrivial residue class rings of integers $\mathbb{Z}$. Note that the numbers of levels corresponds to the number of prime factors of n with multiplicities. The number of states at the top hierarchical level (rightmost) equals the sum of primes dividing n.

Ring	Levels	$\lvert\mathcal{I}\rvert$	Holonomy Decomposition
R_2	1	3	$(\mathbf{2}, \overline{S_2})$
R_3	1	4	$(\mathbf{3}, \overline{S_3})$
R_4	2	7	$(\mathbf{2}, \overline{C_2}) \wr (\mathbf{2}, \overline{C_2})$
R_5	1	6	$(\mathbf{5}, \overline{G_{5:4}})$
R_6	2	12	$\overline{((\mathbf{2}, C_2) \times (\mathbf{3}, S_3))} \wr (\mathbf{5}, \overline{D_{12}})$
R_7	1	8	$(\mathbf{7}, \overline{G_{7:6}})$
R_8	3	15	$(\mathbf{2}, \overline{S_2}) \wr (\mathbf{2}, \overline{S_2}) \wr (\mathbf{2}, \overline{S_2})$
R_9	2	13	$(\mathbf{3}, \overline{S_3}) \wr (\mathbf{3}, \overline{S_3})$
R_{10}	2	18	$\overline{((\mathbf{2}, C_2) \times (\mathbf{5}, G_{5:4}))} \wr (\mathbf{7}, \overline{G_{40}})$
R_{11}	1	12	$(\mathbf{11}, \overline{G_{11:10}})$
R_{12}	3	28	$\overline{((\mathbf{2}, C_2) \times (\mathbf{3}, S_3))} \wr \overline{((\mathbf{2}, C_2) \times (\mathbf{5}, D_{12}))} \wr (\mathbf{5}, \overline{D_{12}})$
R_{13}	1	14	$(\mathbf{13}, \overline{G_{13:12}})$
R_{14}	2	24	$\overline{((\mathbf{2}, C_2) \times (\mathbf{7}, G_{7:6}))} \wr (\mathbf{9}, \overline{G_{84}})$
R_{15}	2	24	$\overline{((\mathbf{3}, S_3) \times (\mathbf{5}, G_{5:4}))} \wr (\mathbf{8}, \overline{G_{120}})$
R_{16}	4	31	$(\mathbf{2}, \overline{S_2}) \wr (\mathbf{2}, \overline{S_2}) \wr (\mathbf{2}, \overline{S_2}) \wr (\mathbf{2}, \overline{S_2})$
R_{17}	1	18	$(\mathbf{17}, \overline{G_{17:16}})$
R_{18}	3	39	$\overline{((\mathbf{2}, C_2) \times (\mathbf{3}, S_3))} \wr \overline{((\mathbf{3}, S_3) \times (\mathbf{5}, D_{12}))} \wr (\mathbf{5}, \overline{D_{12}})$
R_{19}	1	20	$(\mathbf{19}, \overline{G_{19:18}})$
R_{20}	3	42	$\overline{((\mathbf{2}, C_2) \times (\mathbf{5}, G_{5:4}))} \wr \overline{((\mathbf{2}, C_2) \times (\mathbf{7}, G_{40}))} \wr (\mathbf{7}, \overline{G_{40}})$

[b]For identifying certain groups in our automated decompositions we used the Small Groups data library for `GAP` [1].

One of the main advantages of using a software tool is that it enables automatic visualizations for complicated structures. For example, Figure 2 shows the tiling picture of $S(R_6)$.

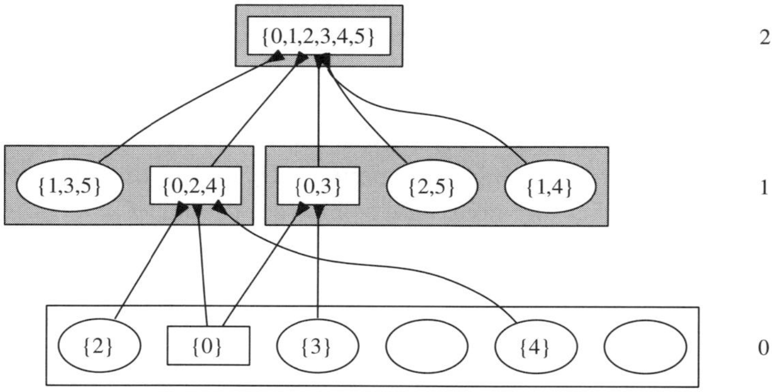

Figure 2. The tiling picture of $S(R_6)$. Nodes are the elements of the extended set of images. Large rectangles denote equivalence classes and the arrows show the tile-of relation. Small rectangles and ovals indicate equivalence class representatives and other members, respectively. Grey shading indicates classes having a nontrivial holonomy group. The numbers on the right indicate the height, i.e. the hierarchical level. Note that only those subsets are displayed that are needed for constructing the cascaded product.

6. Conclusion and Future Work

We described the holonomy decomposition of finite residue class of integers modulo n culminating in the presentation of how the decompositions are put together by using the unique top level components in the holonomy decompositions of smaller rings. The reason why we chose these rings for studying their decompositions is that they are quite regular (hence the nice structure theorem), yet they are also far from trivial. In this way we were able to demonstrate the usability of a computational tool for research in Krohn-Rhodes Theory. We suspect that the results presented here may be generalizable to other interesting classes of semigroups and automata, including other constructions based on rings and fields (e.g. affine transformation semigroups of higher dimension). We also suggest to use the findings here as a guiding metaphor in the next steps of research toward efficient calculation of Krohn-Rhodes decompositions. The distinction between the additive and the multiplicative factor was important here in building the skeleton for the decomposition. Roughly speaking, the multiplicative factors determine the poset of equivalence classes, while the additive factors fill the equivalence classes. The question is to what extent this distinction can be generalized for the permutations and collapsing transformations of an arbitrary ts.

References

1. GAP – Groups, Algorithms and Programming, Version 4.3, (`http://www.gap-system.org`), Centre for Interdisciplinary Research in Computational Algebra, University of St. Andrews, 2002.
2. A. H. Clifford and G. B. Preston, The Algebraic Theory of Semigroups, *AMS Mathematical Surveys* **7(1)**, 2nd edition (1967).
3. Pál Dömösi and Chrystopher L. Nehaniv, *Algebraic Theory of Automata Networks: An Introduction*, SIAM Series on Discrete Mathematics and Applications, chapter 3, The Krohn-Rhodes and Holonomy Decomposition Theorems (2005).
4. Attila Egri-Nagy and Chrystopher L. Nehaniv, *GrasperMachine, Computational Semigroup Theory for Formal Models of Understanding*, (`http://graspermachine.sf.net`) (2003).
5. Attila Egri-Nagy and Chrystopher L. Nehaniv, Algebraic Hierarchical Decomposition of Finite State Automata: Comparison of Implementations for Krohn-Rhodes Theory, Conference on Implementations and Applications of Automata CIAA 2004, *Lecture Notes in Computer Science* **3317**, 315-316 (2004).
6. Samuel Eilenberg, *Automata, Languages and Machines*, volume B, Academic Press (1976).
7. Abraham Ginzburg, *Algebraic Theory of Automata*, Academic Press (1968).
8. Marshall Hall, *The Theory of Groups*, The Macmillan Company, New York (1959).
9. Kenneth Krohn, John L. Rhodes and Bret R. Tilson, *Algebraic Theory of Machines, Languages and Semigroups*, (M. A. Arbib, ed.), chapter 5, The Prime Decomposition Theorem of the Algebraic Theory of Machines, Academic Press, 81-125 (1968).
10. Derek J. S. Robinson, A Course in the Theory of Groups, *Graduate Texts in Mathematics* **80**, Springer Verlag (1993).
11. H. Paul Zeiger, Cascade synthesis of finite state machines, *Information and Control* **10**, 419-433 (1967). Plus erratum.

A P–THEOREM FOR ORDERED GROUPOIDS

N. D. GILBERT

School of Mathematical and Computer Sciences,
Heriot-Watt University,
Riccarton, Edinburgh EH14 4AS, U.K.
E-mail: ndg@ma.hw.ac.uk

McAlister's P-theorem for E–unitary inverse semigroups is one of the most significant components of the structure theory for inverse semigroups: since its first appearance in 1974 several different proofs have been given and the scope of the theorem has been extended to strongly E^*–unitary and strongly categorical inverse semigroups. In this paper, we prove a P–theorem for the wider class of incompressible ordered groupoids, which encompasses previous versions of the P–theorem and offers a unified proof. Moreover, the class of incompressible ordered groupoids may be of interest in its own right, and we look at examples related to Bass-Serre theory for groups.

1. Introduction

This paper generalises some of the structure theory of inverse semigroups to the wider class of ordered groupoids. The use of ordered groupoids as a conceptual framework for proving results in semigroup theory is well-established (see for example [2, 5, 15]): here it offers a uniform approach to the structure of inverse semigroups with or without a zero.

Recall that the the natural partial order on an inverse semigroup S is given by setting $x \leqslant y$ if and only if there exists an idempotent $e \in S$ such that $x = ey$. An inverse semigroup S is said to be *E–unitary* if each element above an idempotent in the natural partial order is also an idempotent.

A pivotal component of the structure theory of inverse semigroups is McAlister's P–theorem for E–unitary inverse semigroups, first proved in [11]. As McAlister himself explains in [12], the theorem was in part motivated by the new approach to the structure of free inverse semigroups provided by the work of Schleiblich [16]. The P–theorem gives a uniform description of E–unitary inverse semigroups in terms of groups acting on posets. Complementary accounts of the P–theorem and its impact on inverse semigroup theory are given in the surveys [12] and [8].

For inverse semigroups with zero, the P–theorem approach needs modification, since an inverse semigroup S with zero is E–unitary if and only if it is a semilattice. We are led either to Szendrei's notion of E^**–unitary* inverse semigroups [18], or to the class of *strongly* E^*–unitary inverse semigroups introduced by Bulman-Flemimg, Fountain, and Gould [1], which coincides with the class of Rees quotients of E–unitary inverse semigroups. A P–theorem for strongly E^*–unitary inverse semigroups can be obtained by describing the Rees quotients of E–unitary inverse semigroups as given by the P–theorem. An alternative approach to a P–theorem for inverse semigroups with zero was given by Gomes and Howie [3], who prove a structure theorem for the class of *strongly categorical* inverse semigroups with zero in terms of a Brandt semigroup acting on a poset.

An inverse semigroup S can be regarded as a groupoid $\mathbb{G}(S)$, whose set of identities is the set of idempotents $E(S)$ of S, and which is ordered by the natural partial order on S. Since $E(S)$ is a semilattice, $\mathbb{G}(S)$ is by definition an *inductive* groupoid. If S has a zero, then 0 is an isolated identity of $\mathbb{G}(S)$ and its removal leaves a $*$*–inductive* groupoid $\mathbb{G}(S)^*$ in which any two identities that have a lower bound have a greatest lower bound. In this way, inverse semigroups with or without zero lead to special types of ordered groupoids determined by order conditions on the set of identities.

In this paper we look at ordered groupoids in general and construct a means to view their structure that encompasses the P–theorem for inverse semigroups, its analogue for strongly E^*–unitary inverse semigroups, and the Gomes-Howie P–theorem for strongly categorical inverse semigroups with zero. Our approach is in the same spirit as that of [1] and [10], in that we make essential use of category structures. However, the details are modelled on Munn's proof of the P–theorem [14], and we now briefly describe the various structural components that we use.

From an ordered groupoid G we construct a groupoid $G_{\updownarrow}$ and a universal *levelling functor* $\lambda : G \to G_{\updownarrow}$: that is, if $g \leqslant h$ then $g\lambda = h\lambda$. We say that G is *incompressible* if λ is star-injective: this notion corresponds to that of an inverse semigroup being E–unitary or strongly E^*–unitary. Our P–theorem describes the structure of an incompressible ordered groupoid G in terms of an action of $G_{\updownarrow}$ on a poset X.

To construct X we use the category $\mathcal{C}(G)$ derived from an ordered groupoid G as in [7] (and generalising the construction of [9] for inverse semigroups) acting on a pullback of the set of identities G_o and $G_{\updownarrow}$. This action determines a quotient set X which is partially ordered and admits an action of $G_\updownarrow$. If G is incompressible, then we show that X has an order ideal Y isomorphic to G_o and such that X is the orbit of Y under the action of $G_\updownarrow$. Finally, from X, Y and $G_\updownarrow$ we construct an ordered groupoid that is isomorphic to G.

For E–unitary inverse semigroups, and strongly E^*-unitary semigroups with zero, the construction just outlined recovers the McAlister P–theorem.

If S is an inverse semigroup that is categorical at zero, then $(\mathbb{G}(S)^*)_\updownarrow = \mathbb{G}(S)^*/\beta$ for a certain congruence β on S. In this case our P–theorem for $\mathbb{G}(S)^*$ gives its structure in terms of the action of $\mathbb{G}(S)^*/\beta$ on the poset X, and this recovers the P–theorem for inverse semigroups with zero in [3].

The generalisation of the P–theorem to ordered groupoids captures other interesting cases. For example, we can regard amalgamated free products and HNN extensions of groups as being derived from ordered groupoids, and in these cases the poset X is the canonical tree on which the group acts and the subset Y is the fundamental domain for the action.

2. Levelling ordered groupoids

A *groupoid* G is a small category in which every morphism is invertible. We think of a groupoid as an algebraic structure (as in [4, 5]): the elements are the morphisms, and composition is an associative partial binary operation on morphisms. The set of identities in G is denoted G_o, and an element $g \in G$ has domain $\mathbf{d}(g) = gg^{-1}$ and range $\mathbf{r}(g) = g^{-1}g$. The structure-preserving maps between groupoids are functors.

Let $e \in G_o$. Then the *star* of e in G is the set $\text{star}_G(e) = \{g \in G : \mathbf{d}(g) = e\}$. A functor $\phi : G \to H$ is said to be *star injective* if, for each $e \in G_o$, the restriction $\phi : \text{star}_G(e) \to \text{star}_H(e\phi)$ is injective. *Star surjective* and *star bijective* functors are defined similarly. A star bijective functor is a *covering*.

An *ordered groupoid* $(G, \leqslant)$ is a groupoid G with a partial order $\leqslant$ satisfying the following axioms:

- OG1 for all $x, y \in G$, if $x \leqslant y$ then $x^{-1} \leqslant y^{-1}$,
- OG2 if $x_1 \leqslant x_2$, $y_1 \leqslant y_2$ and if the compositions x_1y_1 and x_2y_2 are defined, then $x_1y_1 \leqslant x_2y_2$,
- OG3 if $x \in G$ and f is an identity of G with $f \leqslant \mathbf{d}(x)$, there exists a unique element $(f|x)$, called the *restriction* of x to f, such that $\mathbf{d}(f|x) = f$ and $(f|x) \leqslant x$,
- OG3* if $x \in G$ and f is an identity of G with $f \leqslant \mathbf{r}(x)$, there exists a unique element $(x|f)$, called the *corestriction* of x to f, such that

$\mathbf{r}((x|f)) = f$ and $(x|f) \leqslant x$.

It is shown in [5] that OG3* is a consequence of OG1 and OG3: the corestriction of x to f may be defined as $(f|x^{-1})^{-1}$.

Let G be an ordered groupoid and let $a, b \in G$. Suppose that $\mathbf{r}(a)$ and $\mathbf{d}(b)$ have a greatest lower bound – which we write as $\mathbf{r}(a)\mathbf{d}(b)$ – in G_o. Then we may define the *pseudoproduct* of a and b in G as

$$a * b = (a|\mathbf{r}(a)\mathbf{d}(b))(\mathbf{r}(a)\mathbf{d}(b)|b),$$

where the right-hand side is a composition in the groupoid G. As Lawson shows in Lemma 4.1.6 of [5], this is a partially defined associative operation on G.

Our discussion of ordered groupoids follows the terminology and notation of [5], except that we have interchanged the use of domain and range and that we use $*$ to denote the pseudoproduct, rather than $\otimes$.

An ordered groupoid $(G, \leqslant)$ is *inductive* if the set of identities is a meet semilattice under $\leqslant$. In an inductive groupoid G, the pseudoproduct is everywhere-defined and $(G, *)$ is an inverse semigroup (see Proposition 4.1.7 of [5]). This correspondence is one half of the Ehresmann-Schein-Nambooripad theorem, establishing a isomorphism between the categories of inductive groupoids and inverse semigroups (see Theorem 4.1.8 of [5]). The inverse correspondence is easy to describe: an inverse semigroup S is regarded as a groupoid $\mathbb{G}(S)$ by restricting the multiplication $(s,t) \mapsto st$ of S to be defined only if $s^{-1}s = tt^{-1}$. This gives a groupoid with $\mathbb{G}(S)_o$ equal to the set of idempotents $E(S)$ of S and ordered by the natural partial order on S. Since $E(S)$ is a semilattice, $\mathbb{G}(S)$ is inductive. If S has a zero, then $\{0\}$ is an isolated identity of $\mathbb{G}(S)$, and $\mathbb{G}(S)^* = \mathbb{G}(S) \setminus \{0\}$ is a groupoid whose set of identities $E(S)^*$ has the property that if two elements have a lower bound then they have a greatest lower bound. Such an ordered groupoid is called $*$–inductive. It is easy to see that every inverse semigroup with zero arises by adjoining 0 to a $*$–inductive groupoid G, and setting all products not defined in G to be equal to 0.

Let G be an ordered groupoid. A functor $\theta : G \to H$ is said to be *levelling* if $g \leqslant h$ in G implies that $g\theta = h\theta$ in H. We shall construct a groupoid $G_\updownarrow$ that is universal for levelling functors from G.

We first recall Higgins' construction [4] of *universal* groupoids. Let G be a groupoid, and $\sigma : G_o \to V$ some function. To construct the universal groupoid $U_\sigma(G)$, we first define a graph G^σ as follows. Its vertex set is V and its edges are the non-identity arrows of G with incidence maps $\mathbf{d}^\sigma : a \mapsto (\mathbf{d}(a))\sigma$ and $\mathbf{r}^\sigma : a \mapsto (\mathbf{r}(a))\sigma$. Let $p = a_1 a_2 \cdots a_n$ be a path of length

$n > 0$ in G^σ. An *elementary reduction* of p is either the deletion of some a_j where $a_j \in G_o$, or the replacement of $a_j a_{j+1}$ by a when $a_j a_{j+1} = a$ in G. Modulo the equivalence relation $\simeq$ generated by elementary reduction, the path category $\mathcal{P}(G^\sigma)$ becomes a groupoid $U_\sigma(G) = \mathcal{P}(G^\sigma)/\simeq$. We note here that if $V = \{*\}$ is a singleton set, then the universal groupoid $U_\sigma(G)$ is the universal group $U_*(G)$.

Now let $\updownarrow$ be the equivalence relation on G generated by the partial order $\leqslant$, and consider the restriction of $\updownarrow$ to the set G_o of identities. Let $\lambda : G_o \to G_o/\updownarrow$ be the quotient map, and construct the universal groupoid $U_\lambda(G)$. Now suppose that $a, b \in G$ have pseudoproduct $a*b$. Then $\mathbf{d}(a*b) \updownarrow \mathbf{d}(a)$ and $\mathbf{r}(a*b) \updownarrow \mathbf{r}(b)$, and therefore $a^{-1}b^{-1}(a*b)$ is in the local group at $(\mathbf{r}(b))\lambda$ in $U_\lambda(G)$. Let N be the normal subgroupoid of $U_\lambda(G)$ generated by all such elements, and set $G_\updownarrow = U_\lambda(G)/N$. The groupoid $G_\updownarrow$ is called the *level* groupoid of G. We also use λ to denote the quotient map $G \to G_\updownarrow$. The universality of this construction is expressed in the following lemma.

Lemma 2.1. *(i) Let G be an ordered groupoid. Then $\lambda : G \to G_\updownarrow$ is a levelling functor.*

(ii) If $\mu : G \to H$ is levelling, there exists a unique functor $\mu_\updownarrow : G_\updownarrow \to H$ such that $\mu = \lambda\mu_\updownarrow$.

Proof. (i) Suppose that $g \leqslant h$ in G. Then $\mathbf{d}(g) \leqslant \mathbf{d}(h)$ and g^{-1}, h have pseudoproduct $g^{-1} * h = g^{-1}(\mathbf{d}(g)|h) = g^{-1}g = \mathbf{r}(g)$. Hence in $G_\updownarrow$, $((g^{-1})\lambda)(h\lambda) = (\mathbf{r}(g))\lambda$ and therefore $g\lambda = h\lambda$.

(ii) Let $e, f \in G_o$. If $e \updownarrow f$ then, since μ is levelling, we have $e\mu = f\mu$ and it follows that μ induces a functor $\mu' : U_\updownarrow(G) \to H$. Now suppose that $a, b \in G$ have pseudoproduct $a * b = \tilde{a}\tilde{b}$ in G, where $\tilde{a} = (a|\mathbf{r}(a)\mathbf{d}(b))$ and $\tilde{b} = (\mathbf{r}(a)\mathbf{d}(b)|b)$. Then

$$
\begin{aligned}
(a*b)\mu' &= (\tilde{a}\tilde{b})\mu = (\tilde{a}\mu)(\tilde{b}\mu) \\
&= (a\mu)(b\mu) = (a\mu')(b\mu')
\end{aligned}
$$

since μ is levelling. It follows that μ' induces a functor $\mu_\updownarrow : G_\updownarrow \to H$ satisfying $a\lambda = a\mu$ and which is therefore uniquely defined on $G_\updownarrow$. $\square$

Examples

1. Let Γ be a group and E a poset. Then $G = E \times \Gamma$ is an ordered groupoid with the product of (e, g) and (e', g') defined when $e = e'$ and with $(e, g)(e, g') = (e, gg')$, and ordered by $(e, g) \leqslant (e', g')$ if and only if

$g = g'$ and $e \leqslant e'$. The level groupoid $G_{\updownarrow}$ is the disjoint union of copies of Γ indexed by the $\updownarrow$–classes in E.

2. Let I be a connected poset, regarded as a category with a unique arrow $i \to j$ whenever $i \geqslant j$, and let $\mathcal{G}$ be a functor from I to the category of groups, associating to each $i \in I$ a group G_i and a unique homomorphism $\alpha_{ij} : G_i \to G_j$ whenever $i \geqslant j$. The disjoint union $G = \bigsqcup_{i \in I} G_i$ becomes an ordered groupoid if we define $g \geqslant h$ whenever $g \in G_i$ and $h \in G_j$ with $g\alpha_{ij} = h$. The level groupoid $G_{\updownarrow}$ is the colimit $\operatorname{colim} \mathcal{G}$.

3. If G is an inductive groupoid, so that G_o is a semilattice, then $\updownarrow$ is the universal relation on G_o and $U_\lambda(G)$ is a group, the *universal group* of the groupoid G. The level groupoid $G_{\updownarrow}$ is of course also a group, and is the maximum group homomorphic image of the inverse semigroup associated to G.

4. If G is $*$–inductive then $G_{\updownarrow}$ need not be a group. However, $G_{\updownarrow}$ has a universal group $U_*(G_{\updownarrow})$, and this is isomorphic to the universal group $\underline{G}(S)$ of the inverse semigroup with zero $S = G \cup \{0\}$ discussed in [6].

5. A construction due to Lawson [7] associates to any right cancellative category $\mathcal{C}$ (with the conventions of the present paper) an ordered groupoid $\mathbf{G}(\mathcal{C})$. It can be shown that the level groupoid $\mathbf{G}(\mathcal{C})_{\updownarrow}$ is the groupoid completion $\widehat{\mathcal{C}}$ of $\mathcal{C}$ – that is, the universal groupoid with a functor from $\mathcal{C}$. In the special case that $\mathcal{C}$ is the path category $\mathcal{P}(D)$ of a directed graph D, we obtain an isomorphism between $\mathbf{G}(\mathcal{P}(D))_{\updownarrow}$ and the fundamental groupoid $\pi(D)$.

2.1. *Incompressible ordered groupoids*

An ordered groupoid G is *filtered* if its set of identities G_o is an order filter: that is, whenever $a \in G$ and $a \geqslant e$ for some $e \in G_o$ then $a \in G_o$. We say that G is *incompressible* if $\lambda : G \to G_{\updownarrow}$ is star-injective.

Lemma 2.2. *G is filtered if and only if, whenever $a \updownarrow e$ with $e \in G_o$ then $a \in G_o$.*

Proof. Suppose that G is filtered and that $a \updownarrow e$ with $e \in G_o$. Then there exists a finite sequence $(a_0, a_1, \ldots, a_m)$ of elements of G such that $a_0 = a, a_m = e$ and for each i, either $a_i \leqslant a_{i+1}$ or $a_i \geqslant a_{i+1}$. We argue that $a \in G_o$ by induction on the length m of such a sequence. If $m = 0$ then $a = e$. Otherwise, if $m > 0$ then either $a_{m-1} \leqslant e$, in which case $a_{m-1} \in G_o$, or $a_{m-1} \geqslant e$ and again $a_{m-1} \in G_o$ since G is filtered. Hence $a \updownarrow a_{m-1} \in G_o$, and we are done. $\square$

Proposition 2.1. *(i) If G is an incompressible ordered groupoid then it is filtered.*

(ii) If G admits any star-injective levelling functor $\mu : G \to H$ then G is incompressible.

(iii) G is incompressible if and only it it admits a star-injective levelling functor to a group.

Proof. (i) Suppose that $a \updownarrow e$ with $e \in G_o$. By Lemma 2.1 we have $a\lambda = e\lambda$. But also $\mathbf{d}(a) \updownarrow e$ and therefore $\mathbf{d}(a)\lambda = e\lambda$. Since λ is star-injective, $a = \mathbf{d}(a)$.

(ii) This follows from the universality of λ as in part (ii) of Lemma 2.1.

(iii) If $\lambda : G \to G_{\updownarrow}$ is star-injective then so is the composite functor $G \to G_{\updownarrow} \to U_*(G_{\updownarrow})$ to the universal group of the level groupoid $G_{\updownarrow}$. The converse follows from part (ii). □

2.1.1. *E–unitary inverse semigroups*

Let S be an inverse semigroup and $\mathbb{G}(S)$ its associated inductive groupoid. Since the ordering on S and $\mathbb{G}(S)$ is the same, we have that S is E–unitary if and only if $\mathbb{G}(S)$ is filtered. Let $\widehat{S}$ be the maximum group image of S. Then S is E–unitary if and only if the canonical mapping $\sigma : S \to \widehat{S}$ is $\mathcal{R}$–injective. Since $\widehat{S} = \mathbb{G}(S)_{\updownarrow}$, the map σ induces $\lambda : \mathbb{G}(S) \to \mathbb{G}(S)_{\updownarrow}$ and we see that S is E–unitary if and only if $\mathbb{G}(S)$ is incompressible. Rephrasing these remarks solely in terms of inductive groupoids, we have:

Lemma 2.3. *Let G be an inductive groupoid. Then G is filtered if and only if G is incompressible.*

If S is a semigroup with a zero, then S is E–unitary if and only if S is a semilattice, and so instead we look at the E^*–unitary property of [18]: S is E^*–unitary if each element above a non-zero idempotent in the natural partial order is also an idempotent. If we delete the 0 from $\mathbb{G}(S)$ we get the ordered groupoid $\mathbb{G}(S)^*$, and it is easy to see that S is E^*–unitary if and only if $\mathbb{G}(S)^*$ is filtered. Of more interest is the class of strongly E^*–unitary inverse semigroups introduced in [1]. An inverse semigroup with zero S is *strongly E^*–unitary* if there exists a group G and a map $\theta : S \to G \cup \{0\}$ such that $s\theta = 0$ if and only if $s = 0$, $s\theta = 1$ if and only if s is a non-zero idempotent, and such that if $s, t \in S$ with $st \neq 0$ then $(st)\theta = (s\theta)(t\theta)$. It is easy to see that a strongly E^*–unitary inverse semigroup is E^*–unitary.

Lemma 2.4. *The inverse semigroup S is strongly E^*–unitary if and only if $\mathbb{G}(S)^*$ is incompressible.*

Proof. If S is strongly E^*–unitary, there exists a group G and a function $\theta : S \to G^0$ such that $s\theta = 0$ if and only if $s = 0$, $s\theta = 1$ if and only if $0 \neq s \in E(S)$, and $(st)\theta = (s\theta)(t\theta)$ if $st \neq 0$. We can regard θ as a functor $\mathbb{G}(S)^* \to G$ to G that is necessarily levelling, and is star-injective. Hence $\mathbb{G}(S)^*$ is incompressible.

Conversely, suppose that $\mathbb{G}(S)^*$ is incompressible, and let $\underline{G}(S)$ be the universal group of the level groupoid $(\mathbb{G}(S)^*)_\updownarrow$, with $\pi : (\mathbb{G}(S)^*)_\updownarrow \to \underline{G}(S)$ the canonical map. Let $\theta = \lambda\pi$.

Now if $s, t \in S$ with $st \neq 0$, then $s^{-1}s \updownarrow tt^{-1}$ in $\mathbb{G}(S)$ and in $(\mathbb{G}(S)^*)_\updownarrow$ we have $(st)\lambda = (s\lambda)(t\lambda)$. Hence $(st)\theta = (s\theta)(t\theta)$. Suppose that $s \neq 0$ and that $s\theta = 1$. Using Higgins' solution of the word problem in $\underline{G}(S)$ given in Corollary 1 to Theorem 4 of [4], $s\lambda$ must be an identity in $(\mathbb{G}(S)^*)_\updownarrow$. Since $\mathbb{G}(S)$ is incompressible, s must be an identity of $\mathbb{G}(S)^*$, that is $s \in E(S)$. □

As mentioned in example **4** above, the group $\underline{G}(S)$ is the *universal group* of the inverse semigroup S as defined in [6].

2.2. *β–transitive ordered groupoids*

Let G be an ordered groupoid, and define the relation β on G by

$$\beta = \{(a, b) : \text{ there exists } c \in G \text{ with } c \leqslant a, b\}.$$

Then β is reflexive and symmetric: we say that G is *β-transitive* if β is also a transitive relation on G. Equivalently, each principal order ideal

$$(g)_\downarrow = \{a \in G : a \leqslant g\}$$

is a directed set in G. An inductive groupoid is always β–transitive, and interpreted on the associated inverse semigroup, β is the minimum group congruence.

The relation β was studied for inverse semigroups S with zero by Munn [13] and his work is the basis for the P–theorem of Gomes and Howie [3]. We shall return briefly to these ideas in section 4.1 below. Recall that S is *categorical at zero* if whenever $a, b, c \in S$ and $abc = 0$ then either $ab = 0$ or $bc = 0$.

Lemma 2.5. *Let S be an inverse semigroup with zero. Then $\mathbb{G}(S)^*$ is β–transitive if and only if S is categorical at zero.*

Proof. Suppose that $a, b, c \in S^*$ and that $ab, bc \neq 0$. Then $a^{-1}ab, bcc^{-1} \neq 0$ and each is $\leqslant b$. If β is transitive then $a^{-1}ab\beta bcc^{-1}$ and so there exist $e, f \in E(S)$ such that $0 \neq ea^{-1}ab = bcc^{-1}f$, and therefore $0 \neq ea^{-1}abb^{-1} = bcc^{-1}fb^{-1}$. Noting that $ea^{-1}abb^{-1}$ is an idempotent, we have

$$\begin{aligned} 0 \neq ea^{-1}abb^{-1} &= (ea^{-1}abb^{-1})^2 \\ &= (ea^{-1}abb^{-1})(bcc^{-1}fb^{-1}) = ea^{-1}(abc)c^{-1}fb^{-1} \end{aligned}$$

and therefore $abc \neq 0$.

The converse is well known, but we give a proof for completeness. Suppose that $r, s, t \neq 0$ with $r\beta s$ and that $s\beta t$. Then there exist non-zero u, v with $u \leqslant r, s$ and $v \leqslant s, t$, and so $u = es$ and $v = fs$ for some $e, f \in E(S)$. Let $c = efs = es(s^{-1}fs)$: then $c \leqslant r, t$. Further, $es = u \neq 0$ and $s(s^{-1}fs) = fs = v \neq 0$. Since S is categorical at zero, it follows that $c \neq 0$ and therefore that $r\beta t$. □

Since $a \leqslant b \Longrightarrow a\beta b \Longrightarrow a \updownarrow b$, if β is transitive then $\beta = \updownarrow$.

Proposition 2.2. *If G is β–transitive, then G/β is a groupoid isomorphic to $G_{\updownarrow}$.*

Proof. We set $(G/\beta)_o = \{\beta(e) : e \in G_o\}$ and define $\mathbf{d}(\beta(a)) = \beta(\mathbf{d}(a))\,, \mathbf{r}(\beta(a)) = \beta(\mathbf{r}(a))$. Now suppose that $\mathbf{r}(a)\beta\mathbf{d}(b)$: then there exists $f \leqslant \mathbf{r}(a), \mathbf{d}(b)$ and we set

$$\beta(a)\beta(b) = \beta((a|f)(f|b)).$$

This is independent of the choice of f, for if also $f' \leqslant \mathbf{r}(a), \mathbf{d}(b)$ then since G is β–transitive, $f\beta f'$ and so there exists $x \leqslant f, f'$. Now $(a|x)(x|a) \leqslant (a|f)(f|b)$ and so

$$\beta((a|f)(f|b)) = \beta((a|x)(x|b)) = \beta((a|f')(f'|b)).$$

The product is also well-defined on β–classes, for suppose that $a\beta a'$ and $b\beta b'$, with $a'' \leqslant a, a'$ and $b'' \leqslant b, b'$. Then

$$\mathbf{r}(a'')\ \beta\ \mathbf{r}(a')\ \beta\ \mathbf{r}(a)\ \beta\ \mathbf{d}(b)\ \beta\ \mathbf{d}(b'').$$

Then for any $f \leqslant \mathbf{r}(a''), \mathbf{d}(b'')$ we have

$$\beta(a)\beta(b) = \beta((a|f)(f|b)) = \beta((a''|f)(f|b'') = \beta((a'|f)(f|b')) = \beta(a')\beta(b')\,.$$

We now have a well-defined mapping

$$\lambda' : G/\beta \to G_{\updownarrow}\,, \beta(a) \mapsto a\lambda$$

between groupoids with the same set of identities $G_o/\beta = G_o/\updownarrow$. This is a functor, since if $a, b \in G$ and $\mathbf{r}(a)\beta\mathbf{d}(b)$, then for any $f \leqslant \mathbf{r}(a), \mathbf{d}(b)$,

$$\beta(a)\lambda'\beta(b)\lambda' = (a\lambda)(b\lambda) = (a|f)\lambda(f|b)\lambda = ((a|f)(f|b))\lambda = (\beta(a)\beta(b))\lambda' .$$

The canonical map $\pi : G \to G/\beta$ is a levelling functor, and so induces $\pi' : U_\updownarrow(G) \to G/\beta$. Now if $a * b$ is defined in G then $\beta(a * b) = \beta(a)\beta(b)$ and hence π' induces $\pi_\updownarrow : G_\updownarrow \to G/\beta$ which is inverse to λ. □

Generalising Lemma 2.3 we have:

Corollary 2.1. *If G is β–transitive then G is incompressible if and only if G is filtered.*

Proof. By Proposition 2.1 (i) we need only prove that a filtered, β–transitive ordered groupoid is incompresssible, and by Proposition 2.2 we can identify $\lambda : G \to G_\updownarrow$ with $\pi : G \to G/\beta$. Suppose that $a, b \in G$ with $\mathbf{d}(a) = \mathbf{d}(b)$. If $\beta(a) = \beta(b)$ then there exists $c \in G$ with $c \leqslant a, b$, and so $c^{-1}c \leqslant a^{-1}b$. Since G is filtered we conclude that $a^{-1}b \in G_o$ and so that $a = b$. □

This corollary is closely related to Proposition 3.1 of [1] which shows that E^*–unitary strongly categorical inverse semigroups are strongly E^*–unitary.

3. Groupoids acting on posets

Consider a groupoid G and a functor from G to the category of posets. For each $e \in G_o$ we have a poset X_e, and each arrow $g \in G$ determines an isomorphism $X_{\mathbf{d}(g)} \to X_{\mathbf{r}(g)}$: we say that G acts on the disjoint union

$$X = \bigsqcup_{e \in G_o} X_e .$$

Fix an order ideal Y in X, and set $Y_e = Y \cap X_e$. We now define an ordered groupoid $P = P(X, Y, G)$ as follows. As a set,

$$P = \{(a, y) \in G \times Y : y \in Y_{\mathbf{r}(a)} , ya^{-1} \in Y\}.$$

The object set of P is $P_o = \{(e, y) \in G_o \times Y : y \in Y_e\}$ and the domain and range maps are

$$\mathbf{d}(a, y) = (\mathbf{d}(a), ya^{-1}) , \mathbf{r}(a, y) = (\mathbf{r}(a), y) .$$

The composition of (a, x) and (b, z) is defined if and only if $\mathbf{r}(a) = \mathbf{d}(b)$ and $xb = z$, in which case $(a, x)(b, z) = (ab, z)$. If we set $(a, x)^{-1} = (a^{-1}, xa^{-1})$ we have a groupoid structure on P.

A partial ordering on P is given by

$$(a, x) \leqslant (b, y) \text{ if and only if } a = b\,, x \leqslant y\,.$$

If $y \leqslant xa^{-1} \in Y_{\mathbf{d}(a)}$ then the restriction of (a, x) at $(\mathbf{d}(a), y)$ is

$$((\mathbf{d}(a), y)|(a, x)) = (a, ya).$$

Remark. In the construction of P, only elements of the G-invariant poset YG need be considered, and so without loss of generality, we may assume that $X = YG$.

Lemma 3.1. *$(P, \leqslant)$ is an incompressible ordered groupoid, and the functor $\theta : P \to G$ mapping $(a, y) \mapsto a$ is a covering of groupoids that is levelling.*

Proof. We verify the axioms for an ordered groupoid.

(OG1): Suppose that $(a, x) \leqslant (b, y)$. Then $a = b$ and $x \leqslant y$, so that $a^{-1} = b^{-1}$ and $xa^{-1} \leqslant ya^{-1}$. Therefore $(a, x)^{-1} = (a^{-1}, xa^{-1}) \leqslant (a^{-1}, ya^{-1}) = (b, y)^{-1}$.

(OG2): Suppose that (a, x) and (b, y) are composable, that $(a', x') \leqslant (a, x)\,, (b', y') \leqslant (b, y)$ and that (a', x') and (b', y') are composable. Then $a = a'\,, b = b'$ and $x' \leqslant x\,, y' \leqslant y$. It follows that $(a', x')(b', y') = (ab, y') \leqslant (ab, y) = (a, x)(b.y)$.

(OG3): Suppose that $(c, z) \leqslant (a, x)$ with $\mathbf{d}(c, z) = (\mathbf{d}(a), y)$ and $y \leqslant ax^{-1}$. Then $a = c$ and $zc^{-1} = za^{-1} = y$, whence $z = ya$. Hence $(c, z) = (a, ya)$ and the restriction is unique.

It is clear that θ is a levelling functor. If $a \in \text{star}_G(e)$ then for each $z \in Y_e$ we have $a = (a, za)\theta$ with $(a, za) \in \text{star}_P(e, z)$. Moreover, if $(a, x) \in \text{star}_P(e, z)$ then $xa^{-1} = z$ so that $x = za$. Therefore θ is a covering of groupoids. In particular, it is star-injective and so (by Lemma 2.1) P is incompressible. □

4. The P–theorem

In this section we prove our version of the P–theorem for ordered groupoids, namely we show that an incompressible ordered groupoid is isomorphic to the groupoid $P = P(X, Y, G_{\updownarrow})$ constructed from the action of the level groupoid $G_{\updownarrow}$ on a certain poset X. Our approach follows the account given by McAlister in [12] of Munn's proof from [14].

Let G be an ordered groupoid. Following Lawson [7], we construct a right cancellative category $\mathbf{C}(G)$. (The reader should note that the conventions of [7] lead to a left cancellative category.) As a set, we have

$$\mathbf{C}(G) = \{(a, f) \in G \times G_o : \mathbf{r}(a) \leqslant f\}.$$

The identities of $\mathbf{C}(G)$ are the arrows of the form (e, e) with $e \in G_o$, with $\mathbf{d}(a, f) = (\mathbf{d}(a), \mathbf{d}(a))$ and $\mathbf{r}(a, f) = (f, f)$. If $f = \mathbf{d}(b)$ then the composition of (a, f) and (b, f') is defined, with $(a, f)(b, f') = (a(\mathbf{r}(a)|b), f') = (a * b, f')$.

We get a right action of $\mathbf{C}(G)$ on G_o by setting $e(a, f) = f$ when $e = \mathbf{d}(a)$, and a left action of $\mathbf{C}(G)$ on $G_\updownarrow$ by setting $(a, f)w = (a\lambda)w$ whenever this is defined in $G_\updownarrow$. This pair of actions defines a quasiorder $\preceq$ on the pullback

$$G_o \between G_\updownarrow = \{(e, w) \in G_o \times G_\updownarrow : e\lambda = \mathbf{d}(w)\}$$

defined by

$$(e, (a, f)u) \preceq (e(a, f), u)$$

whenever $(a, f) \in \mathbf{C}(G)$ with $\mathbf{d}(a) = e$. Let $e \otimes u$ denote the equivalence class of (e, u) under the equivalence relation determined by $\preceq$, so that

$$e \otimes u = f \otimes v \iff (e, u) \preceq (f, v) \text{ and } (f, v) \preceq (e, u).$$

Let $G_o \otimes G_\updownarrow$ denote the set of equivalence classes. It is easy to check that $e \otimes u = f \otimes v$ if and only if there exists $a, b \in G$ with $\mathbf{d}(a) = e \geqslant \mathbf{r}(b), \mathbf{d}(b) = f \geqslant \mathbf{r}(a)$, such that $u = (a\lambda)v$ and $v = (b\lambda)u$. Note that, in particular, $\mathbf{r}(u) = \mathbf{r}(v)$ in $G_\updownarrow$. Moreover, it follows that $(a\lambda)(b\lambda) = e\lambda$ and $(b\lambda)(a\lambda) = f\lambda$.

If G is incompressible, then the equivalence relation on $G_o \between G_\updownarrow$ has a simpler description.

Lemma 4.1. *If G is incompressible, then $e \otimes u = f \otimes v$ if and only if there exists $a \in G$ with $\mathbf{d}(a) = e, \mathbf{r}(a) = f$ and such that $u = (a\lambda)v$.*

Proof. Suppose that $e \otimes u = f \otimes v$, with a, b as above. Now the pseudoproduct $a * b$ exists, and $(a * b)\lambda = a\lambda b\lambda = e\lambda$. Since G is incompressible, $a * b \in G_o$ and therefore $a * b = \mathbf{d}(a * b) = \mathbf{d}(a) = e$. Then $e = \mathbf{r}(a * b) \leqslant \mathbf{r}(b) \leqslant e$, so that $\mathbf{r}(b) = e$. Then $ba \in G$ and $(ba)\lambda = f\lambda$ so that $ba \in G_o$ and so $ba = f$. Therefore $\mathbf{r}(a) = f$ as claimed (and $b = a^{-1}$).

Conversely, if we are given a as described, then a and $b = a^{-1}$ have the properties required to ensure that $e \otimes u = f \otimes v$. $\square$

Lemma 4.2. *$G_o \otimes G_{\updownarrow}$ is a partially ordered set that admits a right action of the level groupoid $G_{\updownarrow}$, given by $(e \otimes u)w = e \otimes uw$.*

Proof. By construction, the quasiorder $\preceq$ induces a partial order on $G_o \otimes G_{\updownarrow}$. We have $e \otimes u \leqslant f \otimes v$ if and only if $(e, u) \preceq (f, v)$ in $G_o \between G_{\updownarrow}$. It is easy to check that, since $\preceq$ is a quasiorder, that this does not depend on the choice of representatives for $e \otimes u$ and $f \otimes v$. In detail then, $e \otimes u \leqslant f \otimes v$ if and only if there exists $c \in G$ with $\mathbf{d}(c) = e, \mathbf{r}(c) \leqslant f$ such that $u = (c\lambda)v$.

We note that if $e \otimes u \leqslant f \otimes v$ then $\mathbf{r}(u) = \mathbf{r}(v)$. Hence $G_o \otimes G_{\updownarrow}$ is partitioned as the disjoint union of posets

$$(G_o \otimes G_{\updownarrow})_{f\lambda} = \{e \otimes u : \mathbf{r}(u) = f\lambda\}$$

and if $e \otimes u \in (G_o \otimes G_{\updownarrow})_{\mathbf{d}(w)}$ then $e \otimes uw \in (G_o \otimes G_{\updownarrow})_{\mathbf{r}(w)}$. It is clear that if $e \otimes u \leqslant f \otimes v$ then, for all $w \in G_{\updownarrow}$ with $\mathbf{d}(w) = \mathbf{r}(u) = \mathbf{r}(v)$, we have $e \otimes uw \leqslant f \otimes vw$. $\square$

The poset $G_o \otimes G_{\updownarrow}$ gives us the X we need in the P–theorem construction. Now we identify the order ideal Y in X. There is an obvious map $\lambda_* : G_o \to G_o \otimes G_{\updownarrow}$ carrying $e \mapsto e \otimes e\lambda$. Its image may also be characterised by:

Lemma 4.3. *If $a \in G$ with $\mathbf{r}(a) = e$ then*

$$e\lambda_* = e \otimes e\lambda = \mathbf{d}(a) \otimes a\lambda.$$

Lemma 4.4. *The set $G_o\lambda_* = \{e \otimes e\lambda : e \in G_o\}$ is an order ideal in $G_o \otimes G_{\updownarrow}$, and if G is incompressible then $G_o\lambda_*$ is isomorphic as a poset to G_o.*

Proof. Suppose that $k \otimes w \leqslant e \otimes e\lambda$. Then there exists $c \in G$ with $\mathbf{d}(c) = k, \mathbf{r}(c) \leqslant e$ such that $w = (c\lambda)(e\lambda) = (c\lambda)$. That is,

$$k \otimes w = \mathbf{d}(c) \otimes c\lambda = \mathbf{r}(c) \otimes (\mathbf{r}(c))\lambda$$

by Lemma 4.3, and so $\{e \otimes e\lambda : e \in G_o\}$ is an order ideal in $G_o \otimes G_{\updownarrow}$.

We now show that if G is incompressible, the mapping $\lambda_* : e \mapsto e \otimes e\lambda$ is an ordered embedding of G_o into $G_o \otimes G_{\updownarrow}$. So suppose that $e, f \in G_o$ and that $e \otimes e\lambda = f \otimes f\lambda$. Then there exist $a, b \in G$ such that $\mathbf{d}(a) = e \geqslant \mathbf{r}(b)$ and $\mathbf{d}(b) = f \geqslant \mathbf{r}(a)$, with $e\lambda = (a\lambda)(f\lambda)$ and $f\lambda = (b\lambda)(e\lambda)$. Since $e\lambda, f\lambda$ are identities in $G_{\updownarrow}$, we see that $e\lambda = a\lambda$ and $f\lambda = b\lambda$. If G is incompressible, we conclude that $a, b \in G_o$ and it follows that $e = a \geqslant b$ and $f = b \geqslant a$. Therefore $e = a = b = f$.

Now take $e, f \in G_o$ with $e \leqslant f$. Then clearly $e \otimes e\lambda \leqslant f \otimes f\lambda$ since we take $c = e \in G$ with $\mathbf{d}(e) = e, \mathbf{r}(e) \leqslant f$ and $(e\lambda)(e\lambda) = e\lambda = f\lambda$ since λ is levelling.

Finally, to show that λ_* is an isomorphism of posets, we suppose that $e \otimes e\lambda \leqslant f \otimes f\lambda$. Then there exists $c \in G$ with $\mathbf{d}(c) = e, \mathbf{r}(c) \leqslant f$ and $e\lambda = (c\lambda)(f\lambda) = c\lambda$. Since G is incompressible, $c = e$ and therefore $e \leqslant f$. □

We remark here that if $G_o\lambda_*$ is order isomorphic to G_o, then G need not be incompressible. A counterexample is given in Example 1 of section 5 below.

Theorem 4.1. *(The P–theorem for ordered groupoids.) Let G be an incompressible ordered groupoid. Then G is isomorphic to the ordered groupoid*

$$P = P(G_o \otimes G_{\updownarrow}, G_o\lambda_*, G_{\updownarrow}).$$

Proof. Let $\Lambda : G \to P$ map

$$g \mapsto (g\lambda, \mathbf{r}(g)\lambda_*) = (g\lambda, \mathbf{r}(g) \otimes (\mathbf{r}(g))\lambda).$$

If $e \in G_o$ then $e\Lambda = (e\lambda, e\lambda_*) \in P_o$, so that Λ maps identities to identities. Now if $g, h \in G$ with $\mathbf{r}(g) = \mathbf{d}(h)$, then $(\mathbf{r}(g))\lambda = (\mathbf{d}(h))\lambda$ and

$$(\mathbf{r}(g)\lambda_*)h\lambda = \mathbf{r}(g) \otimes h\lambda = \mathbf{d}(h) \otimes h\lambda = (\mathbf{r}(h))\lambda_*.$$

Therefore, $g\Lambda$ and $h\Lambda$ are composable, and moreover

$$(g\Lambda)(h\Lambda) = (g\lambda, \mathbf{r}(g)\lambda_*)(h\lambda, \mathbf{r}(h)\lambda_*) = ((gh)\lambda, \mathbf{r}(h)\lambda_*) = (gh)\Lambda.$$

We have thus shown that Λ is a functor.

Let $(u, y) \in P$. Since G is incompressible, there exists a unique $e \in G_o$ such that $y = e\lambda_*$: we then note that $\mathbf{r}(u) = e\lambda$ and that $e \otimes u^{-1} \in G_o\lambda_*$, and so there exists a unique $f \in G_o$ such that $e \otimes u^{-1} = f \otimes f\lambda_*$. Therefore there exist $p, q \in G$ with

$$\mathbf{d}(p) = e \geqslant \mathbf{r}(q), \mathbf{d}(q) = f \geqslant \mathbf{r}(p), (p\lambda)(f\lambda) = u^{-1}, (q\lambda)u^{-1} = f\lambda.$$

Hence $q\lambda = u$ and $p\lambda = u^{-1}$. The pseudoproduct $p * q$ exists, and $(p * q)\lambda = (p\lambda)(q\lambda) = e\lambda$. Since G is incompressible and $\mathbf{d}(p * q) = \mathbf{d}(p) = e$ we have $p * q = e$. Now $e = \mathbf{r}(p * q) \leqslant \mathbf{r}(q) \leqslant e$: therefore $\mathbf{r}(q) = e$, and

$$(u, e \otimes e\lambda) = (q\lambda, \mathbf{r}(q) \otimes (\mathbf{r}(q))\lambda) = q\Lambda.$$

If we also have $(u, e \otimes e\lambda) = q'\Lambda$ then $u = q'\lambda$ and $e = \mathbf{r}(q')$ (since e is unique), and so $q'q^{-1} \in G$ with $(q'q^{-1})\lambda = \mathbf{d}(u)$, which implies that $q'q^{-1} \in G_o$, and so $q' = q$.

We have shown that Λ is bijective, and hence is an isomorphism $G \to P$. $\square$

Applied to inductive and $*$–inductive groupoids, Theorem 4.1 recovers the classical McAlister P–theorems for E–unitary inverse semigroups and for strongly E^*–unitary inverse semigroups with zero.

4.1. *The Gomes-Howie P–theorem*

Gomes and Howie [3] prove a P–theorem for *strongly categorical* inverse semigroups with zero, that is those inverse semigroups with zero S that are categorical at zero and in which each pair of non-zero ideals has non-zero intersection. Munn [13] had shown that such an S admits a minimum congruence β such that S/β is a Brandt semigroup B, i.e. an inverse semigroup obtained by adding a zero to a connected groupoid. It is not difficult to show that the condition on the intersection of ideals is equivalent to the connectedness of $\mathbb{G}(S/\beta)^*_\updownarrow$. The Gomes-Howie theorem constructs every strongly categorical E^*–unitary semigroup S from a *Brandt triple* (X, Y, B) in which X is a poset with a zero, Y is a subsemilattice and order ideal in X, and $B = S/\beta$ acts on X by partial order isomorphisms. Applied to a β–transitive $*$–inductive ordered groupoid G, Theorem 4.1 recovers this construction for G, but in the more general case in which $G_\updownarrow = G/\beta$ need not be a connected groupoid.

5. Actions on cosets

5.1. *A Clifford semigroup*

Let G be the disjoint union of groups K and L, with a given homomorphism $\alpha : K \to L$. We order G by setting $k \geqslant k\alpha$ for all $k \in K$. Then G is inductive and its associated inverse semigroup is a Clifford semigroup. Let e_K and e_L be the identity elements of K and L so that $G_o = \{e_K, e_L\}$. Now $G_\updownarrow = L$ and G is incompressible if and only if α is injective. We write K^α for the image of K under α.

By Lemma 4.3, for all $x \in L$ we have $e_L \otimes e_L = e_L \otimes x$, and further, if $u, v \in L$ then $e_K \otimes u = e_K \otimes v$ if and only if $v = (k\alpha)u$ for some $k \in K$. It follows that we may identify $e_K \otimes u$ with the coset $K^\alpha u$, and if we identify $e_L \otimes x$ with L then $X = L/K^\alpha \cup \{L\}$, $Y = \{K^\alpha\} \cup \{L\}$, and L acts by right multiplication on these cosets. It then follows that $P(X, Y, L) \cong L \sqcup K^\alpha$ as an ordered groupoid. Y is order isomorphic to G_o, but of course we have $P \cong G$ if and only if α is injective.

5.2. *Free products with amalgamation*

Developing the ideas in 5.1, let G be the disjoint union of groups A, B, C with fixed homomorphisms $\alpha : C \to A$ and $\beta : C \to B$. Order G as follows: for all $c \in C$, we set $c\alpha \leqslant c$ and $c\beta \leqslant c$. The ordering on G_o is therefore $e_A \leqslant e_C \geqslant e_B$. The level groupoid $G_\updownarrow$ is a group, and is the pushout of the maps α and β.

If we assume that α and β are injective, then G is incompressible and $G_\updownarrow$ is the amalgamated free product $A *_C B$. We then have $G_o \between G_\updownarrow = G_o \times G_\updownarrow$ and, by Lemma 4.1, $e \otimes u = f \otimes v$ if and only if $e = f$ and $u = gv \in A *_C B$ for some $g \in G_e$ (where $G_e = A, B$ or C). It follows that we can identify $G_o \otimes G_\updownarrow$ with the set of all right cosets of A, B and C in $G_\updownarrow$,where $e_A \otimes u$ is identified with Au and so on. Under this identification, $G_o\lambda_* = \{A, B, C\}$, and the ordering is given by $Aw \leqslant Cw \geqslant Bw$ for all $w \in A *_C B$. The action of $G_\updownarrow$ is the usual right action on the cosets: $(Au)v = A(uv)$ (and similarly for the cosets of B and C). Thus $(w, A) \in P$ if and only if $Aw^{-1} = A$, that is, if and only if $w \in A$. In this way, we recover a description of P as the disjoint union of A, B and C. If $a \in A$ and $c \in C$ then $(a, A) \leqslant (c, C)$ if and only if $a = c \in A *_C B$, that is if and only if $a = c\alpha$.

There is also a natural identification of $G_o \otimes G_\updownarrow$ with the standard tree on which $G_\updownarrow = A *_C B$ acts according to Bass-Serre theory [17]. Here a graph Y with vertex set VY and edge set EY is identified with the poset $X = VY \sqcup EY$ in which each edge is greater than or equal to its two incident vertices.

5.3. *HNN extensions*

We can treat HNN extensions in a similar way. Let A be a group and let C, D be subgroups of A with an isomorphism $\varphi : C \to D$. Let $\mathbb{I}$ be the groupoid with set of identities $\mathbb{I}_o = \{e_0, e_1\}$ and with two non-identity arrows $\alpha : e_0 \to e_1$ and $\alpha^{-1} : e_1 \to e_0$. The ordered groupoid G is the disjoint union $G = C \sqcup (A \times \mathbb{I})$. Letting e denote the identity of both C and A, we have $G_o = \{e, (e, e_0), (e, e_1)\}$. The ordering on G is $(c\varphi, e_0) \leqslant c$ and $c \geqslant (c, e_1)$ for all $c \in C$. Therefore, in the group $G_\updownarrow$, $(c\varphi, e_0) = (c, e_1)$, and so $(c\varphi, e_0) = (c, e_1) = (e, \alpha^{-1})(c, e_0)(e, \alpha)$. It follows that $G_\updownarrow \cong A*_{C,\varphi}$, with (e, α) mapping to the stable letter $t \in A*_{C,\varphi}$. As in example **2**, we can identify $G_o \otimes G_\updownarrow$ with the set of all right cosets of A and C in $G_\updownarrow$, where $e \otimes u$ is identified with Cu, $(e, e_0) \otimes u$ with Au, and $(e, e_0) \otimes tv = (e, e_1) \otimes v$ by Lemma 4.1. The ordering on $G_o \otimes G_\updownarrow$ is then given by $Aw \leqslant Cw \geqslant Atw$, and we can recover from this ordering the standard tree on which $A*_{C,\varphi}$ acts [17].

References

1. S. Bulman-Fleming, J. B. Fountain and V. A. R. Gould, Inverse semigroups with zero: covers and their structure, *J. Austral. Math. Soc. (Series A)* **67**, 15-30 (1999).
2. N. D. Gilbert, HNN extensions of inverse semigroups and groupoids, *J. Algebra* **272**, 27-45 (2004).
3. G. M. S. Gomes and J. M. Howie, A P–theorem for inverse semigroups with zero, *Portugaliae Mathematica* **53**, 257-278 (1996).
4. P. J. Higgins, *Notes on categories and groupoids*, Van Nostrand Reinhold Math. Stud. 32 (1971). Reprinted electronically at `www.tac.mta.co/tac/reprints/articles/7/7tr7.pdf`.
5. M. V. Lawson, *Inverse Semigroups*, World Scientific (1998).
6. M. V. Lawson, E^*–unitary inverse semigroups,*Semigroups, Automata, Algebra and Languages*, G.M.S. Gomes *et al.* (Eds.), World Scientific (2002).
7. M. V. Lawson, Left cancellative categories and ordered groupoids, *Semigroup Forum* **68**, 458-476 (2004).
8. M. V. Lawson, In McAlister's footsteps: a random ramble around the P–theorem. Preprint (2005).
9. M. Loganathan, Cohomology of inverse semigroups, *J. Algebra* **70**, 375-393 (1981).
10. S. W. Margolis and J.-E. Pin, Inverse semigroups and extensions of groups by semilattices, *J. Algebra* **110**, 277-297 (1987).
11. D. B. McAlister, Groups, semilattices and inverse semigroups II, *Trans. Amer. Math. Soc.* **196**, 251-270 (1974).
12. D. B. McAlister, An introduction to E^*–unitary inverse semigroups – from an old fashioned perspective. In *Proc. Workshop on Semigroups and Languages, Lisbon 2002.* I.M. Araújo *et al.* (Eds.) World Scientific (2004).
13. W. D. Munn, Brandt congruences on inverse semigroups, *Proc. London Math. Soc* **14** (3), 154-164 (1964).
14. W. D. Munn, A note on E–unitary inverse semigroups, *Bull. London Math. Soc.* **8**, 71-76 (1976).
15. K. S. Nambooripad, *The structure of regular semigroups I.* Mem. Amer. Math. Soc. **224** (1979).
16. H. E. Scheiblich, Free inverse semigroups, *Semigroup Forum* **4**, 351-359 (1972).
17. J.-P. Serre, *Trees*, Springer Verlag (1980).
18. M. B. Szendrei, A generalisation of McAlister's P–theorem for E–unitary regular semigroups, *Acta Sci. Math. (Szeged)* **57**, 229-249 (1987).

ON THE FINITE BASIS PROBLEM FOR THE MONOIDS OF EXTENSIVE TRANSFORMATIONS*

I. A. GOLDBERG

Department of Mathematics and Mechanics
Ural State University
620083 Ekaterinburg, Russia
E-mail: Goldberg@dev.kontur-extern.ru

We obtain a description of identities holding in the monoids of extensive, partial extensive and partial order-preserving extensive transformations of a finite chain. Using the description, we partially solve the finite basis problem for the monoids. Namely, we show that "almost all" these monoids are nonfinitely based.

1. Introduction

Let Σ be an alphabet, Σ^* the free monoid over Σ. Recall that a *monoid identity* over the alphabet Σ is merely a pair $(w, w') \in \Sigma^* \times \Sigma^*$ usually written as $w = w'$. A monoid M *satisfies* the identity $w = w'$ if the equality $w\varphi = w'\varphi$ holds in M under all possible morphisms $\varphi : \Sigma^* \to M$. Given any collection I of monoid identities, we say that the identity $w = w'$ *follows* from I if every monoid satisfying all identities of I satisfies the identity $w = w'$ as well. A monoid M is said to be *finitely based* if all identities holding in M follow from a finite set of such identities; otherwise M is called *nonfinitely based.* Given a class $\mathcal{M}$ of finite monoids, the *finite basis problem for* $\mathcal{M}$ consists in determining which monoids in $\mathcal{M}$ are finitely based and which are not. While the finite basis problem for the class of all finite monoids still remains open, it has been solved for many important subclasses, for instance, for the full transformation monoids and the full relation monoids [2, 3]. A comprehensive report on the finite basis problem for various classes of finite monoids can be found in the recent survey [4].

*This work was partially supported by the Russian Foundation for Basic Research (grant No.05-01-00540), by the President Program of Leading Scientific Schools (grant No.2227.2003.01) and by Federal Education Agency of Russia (grants No.49123, No.04.01.437 and No.A04-2.8-928).

In this paper we address some important classes of transformation monoids of the finite chain $\{1,\ldots,n\}$. Recall that a partial transformation α is said to be *order-preserving* if, for all i and j from the domain of α, the condition $i \le j$ implies $i.\alpha \le j.\alpha$ and *extensive* if all i from the domain of α satisfy $i \le i.\alpha$. By $\mathcal{E}_n$, $\mathcal{PE}_n$ and $\mathcal{POE}_n$ we denote the monoids of all full extensive transformation, partial extensive transformations and partial order-preserving extensive transformations of the chain $\{1,\ldots,n\}$ respectively. These series of transformation monoids are known to serve as sort of "canonical" generators for the important pseudovariety of all $\mathcal{R}$-trivial monoids, see the survey [1]. The finite basis problem for these monoids was explicitly mentioned in [4]. A partial solution to the finite basis problem for the monoids $\mathcal{E}_n$ was announced by Ivanov at the 57th Herzen Readings in 2003 in St Petersburg but this result has never been published. In the present paper we show that "almost all" monoids $\mathcal{E}_n$, $\mathcal{PE}_n$ and $\mathcal{POE}_n$ are nonfinitely based, namely the following theorem holds:

Theorem 1.1. *For each positive integer n, the monoids $\mathcal{E}_{n+1}$, $\mathcal{PE}_n$ and $\mathcal{POE}_n$ satisfy the same identities. These monoids are nonfinitely based for all $n \ge 4$.*

The monoid $\mathcal{E}_1$ is trivial and each of the monoids $\mathcal{E}_2$, $\mathcal{PE}_1$, and $\mathcal{POE}_1$ is nothing but the two-element semilattice whence all these monoids obviously are finitely based. Ivanov has announced that the identities

$$xyzx = xyxzx, \quad xaybxy = xaybyx$$

form an identity basis for the 6-element monoid $\mathcal{E}_3$; combining this with our theorem, one concludes that also the monoids $\mathcal{PE}_2$ and $\mathcal{POE}_2$ are finitely based. The finite basis problem for $\mathcal{E}_4$, $\mathcal{PE}_3$ and $\mathcal{POE}_3$ still remains open.

Observe that the situation when in a sequence of finite transformation monoids (naturally indexed by the size n of the base set) all monoids except a few ones at the beginning of the sequence are nonfinitely based is quite common. For instance, the same behavior is demonstrated by the monoids $\mathcal{T}_n$ of all full transformations or by the monoids $\mathcal{C}_n$ of all full order-preserving extensive transformations of $\{1,2,\ldots,n\}$: the monoid $\mathcal{T}_n$ is finitely based if and only if $n \le 2$, cf. [3], and the monoid $\mathcal{C}_n$ is finitely based if and only if $n \le 4$, cf. [5]. It is very tempting to find out some general reason that forces "large enough" transformation monoids to be nonfinitely based. We note that general methods for proving that a monoid is nonfinitely based known so far (see [2, 3]) work well in the case of $\mathcal{T}_n$ but apply to neither $\mathcal{C}_n$ nor the monoids considered in the present paper.

2. Proof of Theorem 1.1

Let w be a word over the alphabet Σ. By $c(w)$ we denote the *alphabet of* w, i.e. the set of all letters appearing in w. We say that a word $v = a_1 \cdots a_m$, where $a_i \in \Sigma$, is a *scattered subword* of w if w can be represented as

$$w = w_0 a_1 w_1 a_2 \cdots w_{m-1} a_m w_m. \tag{1}$$

If for the decomposition (1) the condition $a_i \notin c(w_{i-1})$ holds for all $i = 1, \dots, m$, then this decomposition is called *the first occurrence of the scattered subword* v *in the word* w and the words w_i ($i = 0, \dots, m$) are called *the first occurrence factors for the word* v *in the word* w.

We define the relation $\equiv_n$ on the free monoid Σ^* by letting $w \equiv_n w'$ if and only if the following conditions hold:

1) the words w and w' have the same sets of scattered subwords of length $\leq n$;
2) if $v = a_1 a_2 \cdots a_m$, where $m \leq n$, is an arbitrary common scattered subword of w and w' and

$$w = w_0 a_1 w_1 a_2 \cdots w_{m-1} a_m w_m, \tag{2}$$
$$w' = w'_0 a_1 w'_1 a_2 \cdots w'_{m-1} a_m w'_m, \tag{3}$$

are the first occurrences of v in w and w', then the alphabets of the first occurrence factors coincide, i.e. $c(w_i) = c(w'_i)$ for $i = 0, \dots, m$.

It is easy to see that the relation $\equiv_n$ is a congruence on Σ^*.

Proposition 2.1. *Let $w \equiv_n w'$. Then the words w and w' have the same sets of scattered subwords of length $n+1$.*

Proof. Let $v = a_1 \cdots a_n a_{n+1}$ be a scattered subword of w. From the condition $w \equiv_n w'$ it follows that $a_1 \cdots a_n$ is a scattered subword of w' and the first occurrence factors in the decomposition (1) have the same alphabets. But $a_{n+1} \in c(w)$ whence $a_{n+1} \in c(w')$, and therefore, v is a scattered subword of w'. □

The following proposition describes the identities of the monoid of all full extensive transformations of the finite chain.

Proposition 2.2. *The identity $w = w'$ holds in the monoid $\mathcal{E}_{n+2}$ if and only if $w \equiv_n w'$.*

Proof. Let the identity $w = w'$ hold in $\mathcal{E}_{n+2}$. Consider a scattered subword $v = a_1a_2\cdots a_m$ $(m \leq n)$ of the word w and its first occurrence (2). Now we construct the morphism $\varphi : \Sigma^* \to \mathcal{E}_{n+2}$. For every letter $a \in \Sigma$, we put

$$i.(a\varphi) = \begin{cases} i & \text{if } a \in c(w_{i-1}), \\ i+1 & \text{if } a = a_i, \\ n+2 & \text{otherwise,} \end{cases} \qquad \text{for each } i = 1, \dots, m;$$

$$(m+1).(a\varphi) = \begin{cases} m+1 & \text{if } a \in c(w_m), \\ n+2 & \text{if } a \notin c(w_m); \end{cases}$$

$$j.(a\varphi) = j \qquad \text{for each } j = m+2, \dots, n+2.$$

The transformation $a\varphi$ obviously belongs to the monoid $\mathcal{E}_{n+2}$. By the construction, $1.(w\varphi) = m+1$ whence $1.(w'\varphi) = m+1$. It is easy to check that the latter condition holds only if v is a scattered subword of w' and the first occurrence factors for v in w' satisfy the conditions $c(w'_i) \subseteq c(w_i)$, where $i = 0, 1, \dots, m$. Applying symmetric considerations, we obtain the inverse inclusions.

Now assume that $w \equiv_n w'$. We have to show that, for every morphism $\varphi : \Sigma^* \to \mathcal{E}_{n+2}$ and for every $k = 1, \dots, n+2$, the equality $k.(w\varphi) = k.(w'\varphi)$ holds. We construct a scattered subword $v = a_1 \cdots a_m$ of the word w satisfying the following properties:

$$w = w_0a_1w'_0, \text{ where } k.(w_0\varphi) = k, \ k.((w_0a_1)\varphi) = k_1 \neq k;$$
$$w'_0 = w_1a_2w'_1, \text{ where } k_1.(w_1\varphi) = k_1, \ k_1.((w_1a_2)\varphi) = k_2 \neq k_1;$$
$$\dots$$
$$w'_{m-2} = w_{m-1}a_mw_m, \text{ where } k_{m-1}.(w_{m-1}\varphi) = k_{m-1},$$
$$k_{m-1}.((w_{m-1}a_m)\varphi) = k_m \neq k_{m-1},$$
$$k.(w\varphi) = k_m.$$

Obviously, the length m of the word v does not exceed $n+1$ and the decomposition

$$w = w_0a_1w_1a_2 \cdots w_{m-1}a_mw_m$$

is the first occurrence of v in w. Hence by Proposition 2.1 v is a scattered subword of w'. Take the first occurrence of v in w':

$$w' = w'_0a_1w'_1a_2 \cdots w'_{m-1}a_mw'_m.$$

Now we consider two possible cases.

Case 1: $m \leq n$. In this case, the alphabets of the first occurrence factors for v should coincide, i.e. $c(w_0) = c(w'_0), \dots, c(w_m) = c(w'_m)$. Obviously,

$$k.(w_0\varphi) = k.(w'_0\varphi) = k, \quad k.((w_0a_1w_1)\varphi) = k.((w'_0a_1w'_1)\varphi) = k_1, \dots,$$
$$k.((w_0a_1 \cdots a_mw_m)\varphi) = k.((w'_0a_1 \cdots a_mw'_m)\varphi) = k_m.$$

Case 2: $m = n+1$. In this case we have $k = 1$, $k_m = n+2$. The alphabets of the first occurrence factors for the word $a_1 \cdots a_{m-1}$ in w and w' should coincide whence

$$k.((w'_0a_1w'_1 \dots w'_{m-2}a_{m-1})\varphi) = k.((w_0a_1w_1 \dots w_{m-2}a_{m-1})\varphi) = n+1.$$

Since $(n+1).(a_m\varphi) = n+2$ and $a_m \in c(w'_m)$, we get $k.(w'\varphi) = n+2$. □

Proposition 2.3. *The monoid $\mathcal{E}_{n+4}$ is nonfinitely based for all $n \geq 1$.*

Proof. In order to show that the monoid $\mathcal{E}_{n+4}$ is nonfinitely based, we construct an infinite series of identities of this monoid and verify that "long" identities in this series cannot be deduced from any set of "short" identities of $\mathcal{E}_{n+4}$. Let $x_1, x_2, \dots, x_m$ be arbitrary distinct letters of Σ. By $\overrightarrow{z}$ and $\overleftarrow{z}$ we denote the words $x_1x_2 \cdots x_m$ and $x_mx_{m-1} \cdots x_1$ respectively. Let us consider the following identities:

$$\overrightarrow{z}(\overleftarrow{z})^n x_m^2 = \overrightarrow{z}(\overleftarrow{z})^n x_m^3. \tag{4}$$

Lemma 2.1. *For every positive integer m, the identity* (4) *holds in the monoid $\mathcal{E}_{n+4}$.*

Proof. Let v be an arbitrary scattered subword of $\overrightarrow{z}(\overleftarrow{z})^n x_m^3$ whose length does not exceed $n+2$. It is easy to check that the word v is contained in the word $\overrightarrow{z}(\overleftarrow{z})^n x_m$. Hence v is a scattered subword of $\overrightarrow{z}(\overleftarrow{z})^n x_m^2$. Thus the alphabets of the first occurrence factors for v in the words of the identity (4) coincide. □

Recall that a word w is said to be an *isoterm relative to a monoid M* if M satisfies no non-trivial identity of the form $w = w'$.

Lemma 2.2. *If the word $w \in \Sigma^+$ is a proper factor of the word $\overrightarrow{z}(\overleftarrow{z})^n x_m^2$, then w is an isoterm relative to $\mathcal{E}_{n+4}$.*

Proof. Obviously, it is sufficient to check that the words $x_2 \cdots x_m(\overleftarrow{z})^n x_m^2$ and $\overrightarrow{z}(\overleftarrow{z})^n x_m$ are isoterms relative to $\mathcal{E}_{n+4}$. Let the monoid $\mathcal{E}_{n+4}$ satisfy the identity $x_2 \cdots x_m(\overleftarrow{z})^n x_m^2 = v$. By Proposition 2.2 the factor x_2x_3 is a scattered subword of v. The alphabets of the respective first occurrence factors coincide, hence the word v can be represented in the form $v = x_2x_3v_1$ for some word $v_1 \in \Sigma^*$. Considering the factors $x_3x_4, \dots, x_{m-1}x_m, x_mx_m$ and applying the same arguments we show that $v = x_2x_3 \cdots x_mx_mv_2$. In the same way, we take the factor $x_mx_mx_{m-1}$ and conclude that $v = x_2x_3 \cdots x_mx_mx_{m-1}v_3$. Repeating the above arguments for the words $(x_m)^kx_px_{p-1}$ for all $k = 1, \dots, n$ and $p = m, \dots, 2$ and the word $x_m^nx_1x_m$, we obtain that $v = x_2x_3 \cdots x_m(\overleftarrow{z})^nx_mv_4$ and $c(v_4) = \{x_m\}$. Suppose $v_4 \neq x_m$. Then the word $x_1x_m^{n+2}$ is a scattered subword of v, but it is not a scattered subword of $x_2 \cdots x_m(\overleftarrow{z})^nx_m^2$. The latter fact contradicts Proposition 2.1.

The proof of the fact that $\overrightarrow{z}(\overleftarrow{z})^nx_m$ is an isoterm relative to $\mathcal{E}_{n+4}$ is analogous. □

Lemma 2.3. *Let $\mathcal{E}_{n+4}$ satisfy a non-trivial identity $w = \overrightarrow{z}(\overleftarrow{z})^nx_m^2$. Then the word w can be represented as $w = \overrightarrow{z}(\overleftarrow{z})^nx_m^k$ for some $k > 2$.*

Proof. The proof uses the same arguments as Lemma 2.2. □

To complete the proof of Proposition 2.3 we have to check that the identity $\overrightarrow{z}(\overleftarrow{z})^nx_m^2 = \overrightarrow{z}(\overleftarrow{z})^nx_m^3$ cannot be deduced from the identities of $\mathcal{E}_{n+4}$ containing less than m variables. Let us consider an arbitrary deduction of this identity, i.e. a sequence

$$\overrightarrow{z}(\overleftarrow{z})^nx_m^2 = v_0, v_1, \dots, v_r = \overrightarrow{z}(\overleftarrow{z})^nx_m^3,$$

where, for each i, there exist words $w_i, w_i' \in \Sigma^*$, a morphism $\varphi : \Sigma^+ \to \Sigma^+$ and a non-trivial identity $u_i = u_i'$ of $\mathcal{E}_{n+4}$ such that

$$v_{i-1} = w_i(u_i\varphi)w_i', \quad v_i = w_i(u_i'\varphi)w_i'.$$

We are going to show that the identity $u_1 = u_1'$ contains at least m variables. Applying Lemma 2.2, we may conclude that the factors w_1 and w_1' are empty, therefore, $\overrightarrow{z}(\overleftarrow{z})^nx_m^2 = u_1\varphi$. By Lemma 2.3, it follows that the word $u_1'\varphi$ can be represented as $u_1'\varphi = \overrightarrow{z}(\overleftarrow{z})^nx_m^k$ for some $k > 2$.

The words u_1 and u_1' are not equal and $u_1\varphi$ is a prefix of $u_1'\varphi$. This implies that either the words u_1 and u_1' have at least one distinct letter or u_1 is a prefix of u_1'. Suppose that the first condition holds, i.e.

$$u_1 = pas, \quad u_1' = pbs', \tag{5}$$

where $p \in \Sigma^*$, $a, b \in \Sigma$, $a \neq b$. We may assume that $a\varphi \neq b\varphi$ because otherwise we could apply the following morphism ψ:

$$x\psi = \begin{cases} b \text{ if } x = a, \\ x \text{ otherwise;} \end{cases}$$

and take the words $u_1\psi$ and $u_1'\psi$ instead of u_1 and u_1', thus decreasing the number of variables in the identity.

By t and t' we denote the number of occurrences of the letters a and b respectively in the word p. Obviously, the words $a\varphi$ and $b\varphi$ have the same first letter. We denote this letter by x_j. We define the transformation τ that interchanges the letters a and b: $a\tau = b$, $b\tau = a$. Then the following lemma holds:

Lemma 2.4. *Let for some $\ell \geq 0$ we have*

$$u_1 = pas_0bs_1\cdots(a\tau^\ell)s_\ell, \quad u_1' = pbs_0'as_1'\cdots(b\tau^\ell)s_\ell',$$

where $s_0,\ldots,s_\ell,s_0',\cdots s_\ell' \in \Sigma^*$ $a\tau^i \notin c(s_{i-1})$, $b\tau^i \notin c(s_{i-1}')$ for all values $i = 1,\ldots,\ell$. Then there exist words $\overline{s}_0,\ldots,\overline{s}_\ell,\overline{s}_{\ell+1},\overline{s'}_0,\ldots,\overline{s'}_\ell,\overline{s'}_{\ell+1}$, such that

$$u_1 = pa\overline{s}_0b\overline{s}_1\cdots(a\tau^{\ell+1})\overline{s}_{\ell+1}, \quad u_1' = pb\overline{s'}_0a\overline{s'}_1\cdots(b\tau^{\ell+1})\overline{s'}_{\ell+1},$$

where $a\tau^i \notin c(\overline{s}_{i-1})$, $b\tau^i \notin c(\overline{s'}_{i-1})$ for all values $i = 1,\ldots,\ell+1$.

Proof. The letter x_j appears in the word $(as_0bs_1\cdots(a\tau^\ell)s_\ell)\varphi$ at least $\ell+1$ times. Let q denote the number of occurrences of this letter in the word $u_1\varphi = \overrightarrow{z}(\overleftarrow{z})^n x_m^2$. Obviously, $t+\ell+1 \leq q \leq n+3$, hence $t \leq n+2-\ell$. Similarly, $t' \leq n+2-\ell$.

The word $a^tab\cdots(a\tau^\ell)$ has length $t+\ell+1 \leq n+3$ and it is a scattered subword of u_1. From Proposition 2.1 follows that this word is a scattered subword of u_1', hence $a\tau^\ell \in c(s_\ell')$, therefore there exist words $\overline{s'}_\ell$ and $\overline{s'}_{\ell+1}$ such that $a\tau^\ell = b\tau^{\ell+1} \notin \overline{s'}_\ell$ and

$$u_1' = pbs_0'as_1'\cdots(b\tau^\ell)\overline{s'}_\ell(b\tau^{\ell+1})\overline{s'}_{\ell+1}.$$

Thus, we have found the desired representation of the word u_1'. Moreover, by Proposition 2.2 the alphabets of the words p and pbs_0' coincide. This implies $b \in c(p)$ whence $t' \geq 1$.

Considering the scattered subword $b^{t'}ba\cdots(b\tau^\ell)$ of u_1' and applying the same arguments as above, we construct the desired representation for the word u_1. □

The proof of Lemma 2.1 also yields the following result:

Corollary 2.1. *Both a and b must occur in the word p.*

We have supposed that the words u_1 and u_1' can be represented in the form (5). We may apply Lemma 2.4 for $\ell = 0$ and apply it inductively till $\ell = n+1$. Thus we obtain that the letter x_j appears in the word $(a\overline{s}_0 b\overline{s}_1 \cdots (a\tau^{n+2})\overline{s}_{n+2})\varphi$ at least $n+3$ times, hence it can not appear in the word $p\varphi$. Therefore $a \notin c(p)$ in a contradiction to Corollary 2.1.

Thus we have proved that the word u_1 is a prefix of u_1'. Hence there exist $t \in \Sigma$, $\overline{u}_1 \in \Sigma^*$ such that $u_1' = u_1 t \overline{u}_1$. It is easy to realize that the condition $u_1 \equiv_{n+2} u_1'$ implies the fact that the letter t should appear in the word u_1 at least $n+3$ times. The only subword of the word $\overrightarrow{z}(\overleftarrow{z})^n x_m^2$ appearing at least $n+3$ times is x_m. Hence the letter t should appear in u_1 exactly $n+3$ times and $t\varphi = x_m$. Therefore the words u_1 and u_1' can be represented in the form

$$u_1 = r_0 t t r_2 \cdots r_{n+1} t^2, \quad u_1 = r_0 t t r_2 \cdots r_{n+1} t^3 \overline{u}_1,$$

moreover $r_0\varphi = x_1 \cdots x_{m-1}$ and $r_2\varphi = x_{m-1} \cdots x_1$.

Let us show that $c(r_0) \supseteq c(r_2)$. Arguing by contradiction, we assume that there exists a letter x such that $x \in c(r_2)$ and $x \notin (r_0)$. Then the word xt^{n+2} is a scattered subword of u_1' but is not a scattered subword of u_0, This contradicts Proposition 2.1. Hence $c(r_0) \supseteq c(r_2)$.

Since $r_2\varphi$ is a product of distinct letters, the word r_2 must also be a product of distinct letters. The words $r_0\varphi$ and $r_2\varphi$ have no common subwords of length 2. Hence, for every letter $y \in c(r_2)$, we have $|y\varphi| = 1$. Therefore $|c(r_2)| = m-1$, and the identity $u_1 = u_1'$ contains at least m variables.

Proposition 2.4. *The identity $w = w'$ holds in the monoid $\mathcal{POE}_{n+1}$ if and only if $w \equiv_n w'$.*

Proof. We restrict ourselves by a schema of the proof because it basically repeats the arguments from the proof of Proposition 2.2.

For an arbitrary identity $w = w'$ of $\mathcal{POE}_{n+1}$ and an arbitrary scattered subword $v = a_1 \cdots a_m$ $(m \leq n)$ of w we construct a morphism $\varphi : \Sigma^* \to \mathcal{POE}_{n+1}$ as follows. Let $a \in \Sigma$. We put

$$i.(a\varphi) = \begin{cases} i & \text{if } a \in c(w_{i-1}), \\ i+1 & \text{if } a = a_i, \\ \text{not defined} & \text{otherwise,} \end{cases} \qquad \text{for each } i = 1, \ldots, m;$$

$$(m+1).(a\varphi) = \begin{cases} m+1, & \text{if } a \in c(w_m), \\ \text{not defined}, & \text{if } a \notin c(w_m); \end{cases}$$

$$j.(a\varphi) = j \qquad \text{for each } j = m+2, \ldots, n+1.$$

The argument that shows that $w \equiv_n w'$ is identical to the first part of the proof of Proposition 2.2.

Now let w and w' be the words such that $w \equiv_n w'$, $k = 1, \ldots, n+1$, $\varphi : \Sigma^* \to \mathcal{POE}_{n+1}$. Suppose $k.(w\varphi)$ is defined. As in Proposition 2.2, we construct a scattered subword $a_1 \cdots a_m$ whose length m does not exceed n. Applying the same argument as in the second part of the proof of Proposition 2.2, one can easily show that $k.(w'\varphi)$ is defined and $k.(w\varphi) = k.(w'\varphi)$. □

The following property of the monoid $\mathcal{PE}_m$ is probably known but for the sake of completeness we provide its proof.

Proposition 2.5. *For every number m the monoid $\mathcal{PE}_m$ is isomorphic to the monoid $\mathcal{E}_{m+1}$.*

Proof. We construct the mapping $\varphi : \mathcal{PE}_m \to \mathcal{E}_{m+1}$ as followed: for every transformation $x \in \mathcal{PE}_m$ and every $i = 1, \ldots, m+1$, we put

$$i.(x\varphi) = \begin{cases} i.x & \text{if } i.x \text{ is defined,} \\ m+1 & \text{if } i = m+1 \text{ or } i.x \text{ is not defined.} \end{cases}$$

Obviously, the transformation $x\varphi$ is extensive. We have to prove that φ is an isomorphism. In order to check that transformation φ is surjective we take an arbitrary transformation $y \in \mathcal{E}_{m+1}$. The pre-image of y is the transformation $x \in \mathcal{PE}_m$ such that

$$i.x = \begin{cases} i.y & \text{if } i.y \neq m+1, \\ \text{not defined otherwise.} \end{cases}$$

Let us check that φ is injective. Suppose $x\varphi = z\varphi$ for some $x,\ z \in \mathcal{PE}_m$. Obviously, x and z coincide on their common domains (this follows from the first line of the definition of φ above). Moreover, the domains of x and z coincide (this follows from the second line). Thus φ is one-to-one correspondence between $\mathcal{PE}_m$ and $\mathcal{E}_{m+1}$.

Now we prove that for every $x,\ y \in \mathcal{PE}_m$ the equality $(xy)\varphi = (x\varphi)(y\varphi)$ holds. Fix an arbitrary number $i = 1, 2, \ldots, m$.

Case 1: If both $i.x$ and $i.(xy) = (i.x).y$ are defined, then

$$i.((xy)\varphi) = i.(xy) = (i.x).y = (i.(x\varphi)).(y\varphi) = i.(x\varphi)(y\varphi).$$

Case 2: If at least one of $i.x$ and $i.(xy) = (i.x).y$ is not defined, then $i.((xy)\varphi) = m+1$ and $i.(x\varphi)(y\varphi) = m+1$.

Hence φ is an isomorphism. □

Propositions 2.2, 2.3, 2.4 and 2.5 imply the statement of Theorem 1.1.

References

1. P. M. Higgins, Pseudovarietes generated by transformation semigroups, in S. Kublanovsky, A. Mikhalev, J. Ponizovskii, and P. Higgins (eds.), *Semigroups and Their Applications, Including Semigroup Rings*, St Petersburg State Techn. Univ., St Petersburg, 85–94 (1999).
2. M. V. Sapir, Problems of Burnside type and the finite basis property in varieties of semigroups, *Izv. Akad. Nauk SSSR, Ser. Mat.* **51**, 319–340 (1987) [Russian; Engl. translation *Math. USSR-Izv.* **30**, 295–314 (1987)].
3. M. V. Volkov, On finite basedness of semigroup varieties, *Mat. Zametki* **45** (3), 12–23 (1989) [Russian; Engl. translation *Math. Notes* **45**, 187–194 (1989)].
4. M. V. Volkov, The finite basis problem for finite semigroups, *Sci. Math. Japon.* **53**, 171–199 (2001).
5. M. V. Volkov, Reflexive relations, extensive transformations and piecewise testable languages of a given height, *Int. J. Algebra and Computation* **14**, 817–827 (2004).

A FREIHEITSSATZ FOR SUBSEMIGROUPS OF ONE-RELATOR GROUPS WITH SMALL CANCELLATION CONDITION

ARYE JUHÁSZ

Department of Mathematics
Technion, Israel Institute of Technology
Haifa 32000, Israel
E-mail: arju@tx.technion.ac.il

In this work we define Magnus subsemigroups of one-relator groups as a generalisation of Magnus subgroups. In contrast with Magnus subgroups which by the Dehn-Magnus Freiheitssatz are freely generated, Magnus subsemigroups are not necessarily free. We give a necessary and sufficient condition for a Magnus subsemigroup to be free in terms of the combinatorial structure of the defining relator R, provided that the given presentation satisfies the small cancellation condition $C(6)\&T(4)$ and R is not exceptional. We use word combinatorics with analysis of van Kampen diagrams, as developed in [1].

Introduction

A group G is termed *one-relator* if it has a presentation $gp\langle X \mid R\rangle$ with a single relator R which we assume to be cyclically reduced. The systematic study of one-relator groups started with the works of Wilhelm Magnus, some 70 years ago. Most notably he solved the word problem for one-relator groups [3]. A key ingredient in his treatment was a class of subgroups, the class of Magnus Subgroups, which are generated by a proper subset of X. Magnus proved that if X_0 is a subset of X which misses at least one letter that occurs in R or (R^{-1}), then $gp\langle X_0\rangle$ is free, freely generated by X_0. Inspired by this result we consider the Magnus subsemigroups of G.

Definition 0.1. Let $G = gp\langle X \mid R\rangle$ be a one-relator group. A *Magnus subsemigroup* S is a subsemigroup of G generated by a proper subset of $X \cup X^{-1}$.

Keywords: Word combinatorics, van Kampen diagrams, subsemigroups, one-relator groups, Freiheitssatz.

In the light of the works of Magnus it is natural to ask *are Magnus subsemigroups free*?

Before trying to answer this question let us see a few examples.

Example 0.1.

$$G = gp\langle\ a, b \mid a^{-1}b^{-1}ab\ \rangle,\ H_1 = sg\langle a, b\rangle,\ (\text{“}sg\text{” for semigroup})$$

Since $ab = ba$ and both ab and ba are elements of H_1, H_1 cannot be free.

Example 0.2.

$$G = gp\langle\ a, b, c, d \mid R\ \rangle,\ R = a^{-1}b^{-1}abc^{-1}d^{-1}cd,\ H_2 = sg\langle a, b, c^{-1}, d^{-1}\rangle.$$

Consider the cyclic conjugate $R^* := baR^{-1}a^{-1}b^{-1}$ of R^{-1}. Thus, if we write $\bar{x}$ for x^{-1} then $R^* = ba(\bar{d}\bar{c}dc\bar{b}\bar{a}ba)\bar{a}\bar{b} = ba\bar{d}\bar{c}dc\bar{b}\bar{a}(ba\bar{a}\bar{b}) = ba\bar{d}\bar{c}dc\bar{b}\bar{a}$. Since we can write R^* as UV^{-1} with $U = ba\bar{d}\bar{c}$ and $V = ab\bar{c}\bar{d}$ hence $U = V$ in G. But $U, V \in H_2$, hence H_2 cannot be free.

Example 0.3.

$$G = gp\langle a, b, c, d \mid R\ \rangle,\ R \text{ as in Example 0.2 },\ H_3 = sg\langle a, b, c, d\ \rangle.$$

Since our relation has the structure of an alternating word in pairs, regarding the exponents, it is impossible to find a cyclic conjugate R^* of R or R^{-1} and words U and V in H_3 as in the previous example such that $R^* = UV^{-1}$ with $U, V \in H_3$. But still there is the possibility that R has a consequence UV^{-1} with $U, V \in H_3$. Indeed, this group is known to be a small cancellation group, (see [2], Ch. V) and applying standard results in Small Cancellation Theory to this particular example, it is easy to show that R cannot have a consequence UV^{-1} with $U, V \in H$. Hence H_3 is freely generated by $\{a, b, c, d\}$.

Example 0.4.

$$G = gp\langle a, b, c, d \mid ababa(cd)^2(b\bar{d})^2\rangle \quad H_4 = sgp\langle a, b, c, d\ \rangle.$$

Then it is easy to see that neither R nor R^* may have a cyclic conjugate R^* with $R^* = UV^{-1}$ reduced as written with $U, V \in H_4$. However, in this case standard small cancellation consideration are not able to prove that R has no consequence UV^{-1}, $U, V \in H_4$.

Our main Theorem is addressed to solve this problem, under the assumptions that R satisfies a small cancellation condition.

As examples 0.1 and 0.2 above show, if R has a cyclic conjugate which can be written as UV^{-1}, where $U, V \in S_{Y_0}$, $S_{Y_0} = sg\langle Y_0\rangle$, then clearly S_{Y_0} is not free. Our main theorem shows that under suitable conditions (see below) these are the only cases where S_{Y_0} is not freely generated by Y_0.

Main Theorem. *Let G be a one-relator group given by a one-relator presentation $\mathcal{P}_R = \langle X|R\rangle$, $|X| \geq 2$, and let Y be a proper subset of $X \cup X^{-1}$. Suppose that each of the following holds:*

(a) $\mathcal{P}_R$ satisfies the small cancellation condition $C(6)\&T(4)$;

(b) R has no cyclic conjugate $ABA^{-1}C$, where A and B are pieces and one of the following holds for some $y \in Y$ such that $y^{-1} \notin Y$:

(i) B contains y and y^{-1}, A contains y but not y^{-1} and C contains y^{-1} but not y.

(ii) A contains y and y^{-1}, B contains y but not y^{-1} and C contains y^{-1} but not y.

Then S_Y is a free subsemigroup of G if and only if R has no cyclic conjugate $W_1W_2^{-1}$ reduced as written, such that $W_1, W_2 \in S_Y$.

Remark 1. We call words R which satisfy condition (b) non-exceptional. With considerably more work on word combinatorics the conditions of part (b) can be avoided.

The work is organized as follows: In Section 1 we prove the Main Theorem, relying on Proposition 4.1. In Section 2 we recall the necessary results on diagrams and develop some word combinatorics. In Section 3, using word combinatorics we show for every van Kampen diagram the existence of a special type of boundary regions which contains every letter from $X \cup X^{-1}$. This result together with Theorem 2.1 is the main ingredient of the proof of Proposition 4.1. In Section 4 we prove Proposition 4.1.

1. Proof of the Theorem

Let G be the group defined by $\mathcal{P}$ and denote by "$\underset{G}{=}$" equality in G. Like in the case of Magnus subgroups we may assume without loss of generality that if $Z \subseteq X \cup X^{-1}$ is the set of all the letters that occur in R then

$$Z \cup Z^{-1} = X \cup X^{-1} \tag{0}$$

Let Y be a proper subset of $X \cup X^{-1}$ and let $H = sg\langle Y\rangle$. If $Y \cup Y^{-1} \neq X \cup X^{-1}$ then we may apply the Freiheitssatz for groups, hence we shall assume

$$Y \cup Y^{-1} \neq X \cup X^{-1} \tag{1}$$

Now, Y can be decomposed to $Y_0 \cup Y_1$, where Y_0 is the set of those letters in Y the inverses of which are not in Y. Thus

$$Y = Y_0 \cup Y_1,\ Y_0 = Y_0^{-1} \ \textit{and}\ Y_1 \cap Y_1^{-1} = \varnothing \tag{2}$$

Clearly, if $Y_1 = \varnothing$ then $sg(Y)$ is a subgroup of G, hence again the Freiheitssatz for groups holds. Hence we shall assume

$$Y_1 \neq \varnothing,\ Y_0 \cup Y_1 \cup Y_1^{-1} = X \cup X^{-1} \tag{2'}$$

Suppose the Theorem is false. Then the following two assertions (3) and (4) hold. First, there exist reduced words U and V in H such that $U \underset{G}{=} V$. Since a subword of a word in H is also in H, we may assume without loss of generality that U and V have different first letters and different last letters. Hence

there exist reduced words U and V in H such that UV^{-1} is cyclically reduced and $UV^{-1} \underset{G}{=} 1$ (3)

no cyclic conjugate of R and R^{-1} can be written by $W_1W_2^{-1}$ with $W_1, W_2 \in H$ (4)

Consider now the structure of the word R. Every reduced word W on $X \cup X^{-1}$ can be written as

$A_1B_1 \cdots A_mB_m$, where A_i if non-empty, contains elements of Y_1 but not elements of Y_1^{-1} and B_i if non-empty, contains elements of Y_1^{-1} but not of Y_1 such that $B_i \neq 1$ for $i = 1, \ldots, m-1$ and $A_i \neq 1$ for $i = 2, \ldots, m$ (5)

If in addition we require that the last letter of A_i is from Y_1 and the last letter of B_i is from Y_1^{-1}, and we shall do so, then the expression in (6) for W is unique. Denote by $\mathcal{A}$ the set of all possible words A_i and by $\mathcal{B}$ the set of all possible words B_i then $\mathcal{A}^{-1} = \mathcal{B}$, hence if W has decomposition (5) then W^{-1} has decomposition $A_1'B_1' \cdots A_m'B_m'$ where $A_i' = B_{m-i+1}^{-1}$ and $B_i' = A_{m-i+1}^{-1}$. In particular, if we denote the length of (5) by $\|W\|$ then $\|W\| = \|W^{-1}\|$. Since we consider R as a cyclic word, by taking a cyclic conjugate if necessary we may assume that in the expression of R in (5), $A_1 \neq 1$ and $B_n \neq 1$. Clearly $\|R\| \geq 2$ and $\|R\|$ is even. We claim that if the result of the Theorem is false then $\|R\| \geq 4$. For suppose $\|R\| < 4$. Then $\|R\| = 2$, hence $R = R_1R_2^{-1}$ with $R_1, R_2 \in H$, violating (4). Hence

If the Theorem is false then $\|R\| \geq 4$. (6)

We turn now to the proof of the Theorem. We are going to use van Kampen diagrams with small cancellation condition. See [2], Ch. V.

Assume the Theorem is false. Then due to (3) it follows from van Kampen's Lemma (see [2]) that there exists a connected, simply-connected diagram M with cyclically reduced boundary label UV^{-1} with $U, V \in H$. Due to (6) we may apply Proposition 4.1 which implies that M is a one-layer diagram. Let $M = \langle D_1, \ldots, D_r \rangle$, D_i the regions of M occuring in this order. See Fig. 1.

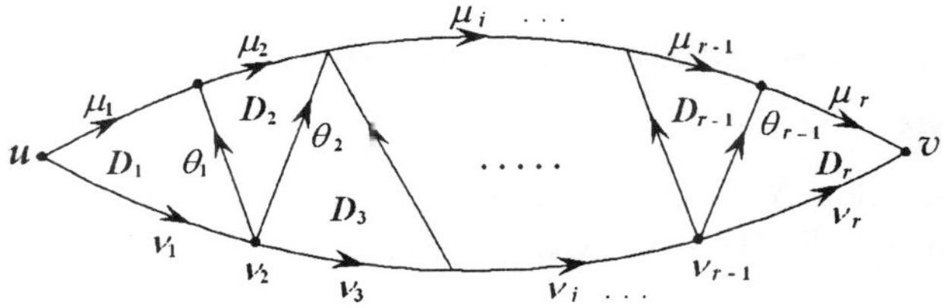

Figure 1.

Let μ and ν be boundary paths of M with $\Phi(\mu) = U$ and $\Phi(\nu) = V$ with corresponding endpoints u and v, where Φ is the labeling function of M. Thus, $u\mu v\nu^{-1}u$ is a boundary cycle of M. Let $u\mu v = u\mu_1 u_1 \mu_2 \ldots \mu_r v$ and let $\nu = u\nu_1 v_1 \nu_2 v_2 \ldots \nu_r v$, where $\mu_i = \partial D_i \cap \mu$ and $\nu_i = \partial D_i \cap \nu$, $i = 1, \ldots, r$, u_i and v_i vertices. Let $\theta_i = \partial D_i \cap \partial D_{i+1}$, $i = 1, \ldots, r-1$. Then u_i and v_i are endpoints of θ_i. The case $\mu_i = \varnothing$ or $\nu_i = \varnothing$ are not excluded, however we cannot have them both, due to the $C(6)$ condition. Let R be given by (5). Then there are non-empty paths $\alpha_1, \ldots, \alpha_m$, $\beta_1, \ldots, \beta_m$ $m \geq 2$ with $\Phi(\alpha_i) = A_i$ and $\Phi(\beta_i) = B_i$, $i = 1, \ldots, m$ such that if ω is a boundary cycle of a region D with $\Phi(\omega) = R$ then $\omega = z_1\alpha_1 s_1 \beta_1 z_2 \alpha_2 \ldots s_m \beta_m$, z_i, s_i vertices and Φ is the labelling function of M (see 2.2). Let $A_i = A_i' A_i''$ such that $A_i' \in gp(Y_0)$ and A_i'' starts with a letter from Y_1 and similarly, let $B_i = B_i' B_i''$ such that $B_i' \in gp(Y_0)$ and B_i'' starts with a letter from Y_1^{-1}, $i = 1, \ldots, m$. Let $\alpha_i = \alpha_i' w_i \alpha_i''$ and $\beta_i = \beta_i' t_i \beta_i''$, w_i, t_i vertices such that $\Phi(\alpha_i') = A_i'$, $\Phi(\alpha_i'') = A_i''$, $\Phi(\beta_i') = B_i'$ and $\Phi(\beta_i'') = B_i''$. We call every vertex on $z_i \alpha_i' w_i$ a *source* and call every vertex on $s_i \beta_i' t_i$ a *sink*. It follows from (4) that

$$u \text{ is a source and } v \text{ is a sink} \tag{7}$$

neither μ_i *nor* ν_i *contains a source and* (8)

neither u_i *nor* v_i *are contained in a source.* (9)

It follows from (7), (8) and (9) that $m-1$ sources are on θ_1. Now $\theta_1^{-1} \subseteq \partial P_2$, hence by the same argument θ_2 contains exactly $m-(m-1)=1$ source. Thus, for odd i, θ_i contains $m-1$ sources and for even i, θ_i contains one source. Hence θ_{r-1} contains either one or $m-1$ sources. But then the complement of θ_{r-1} on D_r, (the last region of M) contains at least one source, violating (7)-(9). This contradiction proves the Theorem.

2. Preliminary results on words and diagrams

2.1. *Words*

In this subsection $X=\{x_1,\ldots,x_n\}, n\geq 2$ and F is the free group on X. All words are in F. We start with the following well known results.

Lemma 2.1. *Let A,B,C be reduced non-empty words such that AB and BC are reduced as written. If $|AB|\geq 2$ and $AB=BC$ then $A=KL$, $C=LK$ and $B=(KL)^\beta K$, $\beta\geq 0$.*

We introduce below majorisation, the key basic notion of the work.

Definitions and notations.

(a) Let $W\in F$, $W=x_{i_1}\ldots x_{i_k}$, $x_{i_j}\in X\cup X^{-1}$ reduced as written. Define $\operatorname{Supp}(W)=\{i_1,\ldots,i_k\}\subseteq\{1,\ldots,n\}$.

(b) Let W_1 and W_2 be reduce words in F. W_2 *majorises* W_1 if $\operatorname{Supp}(W_2)\supseteq\operatorname{Supp}(W_1)$. In this case write $W_2\succ W_1$.

(c) For W_1 and W_2 in part (b) define $W_1\sim W_2$ if $W_1\prec W_2$ and $W_2\prec W_1$. Thus $W_1\sim W_2$ if and only if $\operatorname{Supp}(W_1)=\operatorname{Supp}(W_2)$. Clearly " $\sim$ " is an equivalence relation, which contains the equality of elements in F.

(d) Denote by $\mathcal{H}(W)$ the set of initial subwords of W and by $\mathcal{T}(W)$ the set of terminal subwords of W. Also, for a reduced non-empty word W we denote by $h(W)$ the first letter of W and by $t(W)$ the last letter of W.

The following Lemma is immediate from the definition, hence its proof is omitted.

Lemma 2.2.

(a) *If A is a subword of B then $A \prec B$.*
(b) *If $A \prec B$ then $A^{\pm 1} \prec B^{\pm 1}$.*
(c) *If $A \sim B$ and $A \prec C$ then $B \prec C$.*
(d) *If $A = P_1 \dots P_m$, reduced as written and $P_i \sim Q$ for $i = 1, \dots, m$ then $A \sim Q$.*
(e) *If $A \succ P_1, \dots, P_m$ then $A \succ W(P_1, \dots, P_m)$, for every word W on $P_1, \dots, P_m$.*

Parts (a) and (b) of the following Lemma are immediate corollaries of Lemma 2.1 and Lemma 2.2.

Lemma 2.3.

(a) *Let A, B and C be as in Lemma 2.1. Then $B \prec A \sim C \sim AB \sim BC$.*
(b) *Let A, B, C and K be non-empty words. If $AB = KAC$, reduced as written then $B \succ A$, $B \succ C$ and $K \succ A$.*
(c) *Let K, Q, U, V and F be non-empty words such that KQ, UV, VU and KF are reduced as written. If $KQ = UV$ and $KF = VU$ then $Q \sim F \succ K, U, V$.*
(d) *Let B, Q, L, U and V be non-empty words such that BQ, UV, LB and VU are reduced as written. If $BQ = UV$ and $LB = VU$ then one of the following holds*

   (i) *$B = U$, $Q = L = V$ or*
   (ii) *$Q \succ B, U, V, L$ and $L \succ B, U, V, Q$ (hence $L \sim Q \sim UV$).*

(e) *Let L, K, Q_1, M and N be non-empty reduced words. If $KQ_1 = MN$ and $Q_1 M = LK$ reduced as written, then one of the following holds*

   (i) *$Q_1 = N = L$ and $K = M$ or*
   (ii) *$Q_1 \succ K, L, M, N$.*

Proof. We prove here only part (c), because the proofs of parts (d) and (e) follow the same line of proof. We prove part (c) by induction on $|UV|$. If $|UV| = 2$ then $K = U = F$, $Q = V = K$, hence the claim of part (c) of the Lemma holds true. Assume $|UV| > 2$ and consider the first equation, $KQ = UV$. Then one of the following holds:

Case 1 $K = U$. Then $Q = V$. Subtitution in the second equation gives $UF = VU$, hence the result follows by part (a).

Case 2 $K = UK_1,\ K_1 \neq 1$. Then $V = K_1Q$, hence $UK_1F = K_1QU$. By Lemma 2.1 $K_1F = MN \quad K_1Q = NM,\ U = (MN)^uM,\ u \geq 0$. Since $|MN| < |UV|$ and $K_1 \neq 1$, we may apply the induction hypothesis on these equations to give $F \succ K_1, M, N, \quad Q \succ K_1, M, N$ and $Q \sim F$. Consequently tracing back K and V we get $F \succ K, U, V$.

Case 3 $U = KU_1,\ U_1 \neq 1$. Then $Q = U_1V$. Subtitution in the second equation yields $KF = VKU_1$. The result follows by part (b) and Lemma 2.2.

The Lemma is proved. □

2.2. *Diagrams*

For basic results on diagrams see [2], Ch. V. We denote by Φ_M the labeling function of M over F. If M is fixed we shall write Φ for Φ_M.
We recall the main structure theorem from [1], where it is proved in a more general setting. (Observe that the condition $C(6)\&T(4)$ implies the condition $W(6)$ in [1].)

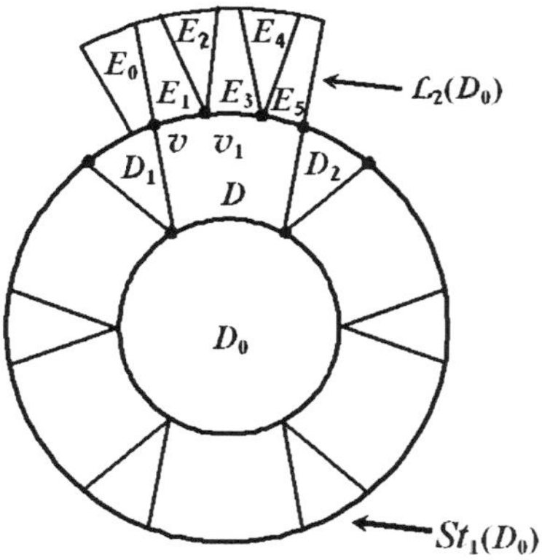

Figure 2.

$$P = \langle E_0, E_1, E_2, E_3, E_4, E_5 \rangle,\ P = 2,$$
$$\mathcal{A}(D) = \{D_0\},\ \mathcal{B}(D) = \{D_1D_2\},\ \mathcal{C}(D) = \{E_1, E_2, E_3, E_4, E_5\},$$
$$d_{St_1(D_0)}(v) = 3, \quad d_{St_1(D_0)}(v_1) = 2$$

Theorem 2.1. (Layer Decomposition,[1]) *See Fig. 2. Let M be a simply connected map (diagram) with connected interior and let D_0 be a region of M. Assume that M satisfies the condition $C(6)\&T(4)$. Define*

$St_0(D_0) = D_0$ and for $i \geq 1$ let $St_i(D_0) = St_{i-1}(D) \cup \mathcal{L}_i(D_0)$ where $\mathcal{L}_0 = \{D_0\}$, $\mathcal{L}_i(D_0) = \langle D$ in $M \backslash St_{i-1}(D_0) | \partial D \cap \partial St_{i-1}(D_0) \neq \varnothing \rangle$. Let p be the smallest number such that $St_p(D_0) = M$ and assume that $p > 0$ (i.e., M contains more than one region).
Then each of the following holds:

(a) Every regular submap of $St_{i+1}(D_0)$ containing $St_i(D_0)$ is simply connected, $0 \leq i \leq p$. (Here a submap is regular if every edge of it is on the boundary of a region.)
(b) Every connected and simply connected submap of $\mathcal{L}_i(D_0)$ is a one-layer map.
(c) For a region $D \in \mathcal{L}_i$, $i \geq 1$, denote by $\mathcal{A}(D)$ the set of regions E in $\mathcal{L}_{i-1}$, which have a non-trivial common edge with D, denote by $\mathcal{B}(D)$ the set of regions F in $\mathcal{L}_i$ with $\partial F \cap \partial D \neq \varnothing$ and let $\mathcal{C}(D)$ be the set of regions K in $\mathcal{L}_{i+1}$ with either $\partial K \cap \partial D$ non-empty or if empty it is an inner vertex of $\partial D \cap \partial St_i(D)$. Also, let $a(D) = |\mathcal{A}(D)|$, $b(D) = |\mathcal{B}(D)|$ and $c(D) = |\mathcal{C}(D)|$. Then $a(D) \leq 1$ and $b(D) \leq 2$. In other words, D has at most two neighbours in $\mathcal{L}_i$ and at most one neighbour in $\mathcal{L}_{i-1}$.
(d) If $v \in \partial St_i(D_0)$ then v has valency at most 3 in $St_i(D_0)$.

When D_0 is fixed, we shall abbreviate $\mathcal{L}_i(D_0)$ by $\mathcal{L}_i$ and call $\Lambda(D_0) = (\mathcal{L}_0, \ldots, \mathcal{L}_p)$ a *layer decomposition* of M. We call D_0 the *center* of the layer decomposition.

Remark 2. Let M be a connected simply connected map (diagram) with connected interior and let D be a region in M. Let $\Lambda(D)$ be a layer decomposition of M with center D. Suppose that D is a boundary region of M with a non-empty edge on ∂M. Then it follows from the above Theorem that $\mathcal{L}_1(D)$ is not annular, hence simply connected. But then due to the simply connectedness of M, $\mathcal{L}_i$ is simply connected for every i.

In the next definition we introduce special subdiagrams and regions the boundaries of which share a large portion with the boundary of M.

Definition 2.1.

(a) Let $\Lambda(D)$ be a layer decomposition of M where D is a boundary region of M with a non-empty edge on ∂M, $\Lambda(D) = (\mathcal{L}_0, \mathcal{L}_1, \ldots, \mathcal{L}_p)$. Let i be an integer, $1 \leq i \leq p$ and let P be a connected component of $\mathcal{L}_i$ with connected interior. We say that P *is a peak relative to* D if $c(E) = 0$ for every region E in P. See Fig. 2. Related to peaks are the following two notions.

(b) A boundary region D of M is a *k-corner region* for $k = 1, 2$ if $\partial D \cap \partial M$ is connected and D has k inner neighbours.

Example 2.1. Let P be the peak, depicted on Fig. 2. Then its right extremal region E_5 is a 2-corner region. If P is a peak consisting of a single region E then E is a 1-corner region.

Remark 3. It is a basic result in small cancellation theory that every $C(4)\&T(4)$ diagram which contains at least two regions, contains at least two k-corner regions, with $k \leq 2$. This is the fundamental "Greendlinger's Lemma". (See [2], Ch. V.)

3. Piece configurations of 1-corner regions and 2-corner regions

We start with the following basic notions.

Definition 3.1.

(a) Let R be a cyclically reduced word in F and let P be a subword of a cyclic conjugate of R. P is a *piece* in R (or relative to the symmetric closure $\mathcal{R}$ of R) if R has distinct elements R_1 and R_2 of $\mathcal{R}$ such that $R_1 = PR_1'$, $R_2 = PR_2'$, reduced as written and $R_1'R_2'^{-1} \neq 1$. We call the two occurrences of P in R_1 and R_2, respectively, a *piece-pair* and denote it by (P, P'), where $P' = P$ is the occurrence of P in R_2.

(b) A piece pair (P, P') as in part (a) of the definition is *right normalized* if $R_1'R_2'^{-1}$ is reduced as written.

Our aim in this section is to show that if D is a k-corner region of M with $k \leq 2$ and the endpoints of $\mu := \partial D \cap \partial M$ have valency 3 in M then μ has a "one-piece neighbourhood" θ the label of which majorises R or R^{-1}. Here, by a one-piece neighbourhood of μ we mean the largest subpath θ of ∂M of the form $\eta\mu\tau$, where η and τ are pieces. To achieve this it is enough to show that θ majorises the label of the complement of μ on ∂D and this is what we show below.

In this section we assume conditions (a) and (b) of the Main Theorem. Let μ be a path in M. For sake of brevity we shall not distinguish between μ and $\Phi(\mu)$. We shall also use the convention that if $\mu = \mu_1 v \mu_2$ and v a

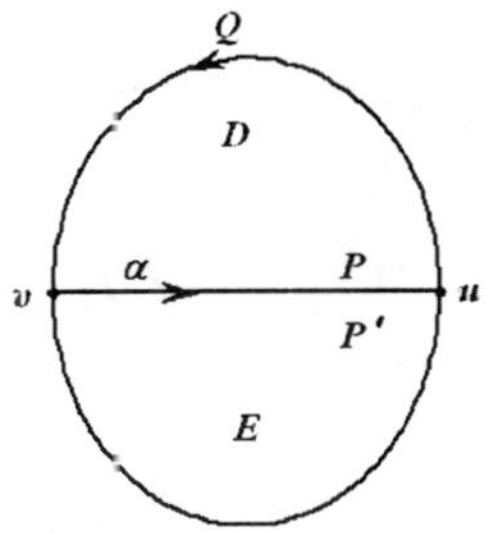

Figure 3.

vertex such that $\Phi(\mu_1) = P_1$, $\Phi(\mu_2) = P_2$ and $\Phi(\mu) = P$ then we shall write $P = P_1 v P_2$. We consider 1-corner regions first. Let D be a 1-corner region in M with inner neighbour E. Denote $\alpha = \partial D \cap \partial E$, $P = \Phi(\alpha)$ and let $Q = \Phi(\partial D \cap \partial M)$. See Fig. 3. Then P is a piece and $vPuQv$ is a boundary label of D, u, v vertices. Let (P, P') be the corresponding piece-pair. Then one of the following holds:

1) P' is a subword of Q;
2) P' contains u as an inner vertex and
3) P' contains v as an inner vertex.

In case 1) $Q \succ P$, by Lemma 2.2 (a). In case 2) we have $P = AX$, $P' = XY$, $Q = YQ_1$, reduced as written, $Q_1 \in \mathcal{H}(Q)$. Applying Lemma 2.3 to the first two of these equations and remembering that P^{-1} cannot overlap P, since F is free, we get $A \sim Y \succ X$ and hence, by Lemma 2.2, $P \sim Y$. Applying Lemma 2.3 to the last equation implies $Q \succ P$. Finally, Case 3 is dual of Case 2. Hence in all the cases $Q \succ P$.

Consider now 2-corner regions. We shall show the existence of the path θ explained above.

Let D be a 2-corner region in M with inner neighbours E_r and E_ℓ. Denote $\alpha_1 = \partial D \cap \partial E_r$ and denote $\alpha_2 = \partial D \cap \partial E_\ell$. Let $v_0 = \alpha_1 \cap \partial M$, let $v_2 = \alpha_2 \cap \partial M$ and let $v_1 = \alpha_1 \cap \alpha_2$. Denote $P_1 = \Phi(\alpha_1)$, $P_2 = \Phi(\alpha_2)$ and $Q = \Phi(\partial D \cap \partial M)$. Then $v_2 Q v_0 P_1 v_1 P_2 v_2$ is a boundary label of D, which we may assume without loss of generality to coincide with R. Thus P_1 and P_2 are pieces. See Fig. 4. Let(P_1, P_1') and (P_2, P_2') be the corresponding piece pairs. Then P_1' and P_2' are subwords of the cyclic word R or R^{-1}, hence one of the following holds for each of P_1' and P_2':

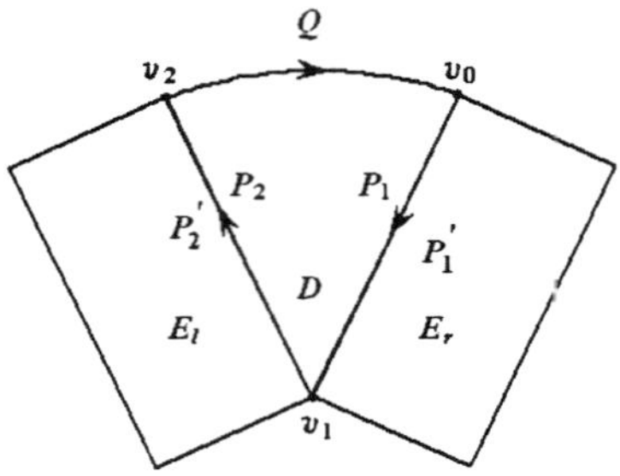

Figure 4.

<u>Case 1</u> v_0 is an inner vertex of P_1'; <u>Case 1′</u> v_2 is an inner vertex of P_2'

<u>Case 2</u> v_1 is an inner vertex of P_1'; <u>Case 2′</u> v_1 is an inner vertex of P_2'

<u>Case 3</u> v_2 is an inner vertex of P_1'; <u>Case 3′</u> v_0 is an inner vertex of P_2'

<u>Case 4</u> P_1' is a subword of P_2; <u>Case 4′</u> P_2' is a subword of P_1

<u>Case 5</u> P_1' is a subword of Q; <u>Case 5′</u> P_2' is a subword of Q

We propose to show that $Q \succ P_1P_2$ in most of the cases. (See precise statement below.) We see that for $i = 1, 2, 3, 4$ and 5, Case i' is the dual of Case i obtained by exchanging P_1 with P_2 and v_0 with v_2. Hence out of the 25 cases (i, j'), $1 \leq j'$, $i \leq 5$, only those with $j' \geq i$ have to be checked, because the rest is obtained by duality.

The following is the main result of this section.

Proposition 3.1. *Let the notation be as above and assume that R satisfies the assumptions of the Main Theorem. Suppose that R has a cyclic conjugate R^* with $\|R^*\| \geq 4$. Assume that the piece pairs (P_1, P_1') and (P_2, P_2') are right normalized. Let $Q_r = \partial E_r \cap \partial M$ and let $Q_\ell = \partial E_\ell \cap \partial M$.*

(a) *In cases (i, j') with $i = 1$, or $j' = 5$, or $i = 2$ and $j = 3$ and $v_2 \notin P_1'$ and $v_0 \notin P_2'$, or $i = 2$ and $j' = 4$ and $v_2 \in P_1'$, or $i = 2$ and $j = 5$ we have $Q \succ P_1P_2$.*

In all remaining cases the following hold

(b) *If $d_M(v_0) = 3$ and Q_r is not a piece, then Q_r has a head Q_ρ which is a piece in $\mathcal{R}$ such that $QQ_\rho \succ P_1^\varepsilon,\ P_2^\delta$ and dually, if $d_M(v_2) = 3$ and Q_ℓ is not a piece, then Q_ℓ has a tail Q_λ which is a piece in $\mathcal{R}$ such that $Q_\lambda Q \succ P_1^\varepsilon,\ P_2^\delta,\ \varepsilon, \delta \in \{1, -1\}$.*

(c) Let notation be as in parts (a) and (b) and suppose $\|Q\| < 2$. If $d_M(v_0) = 3$ then $\|QQ_\rho\| \geq 2$ and if $d_M(v_2) = 3$ then $\|Q_\lambda Q\| \geq 2$.

Proof.

(a) and (b) As explained above it is enough to check the 15 cases (i, j'), $1 \leq i \leq j' \leq 5$. All these cases are rather lengthy and easy to check, using Lemmas 2.2 and 2.3 by considerations similar to those used for 1-corner regions. We give here the details of Case (2,2) as an example. The rest is similar and full details can be obtained from the author.

Case (2, 2)

Consider 4 subcases according to whether v_2 belongs or does not belong to P_1' and dually, v_0 belongs or does not belong to P_2'.

Subcase 1. $v_2 \notin P_1'$ and $v_0 \notin P_2'$. Then $P_1 = XY$, $P_1' = YZ$, $P_2 = ZT$. Assume $d_M(v_0) = 3$, then $X^{-1} \in \mathcal{H}(Q_r)$. Define $Q_\rho = X$. Then we get $P_1 = Q_\rho Y = YZ$, $P_2 = ZT$. Using Lemma 2.2, the first pair of these equations gives

$$Q_\rho^{-1} \sim P_1. \tag{1}$$

But P_2 and P_2' overlap, since v_1 is an inner vertex of P_2', hence $P_2 = LV$, $P_2' = KL$ and $P_1 = UK$. Applying Lemma 2.2 to the first pair of equations gives

$$P_2 \sim K \sim L, \tag{2}$$

while the last equation gives

$$K \prec P_1. \tag{3}$$

From (1), (2) and (3) we get $P_1 \sim Q_\rho^{-1}$, $P_2 \prec Q_\rho$. By the dual argument $P_2 \sim Q_\lambda^{-1}$ and $P_1 \prec Q_\lambda$.

Subcase 2. $v_2 \notin P_1'$ and $v_0 \in P_2'$. Then (1) above still holds, and for P_2 and P_2' we get $P_2' = HP_1L$, $P_2 = LK$, where $H \in \mathcal{T}(Q)$. Thus, by Lemma 2.3, $HP_1 \sim P_2$, i.e., $QQ_\rho \succ P_2, P_1$.

Subcases 3 and 4. $v_2 \in P_1'$. Then $P_1 = XY$, $P_1' = YP_2T$ where $T \in \mathcal{H}(Q)$. Therefore, by Lemma 2.3, $X \sim P_1 \sim P_2T \succ Y$. Since $d_M(v_0) = 3$, hence $X \in \mathcal{H}(Q_r)$. Thus $P_1, P_2 \prec Q_r$. Also, $P_2' = UV$, $P_2 = VW$ $T^{-1}W^{-1} \in \mathcal{T}(Q_\ell)$. Hence $P_1 = YVWT$ and in particular WT is a piece. From the last two equations and Lemma 2.3 we get $U \sim W \sim P_2$, hence $Q_\ell \succ P_2^{-1}$. Let $Q_\lambda = T^{-1}W^{-1}$. Then Q_λ is a piece and $Q_\lambda \succ P_1, P_2$.

(c) Since $\|R\| \geq 4$, if $\|Q\| < 2$ then either $\|P_1\| \geq 2$ or $\|P_2\| \geq 2$. But $\|P_i\| \geq 2$ implies $\|P_i^{-1}\| \geq 2$, $i = 1, 2$ hence it follows from parts (a) and (b) that $\|QQ_\rho\| \geq 2$ and $\|Q_\lambda Q\| \geq 2$ □

4. Proposition 4.1 and its proof

Proposition 4.1. *Suppose that the assumptions of the Main Theorem hold. Let M be a van Kampen $\mathcal{R}$-diagram with boundary label UV^{-1} where $U, V \in S_Y$. If $\|R\| \geq 4$ and $\|UV^{-1}\| = 2$ then M is a one-layer diagram.*

We need Lemmas 4.1 and 4.2 for the proof of Proposition 4.1. In what follows we shall use the notation and rely on the assumptions of Proposition 4.1.

Lemma 4.1. *Let P be a peak of M in $\mathcal{L}_i(D)$ and suppose $\|(\partial P \cup \partial \mathcal{L}_{i-1}(D)) \cap \partial M)\| = 2$. If $|P| > 1$ then $P = \langle D_1, D_2 \rangle$ such that $a(D_1) + a(D_2) = 1$.*

Proof. Suppose $|P| \geq 3$, $P = \langle D_1, \ldots, D_k \rangle$, $k \geq 3$. Consider the extremal regions D_1 and D_k, D_1 first.

(i) If $a(D_1) = a(D_2) = 1$ let $\{E_1\} = \mathcal{A}(D_1)$ and $\{E_2\} = \mathcal{A}(D_2)$. If $E_1 = E_2$ then $v := \partial D_1 \cap \partial D_2 \cap \partial E_1$ is an inner vertex with valency 3, violating the condition $T(4)$. Hence $E_1 \neq E_2$ and since D_1 is extremal in P and D_2 is the only region of P adjacent to D_1, hence $\mathcal{C}(E_1) = \{D_1\}$ and $d_M(E_1) = a(E_1) + b(E_1) + c(E_1) \leq 2 + 1 + 1 = 4$. Consequently, due to the $C(6)$ condition $\partial E_1 \cap \partial M$ is the product of at least two pieces, hence if $u := \partial D_1 \cap \partial E_1 \cap \partial M$ then u is a vertex with valency 3 and every piece starting at u and read anticlockwise is contained in $\partial E_1 \cap \partial M$. Therefore noticing that $d_M(D_1) = 2$, we may apply Proposition 3.1 (c) to D_1 to get

$$\|(\partial E_1 \cap \partial M) \cup (\partial D_1 \cap \partial M)\| \geq 2 \tag{4}$$

(ii) If $a(D_1) = 0$ and $a(D_2) = 1$ then $d(D_1) = 1$, hence by Proposition 3.1 (a)

$$\|(\partial D_1 \cap \partial M)\| \geq 2 \tag{5}$$

(iii) If $a(D_1) = 1$ and $a(D_2) = 0$ then $d(D_1) = 2$ and $d(D_2) \leq 2$, hence by Proposition 3.1 (c)

$$\|(\partial D_1 \cap \partial M) \cup (\partial D_2 \cap \partial M)\| \geq 2 \tag{6}$$

It follows from (4), (5) and (6) that if we define $L = \langle E_1, D_1, D_2\rangle$ when $a(D_1) = 1$ and define $L = \langle D_1, D_2\rangle$ when $a(D_1) = 0$ then $\|\partial L \cap \partial M\| \geq 2$. A similar analysis shows that if $K = \langle E_k, D_k, D_{k-1}\rangle$ when $\mathcal{A}(D_k) = \{E_k\}$ and $K = \langle D_k, D_{k-1}\rangle$ if $a(D_k) = 0$ then $\|\partial K \cap \partial M\| \geq 2$. Consequently, if $k \geq 4$ and $P = \mathcal{L}_p(D)$ then

$$\|(\partial P \cup \partial\mathcal{L}_{p-1}(D)) \cap \partial M)\| \geq 3, \tag{7}$$

violating our supposition, hence $k \leq 3$. If $k = 3$ and one of cases (i) or (ii) above hold for D_1 (or for D_3) then (7) holds true. Assume therefore that $k = 3$ and case (iii) holds for both D_1 and D_3. Then $a(D_1) = 1$, $a(D_2) = 0$ and $a(D_3) = 1$. Now, $d(D_2) = a(D_2) + b(D_2) + c(D_2) = 2 + 1 + 0 = 3$, hence due to the $C(6)$ condition:

$\partial D_2 \cap \partial M$ *is the product of at least three* $(6 - 3 = 3)$ *pieces* $(*)$

Since $d(D_1) = d(D_3) = 2$, we may apply Proposition 3.1 (b) to the pairs (D_1, D_2) and (D_2, D_3), where in the notation of Proposition 3.1 in the first pair $D = D_1$ and $E_r = D_2$ while in the second pair $D = D_3$ and $E_l = D_2$. By their definition Q_ρ and Q_λ are pieces. Since $E_l = E_r = D_2$, Q_ρ is an initial subword of $\Phi(\partial D_2 \cap \partial M)$, which is a piece and Q_λ is a terminal subword of $\Phi(\partial D_2 \cap \partial M)$, which is a piece. Since $(\partial D_2 \cap \partial M)$ is the product of at least three pieces by $(*)$, Q_ρ and Q_λ do not overlap and hence $\|\partial P \cap \partial M\| \geq 3$ violating our supposition. Therefore $|P| = 2$ and if $a(D_1) = a(D_2) = 0$ or $a(D_1) = a(D_2) = 1$ then Proposition 3.1 (a) in the first case and Proposition 3.1 (b) in the second case with the arguments in (i) above imply that $\|(\partial\mathcal{L}_{i-1} \cup \partial\mathcal{L}_i) \cap \partial M\| \geq 3$. Therefore, $a(D_1) + a(D_2) = 1$.
The Lemma is proved. □

Lemma 4.2. *Let* Λ *be a layer structure for* M *and let* P_1 *be a peak of* M *relative to* Λ. *If* P_1 *is (an extremal) component of* $\mathcal{L}_i$ *then*

(a) $\|\partial P_1 \cap \partial M\| \geq 2$. *In particular,* $\|\partial M\| \geq 2$.
(b) *Either* P_1 *contains a region* D *with* $\|\partial D \cap \partial M\| \geq 2$ *or* P_1 *contains adjacent regions* D_1 *and* D_2 *such that* $\|(\partial D_1 \cup \partial D_2) \cap \partial M\| \geq 2$.

Proof.

(a) If $|P_1| = 1$ this follows from Proposition 3.1 (a). Assume $|P_1| \geq 2$. Let $P_1 = \langle D_1, \ldots, D_k\rangle$, $k \geq 2$ and assume P_1 is left-extremal. Then $b(D_1) = 1$ and $c(D_1) = 0$. By Theorem 2.1 (c) $a(D_1) \leq 1$. Consequently, $d(D_1) \leq 1 + 0 + 1 = 2$, hence D_1 is a 2-corner region of M. Let $v = \partial D_1 \cap \partial D_2 \cap \partial M$. Then by Theorem 2.1 (d) $d_{\mathcal{L}_i}(v) = 3$ and since $c(D_1) = c(D_2) = 0$ hence $d_{\mathcal{L}_i}(v) = d_M(v)$. Thus $d_M(v) = 3$ and Proposition 3.1 applies to D_1. Now, in the notation of Proposition 3.1 $D_1 = D$ and $D_2 = E_r$ and $d_M(D_2) = a(D_2) + b(D_2) + c(D_2) \leq 1 + 2 + 0 = 3$, hence Q_r is the product of at least three $(6 - 3 = 3)$ pieces. (Here, as in Proposition 3.1 Q_r is the label of $\partial E_r \cap \partial M$.) Therefore, it follows from Proposition 3.1 that $\|(\partial D_1 \cap \partial M) \cup (\partial D_2 \cap \partial M)\| \geq 2$, as required. Similarly, if P_1 is right-extremal then the above argument applies to D_k.

(b) Immediate from the proof of part (a).

The Lemma is proved. □

Now, it follows from Greendlinger's Lemma (see Remark 3) that due to the $C(4)\&T(4)$ condition (which is implied by the $C(6)\&T(4)$ condition) M contains at least two $k-$corner regions with $k \leq 2$. Consider the layer structure of M with center D_0, where D_0 is a $k-$corner region of M, $k \leq 2$. Since D_0 is a boundary region of M, hence the layer structure of M with center D_0 has a peak P in its last layer. (See Remark 2 following Theorem 2.1.) Hence by Lemma 4.2 (b) either P contains a boundary region D such that $\|\partial D \cap \partial M\| \geq 2$ or contains adjacent regions D and D_1 such that $\|(\partial D \cup \partial D_1) \cap \partial M\| \geq 2$. Consider the layer structure Λ of M with center D. Since $d(D) \leq 3$ for every region in P, all the layers of Λ are simply connected and in particular its last layer $\mathcal{L}_p$. If $\mathcal{L}_p$ has more than one component then $\|\partial \mathcal{L}_p \cap \partial M\| \geq 3$ due to Lemma 4.2 (a), hence $\|\partial M\| \geq 4$, since $\|\partial P \cap \partial M\| \geq 2$ or $\|(\partial D \cup \partial D_1) \cap \partial M\| \geq 2$ and may assume that $D_1, D \not\subseteq \mathcal{L}_p$. (If $D_1 \subseteq \mathcal{L}_p$ or $D \subseteq \mathcal{L}_p$ then $p \leq 1$ and in this case $\|\partial M\| \geq 4$ easily follows.) Thus

If $\mathcal{L}_p$ contains more than one component then $\|\partial M\| \geq 4$. $\quad (\odot)$

Now, we turn to the proof of Proposition 4.1.

Proof. First observe that $\|\partial M\| \geq 2$ due to Lemma 4.2 and if $\|\partial M\| > 2$ then $\|\partial M\| \geq 4$. Suppose by way of contradiction that M is not a one-layer diagram and show that $\|\partial M\| \geq 4$. Let D, Λ and $\mathcal{L}_p$ be as above. Then due to $(\odot)$ we may assume that $\mathcal{L}_p$ has connected interior. It follows that all layers of Λ have connected interior. (See Remark 2 following Theorem 2.1). Let $P = \mathcal{L}_p$. Let $\partial M = \alpha\beta$, where $\alpha = \partial D \cap \partial M$. Since $\|\partial D \cap \partial M\| \geq 2$

or $\|(\partial D \cup \partial D_2) \cap \partial M\| \geq 2$, it is enough to show that $\|\beta\| \geq 3$. Clearly, $\partial P \cap \partial M \subseteq \beta$, hence if $\|\partial P \cap \partial M\| \geq 3$ then $\|\partial M\| \geq 4$. Assume therefore that $\|\partial P \cap \partial M\| = 2$. Then by Lemma 4.1 either $|P| = 1$ or $P = \langle D_1, D_2\rangle$ such that $a(D_1) = a(D_2) = 1$.

Claim. *Let* $(**)$ *be following statement:*

either $|\mathcal{L}_i| = 1$ *or* $\mathcal{L}_i = \langle D_1, D_2\rangle$ *such that* $a(D_1) + a(D_2) = 1$. $(**)$

Then $(**)$ *holds for every* i, $i = 1, ..., p$.

Proof of the Claim. Let $0 \leq t \leq p$ and let $q = p - t$. We prove by induction on q that $\mathcal{L}_t$ satisfies $(**)$. If $q = 0$ then the result follows by Lemma 4.1. Suppose that $q > 0$ and the result holds for every value of q with $q < i$. Suppose $|\mathcal{L}_{i-1}| \geq 2$ and let $\mathcal{L}_{i-1} = \langle E_1, \ldots, E_k\rangle$. Let $\mathcal{A}(D_1) = \{E_j\}$ for some j, $j = 1, \ldots, k$ such that either $j \neq 1$ or $j \neq k$. Suppose first $j \neq 1$. If $a(E_1) = 0$ then $\|\partial E_1 \cap \partial M\| \geq 2$ by Proposition 3.1 (a), hence $\|\beta\| \geq 3$. If $a(E_1) = 1$ and $a(E_2) = 1$ with $\mathcal{A}(E_1) = \{F_1\}$ and $\mathcal{A}(E_2) = \{F_2\}$ then $F_1 \neq F_2$ and $\|(\partial E_1 \cap \partial F_1) \cap \partial M\| \geq 2$. (see proof of part (i) in Lemma 4.1). Therefore,

(i) If $j \neq 1$ then $a(E_1) = 1$ and $a(E_2) = 0$.
Suppose now that $j \neq k$. Then the arguments of the case $j \neq 1$ for E_k apply and yield.

(ii) If $j \neq k$ then $a(E_k) = 1$ and $a(E_{k-1}) = 0$.
Assume now that $k \geq 3$ If $j \neq 2$ and $j \neq 1$ then it follows from (i) and Proposition 3.1 (c) that

$$\|(\partial E_1 \cup \partial E_2) \cap \partial M\| \geq 2 \text{ and hence } |\beta\| \geq 3$$

(iii) If $j \neq k - 1$ and $j \neq k$ then it follows from (ii) and Proposition 3.1 (c) that

$$\|(\partial E_k \cup \partial E_{k-1}) \cap \partial M\| \geq 2 \text{ and hence } |\beta\| \geq 3$$

So assume $j = 2$. We claim that in this case as well, at least one of (6) and (7) in Lemma 4.1 holds. This is clear if $k \geq 4$, hence assume $k = 3$. Since $d(E_2) = 2$ and $\partial E_2 \cap \partial D_1$ is a piece, either $\partial E_2 \cap \partial M$ is connected and is the product of at least three $(6 - (2 + 1) = 3)$ pieces or $\partial E_2 \cap \partial M$ has two connected components γ_1 and γ_3 such that $\partial E_2 \cap \partial St_{i-1} = \gamma_1\gamma_2\gamma_3$ with $\gamma_2 = \partial E_2 \cap \partial D_1$ and either γ_1 is the product of at least two pieces or γ_3 is the product of at least two pieces. Therefore we may apply Proposition 3.1 (c) for E_1 or for E_3 to give $\|(\partial\mathcal{L}_{i-1} \cup \partial\mathcal{L}_i) \cap \partial M\| \geq 3$. Consequently, $|\mathcal{L}_{i-1}| \leq 2$. Now it easily follows by arguments we made several times above that if $|\mathcal{L}_{i-1}| = 2$, then $a(E_1) = 1$ and $a(E_2) = 1$ would imply that either $\|(\partial E_1 \cup \partial F_1) \cap \partial M\| \geq 2$ or $\|(\partial E_2 \cup \partial F_2) \cap \partial M\| \geq 2$, proving the claim.

We show that $(**)$ implies M is a one-layer diagram. Let K be a region of M. Suppose K is in $\mathcal{L}_i$, $0 < i < p$. Then by condition $(**)$ $b(K) = 1$. Also, by Theorem 2.1 (c) $a(K) \leq 1$ and $c(K) \leq 1$. Hence $d(K) \leq 3$. Let $\mathcal{L}_i = \langle K, L \rangle$. If $d(K) = 3$ then it follows from $(**)$ that $c(L) = 0$, $a(L) = 0$ and $b(L) = 1$, hence $d(L) = 1$ and hence $\|\partial L \cap \partial M\| \geq 2$ by Proposition 3.1 (a), implying $\|\beta\| \geq 3$. Therefore $d(K) \leq 2$. Suppose $d(K) = 1$. Then $\|\partial K \cap \partial M\| \geq 2$ by Proposition 3.1 (a) implying again $\|\beta\| \geq 3$. Therefore $d(K) = 2$. Thus, every region K in $\mathcal{L}_i$, $1 < i < p$ has exactly two neighbours. But now, $\mathcal{L}_0 = \{D\}$ and by $(**)$ D has exactly one neighbour (in $\mathcal{L}_1$) and either $\mathcal{L}_p = \{D_1\}$ in which case $d(D_1) = 1$ or $\mathcal{L}_p = \langle D_1, D_2 \rangle$ in which case either $d(D_1) = 2$ and $d(D_2) = 1$ or $d(D_2) = 2$ and $d(D_1) = 1$, by Lemma 4.1. Consequently, M is a one-layer diagram. The Proposition is proved. □

References

1. A. Juhász, Small Cancellation Theory with with a unified small cancellation condition, *J. London Math. Soc. (2)* **40(1)**, 57-80 (1989).
2. R. Lyndon and P. Schupp, *Combinatorial Group Theory*, Springer-Verlag, Berlin-Heidleberg-New York (1977).
3. W. Magnus, Über diskontinuierliche Gruppen mit einer definierenden Relation (Der Freiheitssatz), *J. reine angew. Math.* **163**, 141-165 (1930).

WREATH PRODUCT DECOMPOSITIONS FOR TRIANGULAR MATRIX SEMIGROUPS

MARK KAMBITES

Fachbereich Mathematik / Informatik, Universität Kassel, 34109 Kassel, Germany

E-mail: kambites@theory.informatik.uni-kassel.de

BENJAMIN STEINBERG

School of Mathematics and Statistics, Carleton University, Ottawa, Ontario, K1S 5B6, Canada

E-mail: bsteinbg@math.carleton.ca

We consider wreath product decompositions for semigroups of triangular matrices. We exhibit an explicit wreath product decomposition for the semigroup of all $n \times n$ upper triangular matrices over a given field k, in terms of aperiodic semigroups and affine groups over k. In the case that k is finite this decomposition is optimal, in the sense that the number of group terms is equal to the group complexity of the semigroup. We also obtain some decompositions for semigroups of triangular matrices over more general rings and semirings.

1. Introduction

Some of the most natural and frequently occurring semigroups are those of upper triangular matrices over a given ring or field. For example, such semigroups arise in the study of algebraic semigroups, where Putcha [17] has proven that a connected algebraic monoid with zero over a field has a faithful rational triangular representation if and only if its group of units is solvable [17]. It follows that triangularizable monoids can be thought of as a natural generalisation of solvable groups. More recently, Almeida, Margolis and Volkov [1] have shown that semigroups of triangular matrices over finite fields generate natural pseudovarieties. Almeida, Margolis, Steinberg and Volkov [2, 3] have since considered arbitrary fields and have obtained language-theoretic consequences. Further properties of these semigroups have been described by Okninski [16].

Perhaps the most productive approach to the study of finite semigroups is through coverings by wreath products. In the 1960s, Krohn and Rhodes

[12, 13, 14] showed that every finite semigroup can be expressed as a *divisor* (a homomorphic image of a subsemigroup) of a wreath product of finite groups and finite aperiodic monoids. The *group complexity* of a finite semigroup is the smallest number of group terms in such a decomposition, and is a key concept in finite semigroup theory.

In a previous article [11], the first author computed the group complexity of the semigroup $T_n(k)$ of all $n \times n$ upper triangular matrices over a given finite field k, and of certain related semigroups. However, the methods used did not result in explicit wreath product decompositions. The main objective of this article is to establish an explicit wreath product decomposition for each semigroup of the form $T_n(k)$, and hence for every semigroup of triangular matrices over a finite field. This decomposition is optimal, in the sense that the number of group terms in the decomposition is equal to the group complexity of $T_n(k)$.

In the process, we obtain some results applicable in a more general context. While Krohn-Rhodes theory is traditionally concerned with finite semigroups, there have been numerous attempts to extend it to well-behaved classes of infinite semigroups [4, 6, 8]. Our method for decomposing $T_n(k)$ is fully applicable in the case that the field k, and hence also the semigroup $T_n(k)$, is infinite. We also obtain some wreath product decompositions, although not in terms of groups and aperiodic semigroups, for triangular matrix semigroups over more general rings and semirings with identity.

In addition to this introduction, this paper comprises four sections. In Section 2, we briefly recall the key definitions and results of Krohn-Rhodes theory as applied to abstract monoids, including division, wreath products, the Prime Decomposition Theorem and group complexity. Section 3 introduces triangular matrix semigroups, and briefly describes their structure, before reviewing the results of the first author [11] characterising their group complexity.

Section 4 contains the main original results of the paper; we obtain an explicit decomposition for each semigroup $T_n(k)$, and hence for every semigroup of triangular matrices over a field, as a wreath product of aperiodic monoids and affine groups over k. We also obtain some related decompositions for triangular matrix semigroups over rings and semirings with identity. Finally, in Section 5, we compare our results with those which can be obtained using a standard decomposition method of Eilenberg and Tilson [20]; the latter produces a suboptimal decomposition for $T_n(k)$, but alternative optimal decompositions for certain important divisors.

Throughout this paper, all functions are applied on the right of their arguments. If S and T are sets then we denote by S^T the set of all functions from T to S. We assume familiarity with the standard terminology, notation and foundational results of structural semigroup theory; a detailed introduction to these is given by Howie [10]. By contrast, we assume no prior knowledge which is particular to the study of finite semigroups; we intend that this article should be fully accessible to the reader with experience only of infinite semigroups.

2. Wreath Products, Division and Complexity

In this section, we briefly introduce the basic concepts of wreath products, division and complexity. We restrict ourselves to the special case of abstract monoids (as opposed to transformation semigroups), since this suffices for our purpose. A detailed and more general introduction is given by Eilenberg [5].

Let S and T be semigroups. We say that S *divides* T, and write $S \prec T$, if S is a homomorphic image of some subsemigroup of T. The relation of division is easily verified to be reflexive and transitive.

Let S and T be monoids. Then S^T is a monoid with pointwise product: if $f, g \in S^T$ and $t \in T$, then by definition $t(fg) = (tf)(tg)$. There is also a natural left action of T on S^T defined as follows: if $f \in S^T$, $t_1, t_2 \in T$, then ${}^{t_1}f : T \to S$ is given by

$$t_2{}^{t_1}f = (t_2 t_1)f.$$

Then the *wreath product* of S and T, denoted $S \wr T$, is the monoid with underlying set $S^T \times T$, and multiplication given by

$$(f, a)(g, b) = (f\,{}^a g, ab).$$

The wreath product of monoids is not associative; however, $(S_3 \wr S_2) \wr S_1$ is isomorphic to a submonoid of $S_3 \wr (S_2 \wr S_1)$. For this reason, we define the *iterated wreath product* of a sequence of three or more monoids inductively by

$$S_n \wr S_{n-1} \wr \cdots \wr S_1 \;=\; S_n \wr (S_{n-1} \wr \cdots \wr S_1)$$

so as to obtain the largest monoid possible.

Recall that a semigroup is called *aperiodic* if it has no non-trivial subgroups. In the following proposition we state without proof a few well-known properties of the wreath product which we shall need.

Proposition 2.1. *Let A, B, C and D be finite monoids.*

(i) *If* $A \prec B$ *then* $A \wr C \prec B \wr C$ *and* $C \wr A \prec C \wr B$.
(ii) $A \times B \prec A \wr B$.
(iii) $(A \wr B) \times (C \wr D) \prec (A \times C) \wr (B \times D)$.
(iv) *If* A *and* B *are groups then* $A \wr B$ *is a group.*
(v) *If* A *and* B *are aperiodic then* $A \wr B$ *is aperiodic.*

We shall also need an elementary decomposition that is perhaps not so well known:

Proposition 2.2. *Let* A, B, C *be monoids. Then* $(A \wr B) \times C$ *embeds in* $A \wr (B \times C)$.

Proof. First we define a homomorphism $\alpha : A^B \to A^{B \times C}$ by $(b,c)f\alpha = bf$. Next we define $\psi : (A \wr B) \times C \to A \wr (B \times C)$ by

$$((f,b),c)\psi = (f\alpha, (b,c)).$$

Let us verify that ψ is a homomorphism.

$$\begin{aligned} ((f,b),c)\psi \; ((g,b'),c')\psi &= (f\alpha,(b,c)) \; (g\alpha,(b',c')) \\ &= (f\alpha^{(b,c)}g\alpha, (bb',cc')). \end{aligned}$$

But for $(b_0, c_0) \in B \times C$,

$$(b_0,c_0)^{(b,c)}g\alpha = (b_0b, c_0c)g\alpha = b_0bg = b_0{}^bg.$$

Thus ${}^{(b,c)}(g\alpha) = ({}^bg)\alpha$. Hence we may conclude

$$\begin{aligned} ((f,b),c)\psi \; ((g,b'),c')\psi &= (f\alpha({}^bg)\alpha, (bb',cc')) \\ &= ((f{}^bg)\alpha, (bb',cc')) \\ &= [((f,b),c)((g,b'),c')]\psi \end{aligned}$$

and so ψ is a homomorphism. It is clear that ψ is injective. □

A related, more technical proposition is:

Proposition 2.3. *Let* A, B, C *be monoids. Then* $A \wr (B \times C)$ *embeds in* $(A \wr B) \wr C$.

Proof. Let us begin with some notation. If $f : B \times C \to A$, $b \in B$ and $c \in C$, then we define $f_c : B \to A$ by $b'f_c = (b',c)f$, for $b' \in B$, and we define a function $\langle f \mid b\rangle : C \to A \wr B$ by $c\langle f \mid b\rangle = (f_c, b)$.

We claim that the map $\tau : A \wr (B \times C) \to (A \wr B) \wr C$ given by $(f,(b,c))\tau = (\langle f \mid b\rangle, c)$ is an injective homomorphism. Clearly τ is injective since $f : B \times C \to A$ is determined by $(f_c)_{c\in C}$. Let us check that τ is a homomorphism. We compute

$$(f,(b,c))\tau\ (f',(b',c'))\tau = (\langle f \mid b\rangle, c)\ (\langle f' \mid b'\rangle, c') = (\langle f \mid b\rangle\ {}^{c}\langle f' \mid b'\rangle, cc').$$

Now if we evaluate the leftmost coordinate of the right hand side at $c_0 \in C$ we get

$$[c_0\langle f \mid b\rangle]\ [c_0c\langle f' \mid b'\rangle] = (f_{c_0}, b)\ (f'_{c_0c}, b') = (f_{c_0}\ {}^{b}f'_{c_0c}, bb').$$

The leftmost coordinate of the right hand side then evaluates on $b_0 \in B$ as follows:

$$b_0 f_{c_0}\,{}^{b}f'_{c_0c} = b_0 f_{c_0}(b_0b)f'_{c_0c} = (b_0, c_0)f\ (b_0b, c_0c)f'. \tag{1}$$

On the other hand,

$$\begin{aligned}[(f,(b,c))(f',(b',c'))]\,\tau &= (f^{(b,c)}f', (bb', cc'))\tau \\ &= (\langle f^{(b,c)}f' \mid bb'\rangle, cc').\end{aligned}$$

Evaluating the leftmost coordinate of the right hand side at $c_0 \in C$, we get $((f^{(b,c)}f')_{c_0}, bb')$. Now if, $b_0 \in B$, then

$$b_0(f^{(b,c)}f')_{c_0} = (b_0, c_0)f^{(b,c)}f' = (b_0, c_0)f(b_0b, c_0c)f'. \tag{2}$$

Comparing (1) and (2) completes the proof. □

Let X be a finite set. We denote by $\widetilde{X}$ the monoid consisting of the identity map and all constant maps on X; clearly, $\widetilde{X}$ is an aperiodic monoid. Now if A is a monoid of transformations of X, then the *augmented monoid* $\overline{A}$ of A with respect to its action on X is the monoid generated by transformations in A and those in $\widetilde{X}$. The following proposition, a proof of which can be found in Eilenberg [5], provides a decomposition of an augmented monoid in terms an aperiodic monoid and the underlying monoid.

Proposition 2.4. *Let A be a finite monoid of transformations of a set X. Then $\overline{A} \prec \widetilde{X} \wr A$.*

The importance of wreath products for the study of finite semigroups stems from the following structure theorem of Krohn and Rhodes [12, 14].

Theorem 2.1. *(The Prime Decomposition Theorem, Krohn-Rhodes 1968) Let S be a finite semigroup. Then S divides some iterated wreath product each of whose terms is either a finite simple group which divides S or a finite aperiodic monoid.*

A *Krohn-Rhodes decomposition* for a semigroup S is an expression of S as a divisor of an iterated wreath product of groups and aperiodic monoids. Given such a decomposition for S, Proposition 2.1 tells us that we can combine adjacent groups terms and adjacent aperiodic terms to obtain an *alternating* decomposition of the form:

$$A_n \wr G_n \wr A_{n-1} \wr \cdots \wr A_1 \wr G_1 \wr A_0$$

where each A_i is aperiodic, each G_i is a group, and all terms except possibly A_0 and A_n are non-trivial. (Note, though, that in doing so we may lose the property that the group terms are divisors of S.) The number n, that is, the number of group terms, is called the *group length* of the decomposition. A natural structural constant which can be associated with a finite semigroup S is the minimal group length of a Krohn-Rhodes decomposition for S; this number is called the *group complexity* of S. A decomposition for S is said to be *optimal* if its group length equals the group complexity of S.

Much effort has been put into the study of Krohn-Rhodes decompositions, and in particular of certain algorithmic problems. Various algorithms have been developed for finding wreath product decompositions for semigroups; some of these will be discussed in Section 5 below. A major open question is that of whether group complexity is *decidable*, that is, whether there is an algorithm which, given the multiplication table for a finite semigroup S, determines the group complexity of S.

We remark briefly upon the relationship between these two problems, and in particular on the implications of the latter for the former. In theory, knowing the group complexity of a finite semigroup allows one to compute an optimal decomposition. Indeed, if ones knows that a semigroup S admits a decomposition of group length n, then one can in principle enumerate multiplication tables of divisors of alternating wreath products with n group terms, and test them for isomorphism with S. In practice, of course, this algorithm is completely infeasible – the cardinality of an iterated wreath product grows extremely fast as a function of the cardinalities of the terms, and no sensible upper bounds are known even on the latter. Hence, situations can arise in which the complexity of a semigroup is known, but an explicit optimal decomposition is not. Indeed, the following key result of Rhodes [19] often gives rise to such situations.

Theorem 2.2. (The Fundamental Lemma of Complexity, Rhodes 1974) *Let S and T be finite semigroups, and suppose there exists a surjective morphism $S \to T$ which is injective when restricted to each subgroup of S. Then S and T have the same group complexity.*

The Fundamental Lemma is an extremely powerful tool for computing the group complexity of a semigroup. The proof of the Lemma given by Tilson [21] is constructive in the sense that, given an optimal wreath product decomposition for a semigroup T and a surjective morphism $S \to T$ which is injective on subgroups, it does provide an optimal decomposition for S. However the construction is quite involved and in practice it is hard to see what groups and aperiodic monoids appear.

3. Triangular Matrix Semigroups

Let R be a semiring with identity 1 and zero 0. If x is an $n \times n$ matrix then for $1 \leq i, j \leq n$ we denote by x_{ij} the entry of x in position (i, j), that is, in the ith row and jth column, of x. Recall that the matrix x is *(upper) triangular* if $x_{ij} = 0$ whenever $1 \leq j < i \leq n$. We call an upper triangular matrix *(upper) unitriangular* if, in addition, $x_{ii} = 0$ or $x_{ii} = 1$ for $1 \leq i \leq n$. We call x a *subidentity* if it is unitriangular and $x_{ij} = 0$ whenever $i \neq j$. We denote by $T_n(R)$ and $UT_n(R)$ the semigroups of all $n \times n$ upper triangular matrices and of all $n \times n$ unitriangular matrices respectively, with entries drawn from R, the operation in both cases being usual matrix multiplication. Note that $T_1(R)$ is just the multiplicative semigroup of R.

We shall be especially interested in the case that the semiring R is a field k. In this case, we define a relation σ on each semigroup $T_n(k)$ by $x \,\sigma\, y$ if and only $x = \lambda y$ for some non-zero scalar λ. This relation is easily verified to be a congruence on $T_n(k)$. The *projective triangular semigroup* $PT_n(k)$ is the quotient semigroup $T_n(k)/\sigma$; we denote by $\overline{x}$ the element of $PT_n(k)$ which is the σ-equivalence class of a matrix $x \in T_n(k)$.

The group of units of $T_n(k)$ [respectively, $UT_n(k)$, $PT_n(k)$] is denoted $T_n^*(k)$ [$UT_n^*(k)$, $PT_n^*(k)$]. It consists of those triangular matrices whose diagonal entries are non-zero [respectively, triangular matrices whose diagonal entries are 1, equivalence classes of triangular matrices whose diagonal entries are non-zero]. Note that $T_1^*(k)$ is the multiplicative group of the field k.

We introduce a notion of upper triangular row and column operations on $T_n(R)$ and hence on $UT_n(R)$. By a *row operation* on an upper triangular matrix we shall mean either (i) adding a multiple of one row to a row *above* or (ii) scaling a row by an element of R. There is an obvious analogous definition of *column operations* of different types, a type (i) operation being adding a multiple of one column to a column to the *right*. Row and column operations of type (i) are called *unitriangular*. The following easy proposition characterises Green's relations in $T_n(R)$ and $UT_n(R)$ in terms of these operations.

Proposition 3.1. *Let n be a positive integer and R a semiring with identity. Two matrices in $T_n(R)$ [respectively, $UT_n(R)$] are:*

(i) $\mathcal{L}$-related exactly if each can be obtained from the other by [unitriangular] row operations;
(ii) $\mathcal{R}$-related exactly if each can be obtained from the other by [unitriangular] column operations;
(iii) $\mathcal{J}$-related exactly if each can be obtained from the other by [unitriangular] row and column operations.

We now turn our attention to the case of a finite field k. The following proposition, parts of which go back at least as far as Putcha [17], characterises the regular elements in $T_n(k)$. A proof can be found in a previous article of the first author [11].

Proposition 3.2. *Let n be a positive integer and k a finite field. Let $x \in T_n(k)$ or $x \in UT_n(k)$. Then the following are equivalent:*

(i) x is regular;
(ii) every row in x is a linear combination of rows in x with non-zero diagonal entries;
(iii) every column in x is a linear combination of columns in x with non-zero diagonal entries;
(iv) x is $\mathcal{J}$-related to a subidentity.

Factoring out a monoid by a subgroup of the group of units that is central in the monoid gives rise to a congruence contained in $\mathcal{H}$. The following simple observation is a special case of well-known and elementary facts about congruences contained in $\mathcal{H}$.

Proposition 3.3. *Let n be a positive integer, k a field and $x, y \in T_n(k)$. Then*

(i) x is regular in $T_n(k)$ if and only if $\overline{x}$ is regular in $PT_n(k)$;
(ii) $x \mathrel{\mathcal{L}} y$ in $T_n(k)$ if and only if $\overline{x} \mathrel{\mathcal{L}} \overline{y}$ in $PT_n(k)$;
(iii) $x \mathrel{\mathcal{R}} y$ in $T_n(k)$ if and only if $\overline{x} \mathrel{\mathcal{R}} \overline{y}$ in $PT_n(k)$;
(iv) $x \mathrel{\mathcal{J}} y$ in $T_n(k)$ if and only if $\overline{x} \mathrel{\mathcal{J}} \overline{y}$ in $PT_n(k)$.

We recall the following theorem of the first author [11].

Theorem 3.1. (Kambites 2004) *Let n be a positive integer, and k a finite field. Then $T_n(k)$, $UT_n(k)$ and $PT_n(k)$ have complexity $n-1$, unless $n = 1$ and $k \neq F_2$ in which case $UT_n(k)$ and $PT_n(k)$ have complexity 0 but $T_n(k)$ has complexity 1.*

We remark that the scope of this result has since been extended by Mintz [15]; he observes that triangular matrix semigroups form a special class of *quiver algebra* and that the result extends naturally to cover a somewhat larger class of quiver algebras.

The proof of Theorem 3.1 is somewhat technical, and makes extensive use of the Fundamental Lemma of Complexity, both directly and through the application of a result of Rhodes and Tilson [18]. Consequently, it does not give rise to explicit Krohn-Rhodes decompositions for the semigroups in question. In the next section, we shall show how to obtain such decompositions for semigroups of the form $T_n(k)$, and hence for every triangular matrix semigroup over a field.

4. Decompositions for Triangular Matrix Semigroups

Our main objective in this section is to compute an explicit decomposition for each semigroup of the form $T_n(k)$ with k a field, as a divisor of an alternating wreath product of groups and aperiodic monoids. In the case that k is finite, this decomposition will be optimal, in the sense that its group length equals the group complexity of the semigroup as described by Theorem 3.1. In the process, we also obtain some decompositions for triangular matrix semigroups over more general rings and semirings.

Let R be a semiring and n a positive integer. We consider the R-module R^n of $1 \times n$ row vectors over R. Recall that an *affine transformation* of R^n is a map of the form $v \mapsto vX + c$ for some $n \times n$ matrix X and some vector $c \in R^n$. We say that the transformation is *affine (upper) triangular* if X is upper triangular, and *affine scaling* if X is of the form λI where $\lambda \in R$ and I is the identity matrix.

The *affine monoid* $A_n(R)$ of degree n over R is the monoid of all affine transformations of R^n, with operation composition. It is readily verified that the sets of affine triangular and affine scaling maps form submonoids; these we call the *affine triangular monoid* $AT_n(R)$ and the *affine scaling monoid* $AS_n(R)$ respectively. The *affine group* $A_n^*(R)$, the *affine triangular group* $AT_n^*(R)$ and the *affine scaling group* $AS_n^*(R)$ are the groups of units of $A_n(R)$, $AT_n(R)$ and $AS_n(R)$ respectively. We remark that the various affine groups are semidirect products of the appropriate matrix groups with the additive groups of translations.

There is a natural embedding of an affine triangular monoid of degree $n-1$ into an upper triangular monoid of degree n.

Proposition 4.1. *Let $n \geq 2$ and let R be a semiring. Then $AT_{n-1}(R)$ and $AS_{n-1}(R)$ embed in $T_n(R)$.*

Proof. From the definition, $AS_{n-1}(R)$ is a subsemigroup of $AT_{n-1}(R)$, so it suffices to show that the latter embeds in $T_n(R)$. Given an affine triangular map f given by $v \mapsto vX + c$ we define an $n \times n$ matrix

$$M_f = \begin{pmatrix} 1 & c \\ 0 & X \end{pmatrix}.$$

That the matrix M_f is upper triangular follows from the fact that X is upper triangular. If we identify $v \in R^{n-1}$ with $(1, v)$ then it is routine to verify that $(1, v)M_f = (1, vf)$ and so $f \mapsto M_f$ gives an embedding of $AT_{n-1}(R)$ into $T_n(R)$, as required. □

The following lemma is the main inductive step in our decompositions. If X is a matrix, we write X^T for its transpose.

Lemma 4.1. *Let $n \geq 2$ and R be a semiring with identity. Then*

$$T_n(R) \prec [AS_{n-1}(R) \wr T_{n-1}(R)] \times T_1(R).$$

Proof. We view each $s \in T_n(R)$ as a block matrix

$$s = \begin{pmatrix} M_s & v_s \\ 0 & c_s \end{pmatrix}$$

where M_s is an $(n-1) \times (n-1)$ matrix which clearly lies in $T_{n-1}(R)$, v_s is an $(n-1) \times 1$ column vector and c_s is a 1×1 matrix. Now we define

$$\psi : T_n(R) \to [AS_{n-1}(R) \wr T_{n-1}(R)] \times T_1(R)$$

by

$$s\psi = (f_s, M_s, c_s)$$

where for every $X \in T_{n-1}(R)$, the element $Xf_s \in AS_{n-1}(R)$ is given by

$$w(Xf_s) = (Xv_s + w^T c_s)^T.$$

Clearly, ψ is well-defined; it is also injective, since for any s, we have $v_s = [\underline{0}(If_s)]^T$ where $I \in T_{n-1}(R)$ is the identity matrix and $\underline{0} \in R^{n-1}$ is the zero vector.

To prove the lemma, it will now suffice to show that ψ is a homomorphism. Since

$$\begin{pmatrix} M_s & v_s \\ 0 & c_s \end{pmatrix} \begin{pmatrix} M_t & v_t \\ 0 & c_t \end{pmatrix} = \begin{pmatrix} M_s M_t & M_s v_t + v_s c_t \\ 0 & c_s c_t \end{pmatrix},$$

we have: $M_{st} = M_s M_t$, $c_{st} = c_s c_t$ and $v_{st} = M_s v_t + v_s c_t$. So, recalling the definition of the wreath product, it remains to show that $f_{st} = f_s{}^{M_s} f_t$. That is we must show $w(Xf_{st}) = w[(Xf_s)(XM_s f_t)]$ for all $w \in R^n$ and $X \in T_{n-1}(R)$. But

$$\begin{aligned} w(Xf_{st}) &= (Xv_{st} + w^T c_{st})^T \\ &= (X(M_s v_t + v_s c_t) + w^T c_s c_t)^T \\ &= (XM_s v_t + Xv_s c_t + w^T c_s c_t)^T \\ &= (XM_s v_t + (Xv_s + w^T c_s)c_t)^T \\ &= (XM_s v_t)^T + (Xv_s + w^T c_s)^T c_t \\ &= (XM_s v_t)^T + (wXf_s)c_t \\ &= (XM_s v_t + (w(Xf_s))^T c_t)^T \\ &= w[(Xf_s)(XM_s f_t)] \end{aligned}$$

as required. □

Lemma 4.1 leads easily to the following decomposition for $T_n(R)$ in terms of affine scaling monoids and the multiplicative semigroup of R.

Theorem 4.1. *Let $n \geq 2$ and R be a semiring with identity. Then*

$$T_n(R) \prec AS_{n-1}(R) \wr AS_{n-2}(R) \wr \cdots \wr \left[(AS_1(R) \wr T_1(R)) \times T_1(R)^{n-1}\right].$$

Proof. We use induction on n. When $n = 2$ then using Lemma 4.1 we have

$$T_2(R) \prec [AS_1(R) \wr T_1(R)] \times T_1(R)$$

as required. Now let $n \geq 3$ and assume true for smaller n. Then again using Lemma 4.1 and Proposition 2.2 we obtain

$$\begin{aligned} T_n(R) &\prec [AS_{n-1}(R) \wr T_{n-1}(R)] \times T_1(R) \\ &\prec [AS_{n-1}(R) \wr (AS_{n-2}(R) \wr \cdots \wr \\ &\qquad\qquad [(AS_1(R) \wr T_1(R)) \times T_1(R)^{n-2}])] \times T_1(R) \\ &\prec AS_{n-1}(R) \wr AS_{n-2}(R) \wr \cdots \wr \left[(AS_1(R) \wr T_1(R)) \times T_1(R)^{n-1}\right] \end{aligned}$$

as required. □

As a consequence of Theorem 4.1, we obtain a group length $n-1$ decomposition for each semigroup $T_n(k)$ with k a field.

Theorem 4.2. *Let $n \geq 2$ and k be a field. Then $T_n(k)$ divides*

$$\widetilde{k^{n-1}} \wr AS^*_{n-1}(k) \wr \widetilde{k^{n-2}} \wr AS^*_{n-2}(k) \wr \cdots \wr \widetilde{k} \wr \left[(AS^*_1(k) \wr T^*_1(k)) \times T^*_1(k)^{n-1}\right] \wr U_1^n$$

where U_1 is the two-element semilattice.

Proof. By Theorem 4.1 we have that

$$T_n(k) \prec AS_{n-1}(k) \wr AS_{n-2}(k) \wr \cdots \wr \left[(AS_1(k) \wr T_1(k)) \times T_1(k)^{n-1}\right].$$

For each i, the affine monoid $AS_i(k)$ consists precisely of $AS^*_i(k)$ and the constant maps on k^i. Indeed, given a map $v \mapsto \lambda v + c$ in $AS^*_i(k)$ we see that either $\lambda = 0$, in which case the map is constant, or $\lambda \neq 0$ in which case there exists an inverse map $v \mapsto \lambda^{-1} v - \lambda^{-1} c$. Hence, $AS_i(k)$ is the augmented monoid of $AS^*_i(k)$ with respect to its action on k^i, and so by Proposition 2.4 we have

$$AS_i(k) \prec \widetilde{k^i} \wr AS^*_i(k).$$

Also, it is easy to see that the group with zero $T_1(k)$ divides $T^*_1(k) \times U_1$. It follows that

$$\begin{aligned}
[AS^*_1(k) \wr T_1(k)] \times T_1(k)^{n-1} &\prec [AS^*_1(k) \wr (T^*_1(k) \times U_1)] \times T^*_1(k)^{n-1} \times U_1^{n-1} \\
&\prec [(AS^*_1(k) \wr T^*_1(k)) \wr U_1] \times \left[T^*_1(k)^{n-1} \wr U_1^{n-1}\right] \\
&\prec \left[(AS^*_1(k) \wr T^*_1(k)) \times T^*_1(k)^{n-1}\right] \wr U_1^n
\end{aligned}$$

where the second division uses Proposition 2.3 and the last division uses Proposition 2.1(iii) and Proposition 2.2. The result is now clear. □

In general, a finite semigroup S does not necessarily admit an optimal Krohn-Rhodes decomposition whose group terms are divisors of S (or even wreath products of divisors of S). Here we have succeeded in finding for $T_n(k)$ an optimal Krohn-Rhodes decomposition in which almost every group is a subgroup of the group of units $T_n(k)$, the exception being $(AS^*_1(k) \wr T^*_1(k)) \times T^*_1(k)^{n-1}$, which is a direct product of a subgroup of $T^*_n(k)$ with the wreath product of two subgroups of $T^*_n(k)$. Indeed, Proposition 4.1 implies that each $AS^*_m(k)$ with $1 \leq m \leq n-1$ embeds in $T^*_n(k)$. On the other hand $T^*_1(k)^n$ is just the diagonal subgroup of $T^*_n(k)$ and contains $T^*_1(k)^{n-1}$.

5. Comparison with Depth Decomposition

Considerable thought has been put into algorithmic methods for obtaining explicit Krohn-Rhodes decompositions for finite transformation semigroups. The original proof of Krohn and Rhodes [13] is essentially algorithmic; however, the decompositions it yields are far from optimal. A substantial improvement is the *holonomy method*, which was developed by Eilenberg [5], in conjunction with Tilson, using techniques of Zeiger [23] and Ginzburg [7]; see also Holcombe [9] for a good exposition with a small correction to Eilenberg's definitions.

When attention is restricted to abstract semigroups (as opposed to transformation semigroups), better methods are available. The *depth decomposition* method of Eilenberg and Tilson [20] is known to yield decompositions for abstract semigroups which are at least as short as, and sometimes shorter than, holonomy decompositions. We briefly recall the depth decomposition method; for full details, see Tilson [20].

Recall that a $\mathcal{J}$-class is called *essential* if it contains a non-trivial subgroup. The *depth* of an essential $\mathcal{J}$-class is the length of the longest chain of essential $\mathcal{J}$-classes strictly above it. The *depth of the semigroup* is defined to be the length of the longest chain of essential $\mathcal{J}$-classes in the semigroup, that is, *one more than* the greatest depth of an essential $\mathcal{J}$-class, Let n denote the depth of the semigroup S. For each essential $\mathcal{J}$-class J, let G_J denote the maximal subgroup of J. Now for $0 \leq i < n$, let K_i be the direct product of all subgroups G_J where J is an essential $\mathcal{J}$-class of depth i.

Theorem 5.1. *(The Depth Decomposition Theorem, Eilenberg-Tilson 1976) Let S be a finite semigroup of depth n, and let $K_0, \ldots, K_{n-1}$ be as defined above. Then there exist aperiodic monoids $A_0, \ldots A_n$ such that S divides the wreath product*

$$A_n \wr K_{n-1} \wr A_{n-1} \wr \cdots \wr K_0 \wr A_0.$$

Thus, the depth decomposition theorem gives, for any finite semigroup S, a Krohn-Rhodes decomposition with group length equal to the depth of S. To apply the depth decomposition theorem, we need some information about the $\mathcal{J}$-class structure and maximal subgroups of our semigroups. The following proposition provides a description; various parts of it have been observed before [1, 16, 17, 22] but for completeness we prove the entire statement.

Proposition 5.1. *Let n be a positive integer and k a finite field. Then* *(i)* $T_n(k)$ *has depth* $n-1$ *if* $k = F_2$, *or depth* n *otherwise. For* $0 \le i \le n-2$ *or* $0 \le i \le n-1$ *as appropriate,* $T_n(k)$ *has* $\binom{n}{i}$ *essential* $\mathcal{J}$*-classes of depth* i, *each of which has maximal subgroup isomorphic to* $T^*_{n-i}(k)$;

(ii) $UT_n(k)$ *has depth* $n-1$. *For* $0 \le i < n-1$, $UT_n(k)$ *has* $\binom{n}{i}$ *essential* $\mathcal{J}$*-classes of depth* i, *each of which has maximal subgroup isomorphic to* $UT^*_{n-i}(k)$; *and*

(iii) $PT_n(k)$ *has depth* $n-1$. *For* $0 \le i < n-1$, $PT_n(k)$ *has* $\binom{n}{i}$ *essential* $\mathcal{J}$*-classes of depth* i, *each of which has maximal subgroup isomorphic to* $PT^*_{n-i}(k)$.

Proof. We begin with the case of $T_n(k)$. By Proposition 3.2, the regular $\mathcal{J}$-classes are exactly the $\mathcal{J}$-classes of the subidentites. Moreover, if e and f are two subidentities, it is easily seen (for example, by using Proposition 3.1), that e is $\mathcal{J}$-below f if and only if $ef = fe = e$. Thus, the lattice of regular $\mathcal{J}$-classes is isomorphic to the lattice $\{0,1\}^n$, that is to the subset lattice of the set $\{1,\dots,n\}$. In particular, there are $\binom{n}{i}$ regular $\mathcal{J}$-classes at depth i for $i \in \{0,\dots,n\}$.

Now let $e \in T_n(k)$ be a subidentity at depth i, so that e has rank $n-i$. It is easily seen that $eT_n(k)e$ is isomorphic to $T_{n-i}(k)$ via the map that removes from a matrix all rows and columns for which e has a zero in the corresponding diagonal position. Thus the maximal subgroup at e is isomorphic to $T^*_{n-i}(k)$.

Hence, in the case that $k \neq F_2$, all regular $\mathcal{J}$-classes except for that of 0 are essential, giving the required result. In the case that $k = F_2$, however, $T^*_1(k)$ is trivial and so there are no essential $\mathcal{J}$-classes of depth $n-1$. Thus, in this case, the depth of the semigroup is one less.

The case of the unitriangular semigroup $UT_n(k)$ is exactly the same except that the maximal subgroup of the $\mathcal{J}$-class of a subidentity with $n-i$ diagonal entries is isomorphic to the unitriangular group $UT^*_{n-i}(k)$. However, since $UT^*_1(k)$ is trivial regardless of the field k, there are never essential $\mathcal{J}$-classes of depth $n-1$, so the semigroup has depth $n-1$.

For the projective triangular semigroups $PT_n(k)$, Proposition 3.3 tells us that the lattice of $\mathcal{J}$-classes is the same as that of $T_n(k)$; the maximal subgroup of the $\mathcal{J}$-classes of a subidentity of rank $n-i$ is clearly the projective image $PT^*_{n-1}(k)$ of $T^*_{n-i}(k)$. In particular, $PT^*_1(k)$ is trivial so as in the unitriangular case there are no essential $\mathcal{J}$-classes of depth $n-1$, and the semigroup has depth $n-1$. □

Proposition 5.1 supplies the information needed to apply the Depth Decomposition Theorem to our semigroups. Doing so, we obtain:

$$T_n(k) \prec A_n \wr T_1^*(k)^n \wr A_{n-1} \wr T_2^*(k)^{\binom{n}{2}} \wr \cdots \wr T_n^*(k) \wr A_0$$
$$UT_n(k) \prec B_{n-1} \wr UT_2^*(k)^{\binom{n}{2}} \wr B_{n-2} \wr UT_3^*(k)^{\binom{n}{3}} \wr \cdots \wr UT_n^*(k) \wr B_0$$
$$PT_n(k) \prec C_{n-1} \wr PT_2^*(k)^{\binom{n}{2}} \wr C_{n-2} \wr PT_3^*(k)^{\binom{n}{3}} \wr \cdots \wr PT_n^*(k) \wr C_0$$

for some aperiodic semigroups $A_0, \ldots, A_n, B_0, \ldots, B_{n-1}, C_0, \ldots, C_{n-1}$. Thus, depth decomposition gives alternative (by Theorem 3.1, optimal) decompositions of group length $n-1$ for $UT_n(k)$ and $PT_n(k)$ and a (sub-optimal) group length n decomposition for $T_n(k)$. The theorem as stated does not give an explicit description of the aperiodic terms; however, the interested reader could compute appropriate ones through an analysis of the proof [20].

Acknowledgments

The research of the first author was supported by a Marie Curie Intra-European Fellowship within the 6th European Community Framework Programme. The first author would also like to thank Kirsty for all her support and encouragement. The work of the second author was supported by an NSERC discovery grant.

References

1. J. Almeida, S. W. Margolis, and M. V. Volkov, The pseudovariety of semigroups of triangular matrices over a finite field, *Theor. Inform. Appl.* **39 (1)**, 31–48 (2005).
2. J. Almeida, S. W. Margolis, B. Steinberg and M. V. Volkov, Modular and threshold subword counting and matrix representations of finite monoids, "Words 2005, 5th International Conference on Words, 13-17 September 2005, Acts". Edited by S. Brlek and C. Reutenauer, Publications du Laboratoire de Combinatoire et d' Informatique Mathématique, UQAM 36, 65–78 (2005).
3. J. Almeida, S. W. Margolis, B. Steinberg and M. V. Volkov, Representation theory of finite semigroups, semigroup radicals and formal language theory, Preprint (2005).
4. J.-C. Birget and J. Rhodes, Almost finite expansions of arbitrary semigroups, *J. Pure Appl. Algebra* **32 (3)**, 239–287 (1984).
5. S. Eilenberg, *Automata, languages, and machines. Vol. B*, Academic Press [Harcourt Brace Jovanovich Publishers], New York (1976).
6. G. Z. Elston and C. L. Nehaniv, Holonomy embedding of arbitrary stable semigroups, *Internat. J. Algebra Comput.* **12 (6)**, 791–810 (2002).

7. A. Ginzburg, *Algebraic theory of automata*, Academic Press, New York (1968).
8. K. Henckell, S. Lazarus and J. Rhodes, Prime decomposition theorem for arbitrary semigroups: general holonomy decomposition and synthesis theorem, *J. Pure Appl. Algebra* **55 (1-2)**, 127–172 (1988).
9. W. M. L. Holcombe, Algebraic automata theory, *Cambridge Studies in Advanced Mathematics* **1**, Cambridge University Press, Cambridge (1982).
10. J. M. Howie, *Fundamentals of semigroup theory*, Clarendon Press (1995).
11. M. E. Kambites, On the Krohn-Rhodes complexity of semigroups of upper triangular matrices, to appear in *Internat. J. Algebra Comput.* (2004).
12. K. Krohn and J. Rhodes, Algebraic theory of machines. I. Prime decomposition theorem for finite semigroups and machines, *Trans. Amer. Math. Soc.* **116**, 450–464 (1965).
13. K. Krohn and J. Rhodes, Complexity of finite semigroups, *Ann. of Math. (2)* **88**, 128–160 (1968).
14. K. Krohn, J. Rhodes and B. Tilson, Lectures on the algebraic theory of finite semigroups and finite-state machines, Chapters 1, 5-9 (Chapter 6 with M. A. Arbib), *Algebraic Theory of Machines, Languages, and Semigroups*, (M. A. Arbib, ed.), Academic Press, New York (1968).
15. A. Mintz, Structure and complexity of the multiplicative monoids of path algebras, Talk given at the International Conference on Semigroups and Languages, Lisbon (2005).
16. J. Okniński, Semigroups of matrices, *Series in Algebra* **6**, World Scientific Publishing Co. Inc., River Edge, NJ (1998).
17. M. S. Putcha, Linear algebraic monoids, *London Mathematical Society Lecture Note Series* **133**, Cambridge University Press, Cambridge (1988).
18. J. Rhodes, Algebraic theory of finite semigroups. Structure numbers and structure theorems for finite semigroups (with an appendix by B. R. Tilson), *Semigroups (Proc. Sympos., Wayne State Univ., Detroit, Mich., 1968)*, Academic Press, New York, 125–208 (1969).
19. J. Rhodes, Proof of the fundamental lemma of complexity (strong version) for arbitrary finite semigroups, *J. Combinatorial Theory Ser. A* **16**, 209–214 (1974).
20. B. R. Tilson, Depth decomposition theorem, Chapter XI in Eilenberg [5].
21. B. R. Tilson, Complexity of semigroups and morphisms, Chapter XII in Eilenberg [5].
22. M. V. Volkov and I. A. Goldberg, Identities of semigroups of triangular matrices over finite fields, *Mat. Zametki* **73 (4)**, 502–510 (2003).
23. H. P. Zeiger, Cascade synthesis of finite machines, *Information and Control* **10 (4)**, 419–433 (1967).

IN McALISTER'S FOOTSTEPS: A RANDOM RAMBLE AROUND THE P-THEOREM

MARK V. LAWSON

Department of Mathematics
Heriot-Watt University
Riccarton
Edinburgh EH14 4AS
Scotland, U.K.
E-mail: M.V.Lawson@ma.hw.ac.uk

STUART W. MARGOLIS*

Department of Mathematics
Bar Ilan University
Ramat Gan 52900
Israel
E-mail: margolis@math.biu.ac.il

The work of Don McAlister has been an inspiration to all of us interested in semigroups. In 1974, Don published two papers which had a decisive impact on the subsequent development of semigroup theory. In these papers, two major theorems were proved: the 'covering theorem' and the 'P-theorem'. In this paper, we shall take the latter as the starting point for some excursions through our own and others' work.

1. A primer on categories and inverse semigroups

In this section, we shall review the basic definitions and results about categories and inverse semigroups we shall need, and indicate one way in which inverse semigroups give rise to categories.

There are two equivalent definitions of 'category' in the literature. We shall give the 'arrows only' definition first, and then briefly indicate how the other definition works. Let C be a set equipped with a partially defined binary operation. An *identity* is any element e such that if ea is defined

*Partially funded by the Excellency Center "Group Theoretic Methods in the Study of Algebraic Varieties" of the Israel Science Foundation.

then $ea = a$, and if ae is defined then $ae = a$. If e is an identity and ae is defined then e is called a *right identity of* a, and if ea is defined then e is called a *left identity of* a. A *category* C is a set equipped with a partially defined multiplication such that each element has a unique left identity, called its *source*, and a unique right identity, called its *target*; the product ab is defined iff the source of a is equal to the target of b; the product is associative when it is defined; and finally, the source of ab is equal to the source of b, and the target of ab is equal to the target of a.

The other definition of a category is the 'objects and arrows' one. This starts with a directed graph whose vertices are called *objects* and whose directed edges are called *arrows*. In addition, for each object v there is a unique arrow 1_v which forms a loop at v. We now require a partial multplication to be defined on the arrows in which the loops of the form 1_v are the identities and the above axioms for a category hold.

Apart from these two variations on the definition of a category, the reader should also be aware that the product of two arrows ab is sometimes instead defined iff the target of a is equal to the source of b. Our first definition of multiplication models composition of functions when the arguments are written on the right, whereas our second definition models composition of functions when the arguments are written on the left.

In a category, the set of all arrows with source and target equal to a given identity e is a monoid called the the *local monoid at* e. A category with a single identity is therefore just a monoid. For us categories are algebraic structures in the usual way generalising monoids. The morphisms between categories are called *functors*.

A *groupoid* is a category in which for each element a there is an element b such that ab and ba are identities. Thus a groupoid with one identity is a group. A category is said to be *connected* if for each pair of identities e and f there is an arrow with source e and target f. Connected groupoids can be described in terms of groups. If G is a group and I is a set then $I \times G \times I$ becomes a connected groupoid when the product of triples is defined by $(i, g, j)(j, h, k) = (i, gh, k)$ and undefined otherwise. The identities in this case are the elements of the form $(i, 1, i)$. Groupoids constructed in this way are often known as *Brandt groupoids*. Given a connected groupoid, choose an identity e and fix it, and denote the set of identities of the groupoid by I. For each identity f, pick an arrow a_f from e to f, which is possible because the groupoid is connected. The local monoid at e is in fact a group, which we shall denote by G. Define a function from the connected groupoid to the Brandt groupoid $I \times G \times I$ by g maps to $(j, a_j^{-1} g a_i, i)$ if g has source i

and target j. It is straightforward to check that we have proved that every connected groupoid is isomorphic to a Brandt groupoid. We say that the isomorphism is defined by 'co-ordinatising' the original groupoid.

A semigroup S is said to be *inverse* if for each element s there exists a unique element t such that the following two equations hold:

$$s = sts \text{ and } t = tst.$$

The uniquely defined element t is called the *inverse* of s and is denoted s^{-1}. This is a generalisation of the definition of inverse used in group theory. However, the elements $s^{-1}s$ and ss^{-1} are not identites in general, even if the semigroup is a monoid, but retain one feature possessed of identities in that they are idempotents, where an *idempotent* is an element e such that $e^2 = e$. Remarkably, it can be proved that the product of two idempotents is an idempotent, and so the set of idempotents, $E(S)$, forms a subsemigroup, which is also commutative.

On every inverse semigroup, we can define a relation $\leq$ by

$$s \leq t \text{ iff } s = te \text{ for some idempotent } e.$$

This relation is a partial order, called the *natural partial order*. It intertwines nicely with the algebraic structure of the inverse semigroup in the sense that if $s \leq t$ then $s^{-1} \leq t^{-1}$, and if $s \leq t$ and $s' \leq t'$ then $ss' \leq tt'$. With respect to the natural partial order the set of idempotents is a meet semilattice when we define $e \wedge f = ef$. For this reason, we often refer to the *semilattice of idempotents* of an inverse semigroup.

The natural partial order can be used to define an important congruence on every inverse semigroup. Define the relation σ by

$$s \, \sigma \, s' \Leftrightarrow t \leq s, s' \text{ for some } t.$$

Then σ is a congruence, S/σ is a group, and σ is the smallest group congruence. In this paper, $G(S) = S/\sigma$ will be called the *universal group* of S. The natural map from S to $G(S)$ is denoted by γ.

The natural partial order plays an important role in the structure of inverse semigroups. Define a partial operation $\circ$ on S as follows:

$$s \circ t \text{ is defined iff } s^{-1}s = tt^{-1}$$

in which case set $s \circ t = st$. The structure $(S, \circ)$ is a groupoid. The groupoid product and the natural partial order together determine the semigroup product since

$$st = (se) \circ (et),$$

where $e = s^{-1}stt^{-1}$, and $se \leq s$ and $et \leq t$.

The representation of the multiplication of an inverse semigroup in terms of a groupoid multiplication is the first clue that categories may well play a role in studying inverse semigroups — they do in a number of different ways as we shall see throughout this paper.

Finally, we define one further relation on an inverse semigroup. The *compatibility relation* $\sim$ is defined by

$$s \sim t \text{ iff } s^{-1}t, st^{-1} \text{ are both idempotents.}$$

This relation is reflexive and symmetric, but not transitive in general. More on inverse semigroups can be found in [17].

A good example of an inverse semigroup is the *symmetric inverse monoid* on the set X, denoted $I(X)$, which consists of all bijections between subsets of X (*partial bijections*) with the operation of composition of partial functions. In this inverse semigroup, the idempotents are the identity functions on subsets, the natural partial order is the usual ordering of partial functions, the groupoid product of elements is only defined when the domain of the first matches exactly the image of the second, and a pair of partial bijections are compatible iff their union is another partial bijection. Every inverse semigroup can be embedded in a symmetric inverse semigroup, a result known as the *Vagner-Preston representation theorem.*

2. The P-theorem

The P-theorem centres on the class of E-unitary inverse semigroups. These semigroups were introduced by Saito in 1965 [52] who called them *proper inverse semigroups.* The 'p' of 'proper' explains the 'p' in 'P-theorem'. As we shall see, there are many, equivalent definitions of E-unitary inverse semigroups, each giving a different way of thinking about them. Perhaps the simplest is the following. An inverse semigroup is *E-unitary* if an element above an idempotent, in the natural partial order, is also an idempotent.

Why is this a good class of semigroups to study? Well, mainly because we can find so many interesting examples of them. For example, free inverse monoids are E-unitary, so we are off to a flying start. But here are some other reasons to study them:

(1) The inverse monoid generated by the Möbius transformations on the complex plane is E-unitary. Its universal group is the Möbius group [16].

(2) The inverse monoid of all right ideal isomorphisms between essential finitely generated right ideals of the free monoid on two generators is E-unitary. Its universal group is the Thompson group V [1].
(3) The linear clause monoids over operator domains having a single operation are E-unitary [27]. These monoids are closely related to the group theory of Patrick Dehornoy.
(4) A further connection between groups and E-unitary inverse semigroups is the M&M expansion of X-generated groups to X-generated E-unitary inverse semigroups [30].
(5) Kellendonk's topological groupoid constructed from an inverse semigroup is T_1 in general, but Hausdorff when the inverse semigroup is E-unitary [17].
(6) For every inverse semigroup S there is an E-unitary inverse semigroup T and a surjective homomorphism $\theta\colon T \to S$ which induces an isomorphism between $E(T)$ and $E(S)$. This is the 'covering theorem' [32]. There is sadly no space to say more about this theorem. We can do no more than refer you to McAlister's papers [36, 37, 39, 40, 41, 42] and those of Lawson [8, 10, 11, 12, 13].
(7) One *disadvantage* of E-unitary inverse semigroups is that if they have a zero then they are necessarily semilattices, because every element lies above the zero. This rather disappointing result is rectified in the definition of E^*-unitary and strongly E^*-unitary inverse semigroups. For more on these see [24].

One striking feature of E-unitary inverse semigroups is that they can be characterised in a wide variety of ways. Proofs and references can be found in [17].

Theorem 2.1. *The following are equivalent for an inverse semigroup S.*

(i) S is E-unitary.
(ii) The homomorphism $\gamma\colon S \to G(S)$ is idempotent pure *meaning that $\gamma(s) = 1$ implies that s is an idempotent.*
(iii) The homomorphism $\gamma\colon S \to G(S)$ is $\mathcal{L}$-injective, *meaning that γ restricted to each $\mathcal{L}$-class is injective.*
(iv) The function from S to $E(S) \times G(S)$ that maps s to $(s^{-1}s, \sigma(s))$ is injective.
(v) $\sigma(e) = E(S)$ for each idempotent e.
(vi) The compatibility relation is transitive.

(vii) The compatibility relation is equal to σ.

Both groups and semilattices are E-unitary. So the idea arises that maybe groups and semilattices could be used as building blocks for constructing all E-unitary inverse semigroups. A very nice heuristic which leads to the P-theorem is described in [42].[a] Essentially, one observes that semidirect products of semilattices by groups are E-unitary; that inverse subsemigroups of semidirect products of semilattices by groups needn't be semidirect products of semilattices by groups; and that inverse subsemigroups of E-unitary inverse semigroups are also E-unitary. If you have very good intuitions about inverse semigroups, you then come up with the following construction. Note that partially ordered sets will be abbreviated to *posets*. An *order ideal* in a poset is a subset that contains all elements beneath each element of the subset. A *McAlister triple* (G, X, Y) consists of a group G, a poset X, and an order ideal Y of X that is a meet semilattice under the induced order, such that G acts on X by order automorphisms satisfying the following two conditions:

(MT1) $G \cdot Y = X$.
(MT2) $g \cdot Y \cap Y \neq \emptyset$ for each $g \in G$.

Put

$$P = P(G, X, Y) = \{(y, g) \in Y \times G \colon \ g^{-1} \cdot y \in Y\}.$$

Define a binary operation on P by

$$(y, g)(y', g') = (y \wedge g \cdot y', gg')$$

where the meet always exists, is defined in the poset X, and belongs to Y. It can be checked that P is an E-unitary inverse semigroup with semilattice of idempotents isomorphic to Y and universal group G. Semigroups of the form $P(G, X, Y)$ are called *P-semigroups*. If $Y = X$ then we get back semidirect products of semilattices by groups. But the construction, although it looks like a semidirect product, isn't. The partially ordered set X is a crucial ingredient even though it seems to stand aloof from the proceedings. It also turns out that it cannot in general be replaced by a semilattice. The 'P-theorem' can now be stated.

Theorem 2.2. *Each E-unitary inverse semigroup is isomorphic to a P-semigroup.*

[a] A paper which is neither random, nor rambling, but *is* about inverse semigroups.

There are many different proofs of this theorem. For example:

(1) Don's own proof 1974 [33]. See Section 6.
(2) Boris Schein's proof 1975 [53]. See Section 3.
(3) Norman Reilly and Douglas Munn's proof 1976 (using free inverse semigroups) [51].
(4) Douglas Munn's proof 1976 [49]. See Section 3.
(5) The maverick alternative: 'the Q-theorem' by Mario Petrich and Norman Reilly 1979 [50]. See Section 3.
(6) Stuart Margolis and Jean-Eric Pin's proof 1987 (using the derived category) [29]. See Section 5.
(7) Mark V Lawson's proof of 1990 [6]. See Section 4.
(8) Helen James and Mark V Lawson's proof of 1999 [19]. See Section 6.
(9) Ben Steinberg's proof 2003 (using Schützenberger graphs) [56].

In subsequent sections, we shall look at the proof of the P-theorem from a number of different points of view each of which will provide a partial answer to the following question: what does the P-theorem mean?

3. Partial group actions

We shall begin with the most concrete way of thinking about E-unitary inverse semigroups. By the Vagner-Preston representation theorem, every inverse semigroup is isomorphic to an inverse semigroup of partial bijections. Accordingly, let S be an E-unitary inverse subsemigroup of a symmetric inverse monoid $I(X)$. Looking at our list of characterisations of a semigroup being E-unitary in Theorem 2.1, there are three that we want to highlight now: (v), (vi) and (vii). An element g of the universal group $G(S)$ is a σ-class and so by (vii) a set of pairwise compatible elements of S. We may therefore form the union, f_g, within $I(X)$, of the elements of g to obtain a well-defined partial bijection of X. If $1 \in G$ is the identity element then f_1 is an idempotent by (v). The idempotents in $I(X)$ are the identity functions defined on subsets of X. It is no loss in generality to assume that the domain of definition of f_1 is the whole of X, since if it isn't we can embed S in $I(X')$ where X' is the domain of f_1. If $g \in G$ and $x \in X$ then the element x need not belong to the set $\mathrm{dom}(f_g)$, but if it does then we define

$$g \cdot x = f_g(x).$$

We therefore have a partial function $G \times X \to X$ that defines what we call a 'partial action' of G on X. We now make this precise. A partial function

$G \times X \to X$ which maps (g, x) to $g \cdot x$ is said to define a *partial group action* [25] if the following three axioms hold.

(PGA1) $1 \cdot x$ is always defined and equals x.
(PGA2) If $g \cdot x$ is defined then $g^{-1} \cdot (g \cdot x)$ is defined and equals x.
(PGA3) If $g \cdot (h \cdot x)$ is defined then $(gh) \cdot x$ is defined and they are equal.

Observe that (PGA3) is the crucial difference with 'global' group actions. If S is an E-unitary inverse semigroup of partial bijections of the set X then $G(S)$ acts partially on X. This leads us to think that there may be some connection between the structure of E-unitary inverse semigroups and partial group actions.

Partial group actions can easily be constructed. Let $G \times X \to X$ be a global group action and $X' \subseteq X$. Then G acts partially on X'. In fact, every partial group action arises in this way. The following theorem, known as the 'Globalisation Theorem', although formally proved by Johannes Kellendonk and Mark V Lawson, has been around in one form or another for a long time.

Theorem 3.1. *Let G act partially on the set X. Then there is a set $\bar{X}$, essentially unique, such that G acts on $\bar{X}$, $\bar{X}$ contains X, the restriction of the action of G to X is equal to the original partial action, and $G \cdot X = \bar{X}$.*

We say that the action of G on $\bar{X}$ is the *globalisation* of the partial action of G on X. We already see parallels with the definition of a McAlister triple: indeed, if (G, X, Y) is a McAlister triple then G acts partially on Y and since $G \cdot Y = X$ it follows that the action of G on X is the globalisation of the partial action of G on Y. The only difference is that the sets and actions have extra structure. Can we use the Globalisation Theorem to prove the P-theorem? The answer is 'yes' in two different ways.

Schein's proof Let S be an E-unitary inverse semigroup. Then by Theorem 2.1(iv), there is an embedding $\kappa\colon\ S \to E(S) \times G(S)$ which maps s to $(s^{-1}s, \sigma(s))$. Let A be the image of κ. The semigroup S is isomorphic to an inverse subsemigroup of $I(A)$ by the Vagner-Preston representation theorem. The globalisation of the partial action of $G(S)$ on A turns out to be $B = E(S) \times G(S)$, where $G(S)$ acts on B by left multiplication on the second component. In other words, the globalisation can be explicitly described and has nice properties. The ingredients for the McAlister triple corresponding to S can now be read off from the globalisation we have constructed.

Munn's proof and the 'Q-theorem' Let S be an E-unitary inverse smeigroup. Then by Theorem 2.1(iii), $\gamma\colon\ S \to G(S)$ is $\mathcal{L}$-injective. Thus given $g \in G(S)$ and $e \in E(S)$ there is *at most one* element $s \in S$ such that $s^{-1}s = e$ and $\gamma(s) = g$. Define a partial action of $G = G(S)$ on the set $E = E(S)$ by

$$g \cdot e = ss^{-1}$$

if $s^{-1}s = e$ and $\gamma(s) = g$. The set of elements of E on which G acts forms a non-empty order ideal of E. The partial action is order preserving, when this makes sense. The data of the partial action of G on E is equivalent to the semigroup S (this is the substance of the 'Q-theorem'). The construction of the globalisation will be found to be a poset; the details are exactly Munn's proof of the P-theorem.

Is the P-theorem equivalent to the globalisation theorem? Almost. We shall clarify this later when we talk about 'enlargements'. Globalisations of partial group actions are interesting and widespread in mathematics. For example, the globalisation in the case of the Möbius group is the Riemann sphere [25]. Ben Steinberg has looked at partial group actions on cell complexes and poses a number of interesting questions about their applications [55]. In particular, he suggests that Bass-Serre theory might be developed from the point of view of partial group actions.

4. Ordered groupoids

There are three ways of proving the P-theorem using category theory: Margolis and Pin's proof [29], Lawson's proof [6], and the proof of James and Lawson [19]. In this section, we shall discuss Lawson's proof, in Section 5 the proof of Margolis and Pin, and in the final section the proof due to James and Lawson.

In the 1950's, Charles Ehresmann developed the theory of ordered groupoids motivated by questions in differential geometry. Since inverse semigroups can be regarded as ordered groupoids the question arose of the implications of Ehresmann's work for inverse semigroup theory. The P-theorem provided the key for understanding these implications. The papers [6, 7, 9, 14, 15] developed the ordered groupoid approach to inverse semigroups motivated by the desire to combine Ehresmann's ideas [2] with those of McAlister. In this section, we shall explain how the proof of the P-theorem looks when viewed from the perspective of ordered groupoids. We shall also say something about the connection between ordered groupoids and partial group actions. We shall finish off by sketching out how the P-theorem can be viewed from the vantage-point of a sort of generalised homotopy theory.

An *ordered groupoid* is a groupoid whose set of arrows is also a poset satisfying a number of additional conditions: the order intertwines nicely with respect to taking inverses and with respect to products, just as the natural partial order for inverse semigroups; if e is an identity less than the source of an arrow g then there exists a unique element, denoted $(g|e)$, which is less than g and whose source is e — the element $(g|e)$ is called the *restriction of g to e*. Functors between ordered groupoids are ordered preserving functors and are termed *ordered functors*. The set of identities of an ordered groupoid is a poset. If this poset is also a meet semilattice then the ordered groupoid is called, for historical reasons, an *inductive groupoid*. Inverse semigroups can be regarded as inductive groupoids when one considers them with respect to their groupoid product and their natural partial order. Groupoids are ordered groupoids when ordered by equality, and posets are ordered groupoids when viewed as groupoids of idempotents. The fact that posets can be viewed as ordered groupoids is promising when thinking about the mysterious poset X in the P-theorem.

The starting point for understanding how ordered groupoids can shed light on the P-theorem is Theorem 4.12 of [37]:

Theorem 4.1. *An inverse semigroup is isomorphic to a semidirect product of a semilattice by a group iff the homomorphism γ to its universal group is $\mathcal{L}$-bijective.*

Henceforth, we shall call $\mathcal{L}$-injective maps *immersions*, and $\mathcal{L}$-bijective maps *coverings*. The difference between arbitrary E-unitary inverse semigroups, and those which are semidirect products of semilattices by groups now resides in the difference between immersions and coverings.

Can we convert our immersion to a covering? Well, yes and no. It can be extended to a covering but the covering will be from an ordered groupoid. There are two ideas on which the ordered groupoid proof of the P-theorem is based.

The first idea requires us to think about group actions in terms of groupoids. Let G be a group acting on the set X. This action can be regarded as a groupoid: in fact when we draw pictures of group actions with points representing the elements of X and arrows showing us how elements of G move the points of X around, we are precisely thinking of the action as a groupoid. Now let G be a group acting by order automorphisms on the poset X. Our groupoid becomes an ordered groupoid. Thus given an action of the group G on the poset X by order automorphisms, we define the semidirect product of X by G, denoted $P(G, X)$, to be the ordered groupoid whose arrows are pairs (x, g) which start at $(g^{-1} \cdot x, 1)$ and terminate at $(x, 1)$. We define $(x, g) \leq (x', g')$ iff $g = g'$ and $x \leq x'$. The projection from $P(G, X)$ to G is an ordered covering functor.

Theorem 4.1 can be generalised from inverse semigroups to arbitrary ordered groupoids.

Theorem 4.2. *An ordered groupoid is isomorphic to a semidirect product of a poset by a group iff it admits a surjective ordered covering functor to a group.*

The second idea we need is that of an 'enlargement'. An enlargement is a particular kind of relationship between an inverse semigroup and an inverse subsemigroup or, more generally, between an ordered groupoid and an ordered subgroupoid. The idea, which can be justified, is that an inverse semigroup (or ordered groupoid) and its enlargement are very similar in structure: the enlargement being a sort of expanded version of its substructure. The definition arose by combining ideas to be found in both Ehresmann and McAlister. In particular, in the case of inverse semigroups, it is McAlister's notion of a 'heavy' inverse subsemigroup combined with an extra notion mentioned in a remark on page 208 of [37]. It also recurs in McAlister's work on the local structure of regular semigroups [43, 44, 45, 46, 47, 48]. Enlargements play an important role in [15, 20, 21, 22].

Let G be an ordered subgroupoid of an ordered groupoid H. We say that H is an *enlargement* if the following three conditions hold.

(E1) G is an order ideal of H.

(E2) If the source and target of an arrow of H belong to G then the arrow belongs to G.

(E3) Each idempotent of H is connected by an arrow to an idempotent of G.

The ordered groupoid version of the P-theorem can now be stated.

Theorem 4.3. *The P-theorem is equivalent to the following statement. Let S be an E-unitary inverse semigroup. Then the immersion $\gamma\colon S \to G(S)$ can be factorised $\theta = \iota\Theta$ where $\iota\colon S \to \bar{S}$ is an inclusion of S into an ordered groupoids $\bar{S}$ which is an enlargement of S, and $\Theta\colon \bar{S} \to G$ is a* *covering.*

We shall call this process the 'enlargement of an immersion to a covering'.

This proof of the P-theorem looks quite different from the partial group action proofs described in the previous section. However, as we shall now show, they are really just different sides of the same coin. Recall that a group G acting on a set X can be repackaged as a groupoid equipped with a covering to G. It is easy to show that a group G acting *partially* on a set X can be repackaged as a groupoid equipped with an *immersion* to G. Theorem 4.3 above can be generalised: any surjective immersion from an ordered groupoid to a group can be enlarged to a covering. The above result is equivalent to the globalisation theorem for groups acting partially by order automorphisms on posets.

We therefore arrive at one answer to our question about the meaning of the P-theorem:

> *"Globalisations and the P-theorem are both aspects of one and the same problem."*

The theory of enlargements above has wide-ranging generalisations; more information can be found in [17]. But this is not quite the end of the story. Ben Steinberg looked afresh at the work on the maximum enlargement theorem through Rhodes-tinted spectacles [54]. Steinberg's work in turn led to a fully-fledged homotopy theory of ordered groupoids (and so of inverse semigroups) [23]. If you put all this together, you get another, quite-different looking interpretation of the P-theorem.

> *"The P-theorem is an analogue (in some sense) of the well-known result in topology that states that every continuous function can be factorised into a homotopy equivalence followed by a fibration."*

5. Extensions of semilattices by groups, the derived category and global semigroup theory

In this section, we take a look at the next categorical proof of the P-theorem due to Margolis and Pin [29], chronologically the first using categorical methods. This had ramifications beyond inverse semigroup theory in that it led to the introduction of methods from the algebraic theory of categories into semigroup theory. This has had a major influence on subsequent developments in semigroup theory and also provides a strong connection between the work of Don McAlister and John Rhodes' notion of Global Semigroup Theory.

The starting point for this approach is the important role played in both group theory and semigroup theory by wreath products. We mention just two examples. First, the theorem of Krasner and Kaloujnine [5] states that if $f : G \to H$ is a morphism of groups, then G embeds into the wreath product $N \wr H$ where $N = Ker(f)$. Thus every finite group G embeds into the wreath product of its Jordan-Hölder factors, arising from a composition series for G. Second, the Krohn-Rhodes Theorem [3] shows that every finite semigroup S divides[b] a wreath product of subgroups of S and the three element monoid consisting of two right zeroes and an identity element. By the Krasner and Kaloujnine Theorem we can further decompose the groups into their simple divisors.

How do these ideas relate to the P-theorem? Characterisation (ii) of an E-unitary inverse semigroup in Theorem 2.1, the one characterisation we have yet to use, says that an inverse semigroup S is E-unitary if and only if it is an extension of its semilattice $E(S)$ by its universal group $G(S)$. Suppose we could find a generalisation of the Krasner-Kaloujnine Theorem that would, given a homomorphism $f : S \to T$, obtain S from T via the wreath product and something playing the role of $Ker(f)$. Then we could apply such a construction to the case where S is an E-unitary inverse semigroup and T is its universal group to try to obtain a proof of the P-Theorem as well as to problems arising from the Krohn- Rhodes Theorem.

For this program to work, an appropriate notion of the kernel of a semigroup morphism $f : S \to T$ has to be found. We know that the image of a morphism is the quotient of S by the congruence relation associated to f. The first approach to the problem was given by Bret Tilson [58] who defined the *derived semigroup* of a morphism to be essentially a partial action of S on the collection of congruences classes of f. The innovation was that the derived semigroup was only locally inside S and not a subsemigroup as in the case of groups. This was successful enough to help prove (along with the notion of the Rhodes expansion) the difficult and fundamental theorem of Rhodes complexity theory which states that if S and T are finite semigroups and f is injective on subgroups of S, then the group complexity of S is equal to that of T [58].

However, the derived semigroup fails to do the job precisely in cases like

[b]That is, is a homomorphic image of a subsemigroup.

that of the morphism between an E-unitary semigroup and its universal group. Even a cursory look at the definition of the derived semigroup [58] shows that it has a category like multiplication with an externally adjoined zero to take care of products that are not defined. In the case of the morphism between an E-unitary semigroup and its universal group, the derived semigroup is an inverse semigroup, but its semilattice is a 0-disjoint union of $|G(S)|$ copies of the semilattice of S, whereas we want the "kernel" to have semilattice $E(S)$.

This problem was solved in [29] by discarding the offending zero and treating the derived object as a category. This was motivated by corresponding uses of groupoids in the theory of groups as exposed by Philip Higgins [4] and others. In this way, the *derived category* of the morphism between an E-unitary semigroup and its universal group is a category $D(\gamma)$ such that each local monoid is a semilattice. The following gives an outline of the use of the derived category to prove the P-Theorem. It is convenient to assume that S is a monoid. This presents no problem, as an inverse semigroup S is E-unitary if and only if the monoid S^1 obtained by adjoining an identity to S is E-unitary. Let $\gamma\colon\ S \to G(S)$ be the map from the E-unitary monoid S to its universal group $G(S)$. The derived category $D(\gamma)$ has objects $G(S)$ and for each $g, h \in G(S)$, the set of morphisms from g to h is $\{(g, m, h) | g(\gamma(m)) = h\}$. There is an obvious Brandt-groupoid-like composition that turns this into a category. The local monoid at any object g is a semilattice isomorphic to $E(S)$. The group G(S) acts on the category $D(\gamma)$ by left multiplication. This action is transitive on the objects. Thus the quotient $D(\gamma)/G(S)$ is a one object category, better known as a monoid! The monoid $D(\gamma)/G(S)$ is canonically isomorphic to S and the map from $D(\gamma)$ to $D(\gamma)/G(S)$ is a covering of categories [4]. The P-structure of S can be recovered from $D(\gamma)$. As expected, the group in the McAlister triple is $G(S)$ and the semilattice is $E(S)$. Pleasantly, the "mysterious" partially ordered set turns out to be the partially ordered set of $\mathcal{J}$-classes of $D(\gamma)$. Here we view $D(\gamma)$ as a partial associative structure and then Green's relations have the same definition and analogous properties as in the case of semigroups. In fact, they are the restriction of Green's relations to $D(\gamma)$ on the semigroup obtained by adding an external zero to $D(\gamma)$. As mentioned above, this semigroup is what Tilson defined to be the derived semigroup of γ.

This was just the beginning of the use of the derived category and related algebraic categories in the decomposition theory of semigroups and its applications to various disciplines. In fact, the idea goes back to what is called the Grothendieck construction. Bret Tilson made fundamental and deep contributions to this area and put it at the centre of semigroup theory. In particular, he showed that the construction is, in a precise sense, adjoint to the construction of semidirect products: another connection to the P-Theorem. See his paper [59] and the later joint work with Benjamin Steinberg [57].

Finally we give a connection between the McAlister Theorems and Rhodes' notion of global semigroup theory. We would like to study an arbitrary inverse semigroup S as an extension of its universal group $G(S)$. If S has a zero element, then except for semilattices this approach is doomed from the start, since $G(S)$ is easily seen to be the trivial group. In global semigroup theory, one looks for an "expansion" of S to remove the obstruction that 0 presents. An expansion is intuitively a semigroup that maps onto S, is close to S in its structure and has nicer properties; in our case, it would be E-unitary and thus we could build the expansion from its universal group and its semilattice via the P-Theorem. McAlister's Covering Theorem gives exactly this. Every inverse semigroup is an idempotent separating image of an E-unitary inverse semigroup (finiteness can be preserved). This means that the covering semigroup has the same semilattice as S in particular. Thus the two McAlister Theorems can be considered to be an early example of the methods and philosophy of Global Semigroup Theory.

6. Cancellative categories

In this section, we shall describe the remaining categorical proof of the P-theorem due to James and Lawson [19]. It is appropriate that we describe it last of all because it in fact takes us back to Don's original proof of the P-theorem, as we shall see.

The starting point is the class of bisimple inverse monoids. An inverse semigroup is bisimple if it consists of a single $\mathcal{D}$-class. It is a classical theorem of Clifford that bisimple inverse monoids are determined by the $\mathcal{R}$-class containing the identity: this is a right cancellative monoid in which the intersection of any two principal left ideals is again a principal left ideal. We shall call such monoids *division monoids* for short. Furthermore, from each (abstract) division monoid we can construct a bisimple inverse monoid. Recall that a (right) cancellative monoid is *right reversible* if for all elements s and t in the monoid, elements p and q can be found such that $ps = qt$. It is a classical theorem of Ore, that each right reversible cancellative monoid can be embedded in a group in such a way that each element of the group is of the form $a^{-1}b$ where a and b are elements of the monoid. It is interesting that Rees proved Ore's embedding theorem using, once again, E-unitary inverse semigroups. You can find his proof on pages 68 and 69 of [17]. Don McAlister and Bob McFadden proved the following result in Section 3 of [35].

Theorem 6.1. *Let S be a bisimple inverse monoid. Then S is E-unitary if and only if its associated division monoid is cancellative. In the E-unitary case, the McAlister triple describing S can be recovered easily from the embedding of the division monoid into its group of fractions.*

In [19], this argument is generalised to an arbitrary E-unitary inverse monoid. The role of the $\mathcal{R}$-class containing the identity is taken by the division category $C(S)$ of S in the sense of Leech [28]. This is a right cancellative category. The existence of least common left multiples is generalised to the condition that each pair of arrows with a common source has a pushout. It's proved that an inverse monoid S is E-unitary if and only if its division category is cancellative. When S is E-unitary the division catgeory $C(S)$ can be embedded in a connected groupoid, its groupoid of fractions, in a way that directly generalises the Ore embedding theorem. The McAlister triple describing S can be recovered easily from this embedding.

McAlister's original proof of the P-theorem can be obtained from the above proof by 'choosing coordinates'. The starting point is the result we mentioned in Section 1: connected groupoids are isomorphic to Brandt groupoids, the isomorphism being defined by 'co-ordinatising' the groupoid. If this result is applied to the proof above then we get *exactly* the first ever proof of the P-theorem.

Again, this is not the end of the story. Leech's categorical description applies to inverse monoids. How can it be generalised to inverse semigroups? McAlister's papers [31, 34, 38] provided the clues, and the theory was worked out in [18, 26].

References

1. J.-C. Birget, The groups of Richard Thompson and complexity, *IJAC* **14**, 569–626 (2004).
2. Ch. Ehresmann, *Oeuvres complètes et commentées*, (ed. A. C. Ehresmann), Suppl. *Cahiers Topologie Géom. Diff.* (Amiens, 1980–1983).
3. S. Eilenberg, *Automata, Languages and Machines*, Academic Press, New York, Vol A, 1974; Vol B, 1976.

4. Ph. Higgins, *Categories and groupoids*, Van Nostrand Reinhold Company, 1971.
5. M. Krasner and L. Kaloujnine, Produit complet des groupes de permutations et problème d'extension de groupes. III, *Acta Sci. Math. Szeged* **14**, 69–82 (1951).
6. M. V. Lawson, The geometric theory of inverse semigroups I: E-unitary inverse semigroups, *J. Pure and Appl. Algebra* **67**, 151–177 (1990).
7. M. V. Lawson, The geometric theory of inverse semigroups II: E-unitary covers of inverse semigroups, *J. Pure and Appl. Algebra* **83**, 121–139 (1992).
8. M. V. Lawson, An equivalence theorem for inverse semigroups, *Semigroup Forum* **47**, 7–14 (1993).
9. M. V. Lawson, Congruences on ordered groupoids, *Semigroup Forum* **47**, 150–167 (1993).
10. M. V. Lawson, Coverings and embeddings of inverse semigroups, *Proc. Edinburgh Math. Soc.* **36**, 399–419 (1993).
11. M. V. Lawson, A note on a paper of Joubert, *Semigroup Forum* **47**, 389–392 (1993).
12. M. V. Lawson, Extending partial isomorphisms and McAlister's covering theorem, *Semigroup Forum* **48** 18–27 (1994).
13. M. V. Lawson, Almost factorisable inverse semigroups, *Glasgow Mathematical J.* **36**, 97–111 (1994).
14. M. V. Lawson, A class of actions of inverse semigroups, *J. of Algebra* **179**, 570–598 (1996).
15. M. V. Lawson, Enlargements of regular semigroups, *Proc. Edin. Math. Soc.* **39**, 425–460 (1996).
16. M. V. Lawson, The Möbius inverse monoid, *J. of Algebra* **200**, 428–438 (1998).
17. M. V. Lawson, *Inverse semigroups: the theory of partial symmetries*, World Scientific, 1998.
18. M. V. Lawson, Constructing inverse semigroups from category actions, *J. of Pure and Applied Algebra* **137**, 57–101 (1999).
19. M. V. Lawson and H. James, An application of groupoids of fractions to inverse semigroups, *Periodica Mathematica Hungarica* **38**, 43–54 (1999).
20. M. V. Lawson and L. Márki, Enlargements and coverings by Rees matrix semigroups, *Monatsh. Math.* **129**, 191–195 (2000).
21. M. V. Lawson and T. Khan, A characterisation of a class of semigroups with locally commuting idempotents, *Periodica Mathematica Hungarica* **40**, 85–107 (2000).
22. M. V. Lawson and T. Khan, Rees matrix covers for a class of semigroups with locally commuting idempotents, *Proc. Edin. Math. Soc.* **44**, 173–186 (2001).
23. M. V. Lawson, J. Matthews and T. Porter, The homotopy theory of inverse semigroups, *IJAC* **12**, 755–790 (2002).
24. M. V. Lawson, E^*-unitary inverse semigroups, in *Semigroups, algorithms, automata and languages* (eds. G. M. S. Gomes, J-E. Pin, P. V. Silva) World Scientific, 195–214 (2002).

25. M. V. Lawson and J. Kellendonk, Partial actions of groups, *IJAC* **14**, 87–114 (2004).
26. M. V. Lawson, Constructing ordered groupoids, *Cahiers de topologie et géométrie différentielle catégoriques* **46**, 123–138 (2005).
27. M. V. Lawson, A correspondence between balanced varieties and inverse monoids, Accepted by *IJAC*.
28. J. Leech, Constructing inverse monoids from small categories, *Semigroup Forum* **36**, 89–116 (1987).
29. S. W. Margolis and J.-E. Pin, Inverse semigroups and extensions of groups by semilattices, *J. of Algebra* **110**, 277–297 (1987).
30. S. W. Margolis and J. C. Meakin, E-unitary inverse monoids and the Cayley graph of a group presentation, *J. of Algebra* **58**, 45–76 (1989).
31. D. B. McAlister, 0-bisimple inverse semigroups, *Proc. London Math. Soc.* (3) **28**, 193–221 (1974).
32. D. B. McAlister, Groups, semilattices and inverse semigroups, *Trans. Amer. Math. Soc.* **192**, 227–244 (1974).
33. D. B. McAlister, Groups, semilattices and inverse semigroups II, *Trans. Amer. Math. Soc.* **196**, 351–370 (1974).
34. D. B. McAlister, On 0-simple inverse semigroups, *Semigroup Forum* **8**, 347–360 (1974).
35. D. B. McAlister and R. McFadden, Zig-zag representations and inverse semigroups, *J. of Algebra* **32**, 178–206 (1974).
36. D. B. McAlister, $\vee$-prehomomorphisms on inverse semigroups, *Pac. J. Math.* **67**, 215–231 (1976).
37. D. B. McAlister, Some covering and embedding theorems for inverse semigroups, *J. Austral. Math. Soc.* **22** (Series A), 188–211 (1976).
38. D. B. McAlister, One-to-one partial right translations of a right cancellative semigroup, *J. of Algebra* **43**, 231–251 (1976).
39. D. B. McAlister and N. R. Reilly, E-unitary covers for inverse semigroups, *Pac. J. Math.* **68**, 161–174 (1977).
40. D. B. McAlister, E-unitary inverse semigroups over a semilattice, *Glas. Math. J.* **19**, 1–12 (1978).
41. D. B. McAlister, Embedding semigroups in coset semigroups, *Semigroup Forum* **20**, 255–267 (1980).
42. D. B. McAlister, A random ramble through inverse semigroups, in *Semigroups* (eds Hall, Jones, Preston), Academic Press, 1–19 (1980).
43. D. B. McAlister, Regular Rees matrix semigroups and regular Dubreil-Jacotin semigroups, *J. Austral. Math. Soc.* (Series A) **31**, 325–336 (1981).
44. D. B. McAlister, Rees matrix covers for locally inverse semigroups, *Trans. Amer. Math. Soc.* **277**, 727–738 (1983).
45. D. B. McAlister, Rees matrix covers for regular semigroups, *J. of Algebra* **89**, 264–279 (1984).
46. D. B. McAlister, Rees matrix covers for regular semigroups, in *Proceedings of the 1984 Marquette Conference on Semigroups* (editors K. Byleen, P. Jones, F. Pastijn), 131–141.
47. D. B. McAlister, Some covering and embedding theorems for locally inverse

semigroups, *J. Austral. Math. Soc.* (Series A) **39**, 63–74 (1985).
48. D. B. McAlister, Quasi-ideal embeddings and Rees matrix covers for regular semigroups, *J. of Algebra* **152**, 166–183 (1992).
49. W. D. Munn, A note on E-unitary inverse semigroups, *Bull. London Math. Soc.* **8**, 71–76 (1976).
50. M. Petrich and N. R. Reilly, A representation of E-unitary inverse semigroups, *Quart. J. Math. Oxford* **30**, 339–350 (1979).
51. N. R. Reilly and W. D. Munn, E-unitary congruences on inverse semigroups, *Glasgow math. J.* **17**, 57–75 (1976).
52. T. Saito, Proper ordered inverse semigroups, *Pac. J. Maths* **15**, 649–666 (1965).
53. B. M. Schein, A new proof for the McAlister "P-theorem", *Semigroup Forum* **10**, 185–188 (1975).
54. B. Steinberg, Factorization theorems for morphisms of ordered groupoids and inverse semigroups, *Proc. Edin. Math. Soc.* **44**, 549–569 (2001).
55. B. Steinberg, Partial actions of groups on cell complexes, *Monatsh. Math.* **138**, 159–170 (2003).
56. B. Steinberg, McAlister's P-theorem via Schützenberger graphs, *Comm. in Algebra* **31**, 4387–4392 (2003).
57. B. Steinberg and B. Tilson, Categories as algebra. II. *IJAC* **13**, 627–703 (2003).
58. B. Tilson, Complexity of Semigroups, Chapter XII in S. Eilenberg, *Automata, Languages and Machines*, Academic Press, New York, Vol B, 1976.
59. B. Tilson, Categories as algebra: An essential ingredient in the theory of monoids, *J. Pure Appl. Algebra* **48**, 83–198 (1987).

ON THE STRUCTURE OF THE LATTICE OF COMBINATORIAL REES–SUSHKEVICH VARIETIES

EDMOND W. H. LEE*

Department of Mathematics
Simon Fraser University
Burnaby, BC V5A 1S6, Canada
E-mail: ewl@sfu.ca

M. V. VOLKOV†

Department of Mathematics and Mechanics
Ural State University
620083 Ekaterinburg, Russia
E-mail: Mikhail.Volkov@usu.ru

We present several recent results on the structure of the lattice of combinatorial Rees–Sushkevich varieties. In particular, we study a natural decomposition of the lattice into a disjoint union of intervals and describe the sublattice generated by the extreme points of these intervals.

Introduction

The lattice $L(\mathbf{S})$ of all semigroup varieties is a complex object but some of its parts are rather well understood by now. For instance, the "upper part" of $L(\mathbf{S})$ consists of *overcommutative* varieties, that is varieties containing the variety of all commutative semigroups. These varieties form a filter in $L(\mathbf{S})$, and it turns out that the filter admits a relatively easy description in terms of congruence lattices of certain unary algebras [39]. In the "lower part" of $L(\mathbf{S})$ consisting of periodic semigroup varieties one can distinguish two important ideals formed by completely regular varieties and by nilsemigroup varieties. The former ideal was investigated in depth by Polák [16, 17, 18] in the 1980s — of course, modulo the lattice of periodic group varieties.

*The first author acknowledges support from the NSERC of Canada, grant A4044.
†The second author acknowledges support from the Russian Foundation for Basic Research, grants 05-01-00540 and 06-01-00613.

The structure of the lattice of nilsemigroup varieties was to some extent clarified by Vernikov and the second author in [35, 36, 37]. Figure 1 shows the relative location of the aforementioned fragments within $L(\mathbf{S})$.

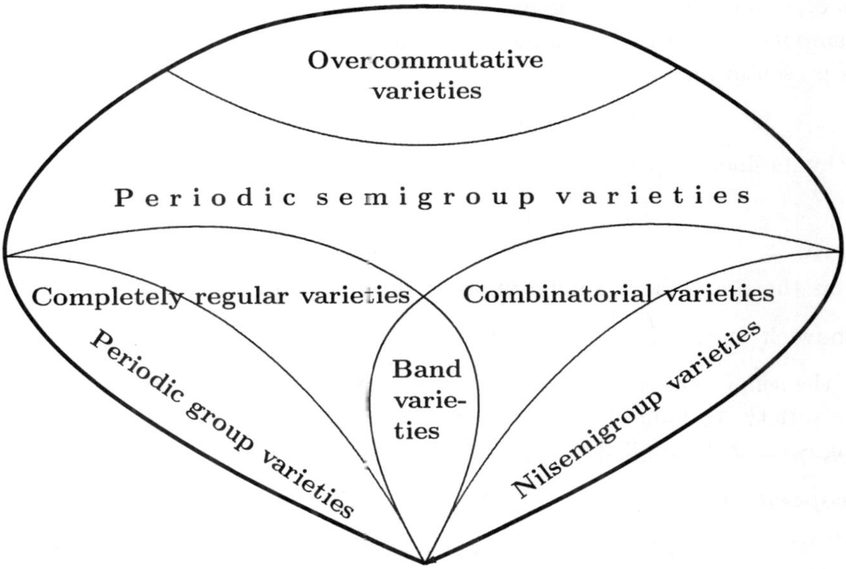

Figure 1. A sketch of the lattice of semigroup varieties

Following Kublanovskii, we say that a semigroup variety $\mathbf{V}$ is a *Rees–Sushkevich variety* if $\mathbf{V}$ is contained in a periodic variety generated by 0-simple semigroups. As it should be clear from Fig.1, any further progress in understanding the lattice of semigroup varieties requires exploring periodic varieties containing 0-simple semigroups with zero divisors. Therefore it is rather natural to start such a quest with investigating the lattice of Rees–Sushkevich varieties. This research program has been initiated by Reilly and the first author, cf. [8, 9, 10, 23, 24, 25], and recently the second author has joined in, cf. [41]. The aim of the present paper is to overview the information gathered so far on the structure of the lattice of combinatorial Rees–Sushkevich varieties. (Recall that a semigroup variety $\mathbf{V}$ is called *combinatorial* if all groups in $\mathbf{V}$ are trivial.) Due to space limitations we have included only proofs that do not require lengthy calculations; a more complete account of our results and their generalizations to the non-combinatorial case will appear elsewhere.

1. The variety $\mathbf{A_2}$ and a decomposition of its subvariety lattice into intervals

We adopt the standard terminology and notation of semigroup theory (see [3, 6, 7]) and universal algebra (cf. [2]). We denote by A_2 the 5-element idempotent-generated 0-simple semigroup that can be described either by its presentation

$$A_2 = \langle a, b \mid aba = a^2 = a,\ bab = b,\ b^2 = 0 \rangle = \{a, b, ab, ba, 0\}$$

or by its linear representation by 2×2-matrices (over any field):

$$A_2 = \left\{ \begin{pmatrix} 0 & 0 \\ 0 & 0 \end{pmatrix}, \begin{pmatrix} 1 & 0 \\ 0 & 0 \end{pmatrix}, \begin{pmatrix} 0 & 1 \\ 0 & 0 \end{pmatrix}, \begin{pmatrix} 1 & 0 \\ 1 & 0 \end{pmatrix}, \begin{pmatrix} 0 & 1 \\ 0 & 1 \end{pmatrix} \right\}$$

or as the Rees matrix semigroup over the trivial group $E = \{1\}$ with the sandwich matrix $\begin{pmatrix} 1 & 1 \\ 0 & 1 \end{pmatrix}$. Let $\mathbf{A_2}$ stand for the variety $\operatorname{var} A_2$ generated by the semigroup A_2. We shall make use of the following identity basis of the variety $\mathbf{A_2}$ found by Trakhtman (see his preprint [31] and its abridged English translation [34]):

Proposition 1.1. *The variety $\mathbf{A_2}$ is defined by the identities*

$$x^2 = x^3, \quad xyx = xyxyx, \quad xyxzx = xzxyx. \tag{1}$$

Now we easily deduce a result that explains the role of the variety $\mathbf{A_2}$:

Proposition 1.2. *The variety $\mathbf{A_2}$ is the largest combinatorial Rees–Sushkevich variety.*

Proof. The variety $\mathbf{A_2}$ is combinatorial as it satisfies the identity $x^2 = x^3$ that cannot hold in a non-trivial group and it is a Rees–Sushkevich variety since it is generated by the 0-simple semigroup A_2.

Now let $\mathbf{V}$ be a combinatorial Rees–Sushkevich variety. As a combinatorial variety, $\mathbf{V}$ must satisfy the identity $x^m = x^{m+1}$ where m is the cardinality of the $\mathbf{V}$-free semigroup on one generator. On the other hand, as a Rees–Sushkevich variety, $\mathbf{V}$ should be contained in a periodic variety $\mathbf{W}$ generated by 0-simple semigroups. Let $n \geq m$ be such that the exponent of every group in $\mathbf{W}$ divides n. Then, by Proposition 3.3 of [5], $\mathbf{W}$ satisfies the identities

$$x^2 = x^{n+2}, \quad xyx = (xy)^{n+1}x, \quad xyx(zx)^n = x(zx)^n yx,$$

and so does its subvariety $\mathbf{V}$. Combining the identity $x^m = x^{m+1}$ with the identity $x^2 = x^{n+2}$, one readily observes that $x^2 = x^{n+2} = x^m \cdot x^{n+2-m} =$

$x^{m+1} \cdot x^{n+2-m} = x \cdot x^{n+2} = x^3$ in $\mathbf{V}$. Clearly, the identities $x^2 = x^3$ and $xyx = (xy)^{n+1}x$ imply the identity $xyx = xyxyx$, and combining the latter identity with $xyx(zx)^n = x(zx)^n yx$ yields $xyxzx = xzxyx$. We have shown that the variety $\mathbf{V}$ satisfies all the identities (1) whence $\mathbf{V} \subseteq \mathbf{A_2}$. □

Thus, the lattice of combinatorial Rees–Sushkevich varieties is nothing but the lattice $L(\mathbf{A_2})$ of all subvarieties of the variety $\mathbf{A_2}$. In spite of the small size and transparent construction of the semigroup A_2, the lattice $L(\mathbf{A_2})$ is known to be quite complex; in particular, it contains an isomorphic copy of every finite lattice, see Proposition 4.1 in [36]. Nevertheless, the structure of $L(\mathbf{A_2})$ can be rather well understood via its decomposition into a union of certain intervals being classes of a complete congruence. In order to describe this decomposition, we need an intermediate observation.

Recall that a semigroup variety $\mathbf{V}$ is said to be *locally finite* if every finitely generated semigroup in $\mathbf{V}$ is finite. A variety generated by a finite semigroup is called *finitely generated*. It is well known that every finitely generated variety is locally finite (see [2], Theorem 10.16). Hence we have

Corollary 1.1. *Combinatorial Rees–Sushkevich varieties are locally finite.*

Clearly, each locally finite variety $\mathbf{V}$ is completely determined by the set $\mathbf{V}_{fin}$ of its finite members. Moreover, if $\mathbf{V}$ is a locally finite variety, then the mapping $\mathbf{W} \mapsto \mathbf{W}_{fin}$, for $\mathbf{W}$ being a subvariety of $\mathbf{V}$, is easily shown to be an isomorphism between the subvariety lattice of $\mathbf{V}$ and the subpseudovariety lattice of $\mathbf{V}_{fin}$. This observation allows us to apply to the lattice $L(\mathbf{A_2})$ a method developed by Auinger [1] for construction of complete congruences in lattices of pseudovarieties. Auinger's method is based on the notion of a *divisor system*. We shall not reproduce this notion in its full generality as we need only one concrete divisor system here (namely, the second of the three systems defined in Sec. 4.4 of [1]). Instead, we directly proceed with some notation and results yielding an adaptation of Auinger's construction to our environment.

Let S be a semigroup from $\mathbf{A_2}$ and J a regular $\mathcal{J}$-class of S. We put $\widetilde{J} = J$ if J is a subsemigroup of S and $\widetilde{J} = J^0$ if J is not a subsemigroup of S (where J^0 is the corresponding principal factor of S, see Sec. 2.6 in [3]). Put $\mathcal{J}(S) = \{\widetilde{J} \mid J \text{ is a regular } \mathcal{J}\text{-class of } S\}$ and, for each variety $\mathbf{V} \subseteq \mathbf{A_2}$, put $\mathcal{J}(\mathbf{V}) = \bigcup_{S \in \mathbf{V}} \mathcal{J}(S)$ and $\mathbf{V}^{\mathcal{J}} = \{S \in \mathbf{A_2} \mid \mathcal{J}(S) \subseteq \mathbf{V}\}$.

Proposition 1.3. *The relation θ defined on the lattice $L(\mathbf{A_2})$ by*

$$\mathbf{V}\ \theta\ \mathbf{W} \iff \mathcal{J}(\mathbf{V}) = \mathcal{J}(\mathbf{W})$$

is a complete congruence on $L(\mathbf{A_2})$*. The congruence classes with respect to this congruence are the intervals* $[\operatorname{var}\mathcal{J}(\mathbf{V}), \mathbf{V}^{\mathcal{J}}]$.

Observe that the bottom elements of the above intervals are precisely the subvarieties of $\mathbf{A_2}$ generated by simple semigroups and/or 0-simple semigroups with zero divisors. All such subvarieties are known; in order to formulate their (folklore) description, we need an extra piece of notation for semigroups and varieties. Let L and R stand respectively for the 2-element left zero and right zero semigroups, and let B_2 denote the 5-element Brandt semigroup that can be described by the presentation

$$B_2 = \langle c, d \mid cdc = c,\ dcd = d,\ c^2 = d^2 = 0 \rangle = \{c, d, cd, dc, 0\}.$$

Alternatively, B_2 can be described by its representation by 2×2-matrices:

$$B_2 = \left\{ \begin{pmatrix} 0 & 0 \\ 0 & 0 \end{pmatrix}, \begin{pmatrix} 1 & 0 \\ 0 & 0 \end{pmatrix}, \begin{pmatrix} 0 & 1 \\ 0 & 0 \end{pmatrix}, \begin{pmatrix} 0 & 0 \\ 1 & 0 \end{pmatrix}, \begin{pmatrix} 0 & 0 \\ 0 & 1 \end{pmatrix} \right\}$$

or as the Rees matrix semigroup over the trivial group $E = \{1\}$ with the sandwich matrix $\begin{pmatrix} 1 & 0 \\ 0 & 1 \end{pmatrix}$. Further, let

$\mathbf{T}$ be the trivial variety;
$\mathbf{L} = \operatorname{var} L$, the variety of left zero semigroups;
$\mathbf{R} = \operatorname{var} R$, the variety of right zero semigroups;
$\mathbf{RB} = \operatorname{var}\{L, R\}$, the variety of rectangular bands;
$\mathbf{B_2} = \operatorname{var} B_2$;
$\mathbf{LB_2} = \operatorname{var}\{L, B_2\}$;
$\mathbf{RB_2} = \operatorname{var}\{R, B_2\}$;
$\mathbf{NB_2} = \operatorname{var}\{L, R, B_2\}$.

Proposition 1.4. *Every subvariety of* $\mathbf{A_2}$ *generated by simple semigroups and/or 0-simple semigroups with zero divisors coincides with one of the nine varieties* $\mathbf{T}$, $\mathbf{L}$, $\mathbf{R}$, $\mathbf{RB}$, $\mathbf{B_2}$, $\mathbf{LB_2}$, $\mathbf{RB_2}$, $\mathbf{NB_2}$, *or* $\mathbf{A_2}$.

Combining Propositions 1.3 and 1.4, we see that the lattice $L(\mathbf{A_2})$ decomposes into a disjoint union of nine intervals. The lattice formed by these intervals (that is, the quotient $L(\mathbf{A_2})/\,\theta$) is shown in Fig. 2.

Now we are in a position to describe the results of this paper in more detail. We have decomposed the lattice $L(\mathbf{A_2})$ into a disjoint union of nine intervals. For future references, we shall call them the *basic intervals* of $L(\mathbf{A_2})$. By the construction, the bottom element of each basic interval is a finitely generated variety; moreover, each such bottom variety has a

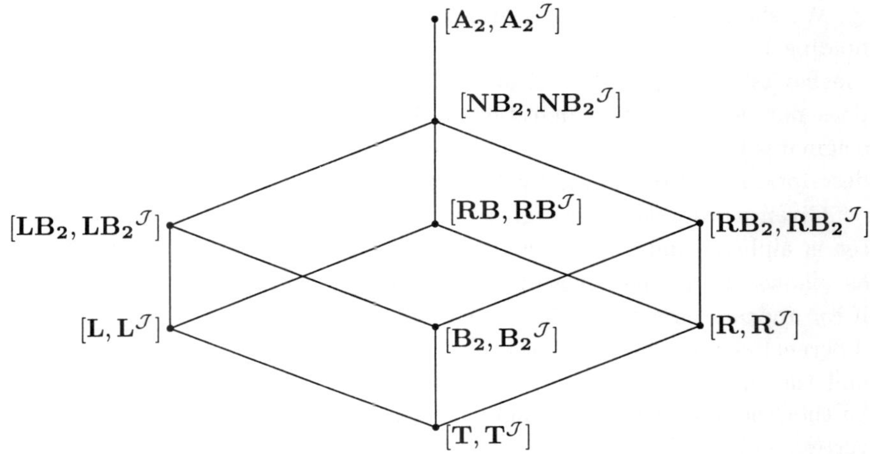

Figure 2. The lattice $L(\mathbf{A_2})/\theta$

finite identity basis (for a majority of them this is known and for the rest this easily follows from known facts). On the other hand, the construction gives only an implicit characterization of the top elements of the basic intervals. We shall show that the top element of each basic interval is a finitely generated variety with a finite identity basis. We shall also describe the sublattice of $L(\mathbf{A_2})$ generated by the extreme elements of the basic intervals; this sublattice forms sort of a skeleton for $L(\mathbf{A_2})$.

Our technique is a mixture of certain tools from the structural theory of periodic semigroups with combinatorial arguments of a graph-theoretical flavor. We shall develop its main ingredients in the next two sections.

2. Regular elements in semigroups from $\mathbf{A_2}$ and their representation by words

First, for the reader's convenience, we recall a few basic definitions related to words. We fix a countably infinite set Σ (the *alphabet*) whose elements are referred to as *letters*. As usual, Σ^+ is the free semigroup over Σ. We call elements of Σ^+ *words* and denote the equality relation on Σ^+ by $\equiv$. If u, v are words, we say that u *occurs* in v or is a *factor* in v whenever $v \equiv u$ or v can be obtained by adding to u some letters on the left and/or on the right. For a word $w \in \Sigma^+$ we denote by $\mathrm{alph}(w)$ the set of letters from Σ that occur in w. If $w \equiv x_1 x_2 \cdots x_n$ where $x_1, x_2, \ldots, x_n$ are letters, then the number n is called the *length* of the word w and is denoted by $|w|$.

We shall make use of a graph-theoretical description of the identities holding in A_2. This description is well known. It is often attributed to Mashevitsky [11] (see, for instance, [9] or [34]) even though the paper [11] does not deal with the identities of A_2 at all. Apparently, this mistake originates from an erroneous reference in the survey paper [27]. In fact, the description has been found by Trakhtman, see his preprint [31].

Given a word w, we assign to it a directed graph $G(w)$ whose vertex set is $\mathrm{alph}(w)$ and whose arrows correspond to factors of length 2 in w as follows: $G(w)$ has an arrow from x to y ($x, y \in \mathrm{alph}(w)$) if and only if xy appears as a factor in w. We will distinguish two (not necessarily different) vertices in $G(w)$: the *initial vertex*, that is the first letter of w, and the *final vertex*, that is the last letter of w. Then the word w can be thought of as a walk through the graph $G(w)$ that starts at the initial vertex, ends at the final vertex and traverses each arrow of $G(w)$ (some of the arrows can be traversed more than once). Figure 3 shows the graph

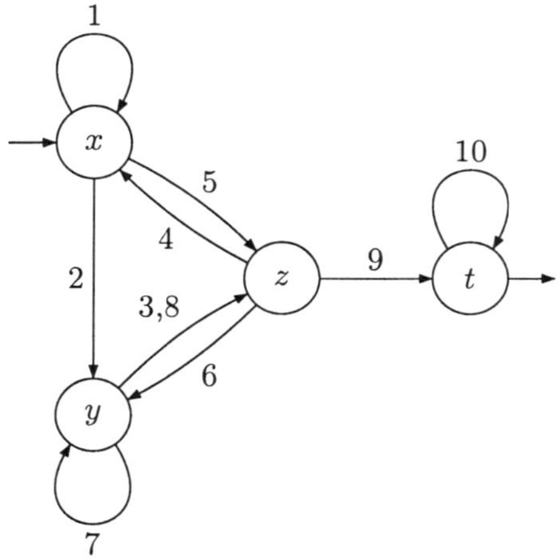

Figure 3. The graph of the word $x^2yzxzy^2zt^2$ and the corresponding walk

$G(w)$ for the word $w \equiv x^2yzxzy^2zt^2$. The ingoing and the outgoing marks show respectively the initial and the final vertices of the graph. In Fig. 3 each arrow of the graph is labelled by the number[s] indicating the order of occurrence[s] of the arrow in the walk induced by the word w. We stress that, in contrast to the vertex names and the ingoing/outgoing marks, these labels are not considered as a part of the data making the graph $G(w)$. Therefore the graph does not determine the word w: for instance, the word $xy^3zyzx^2zyzt^3$ has exactly the same graph (but corresponds to a different walk through it, see Fig. 4).

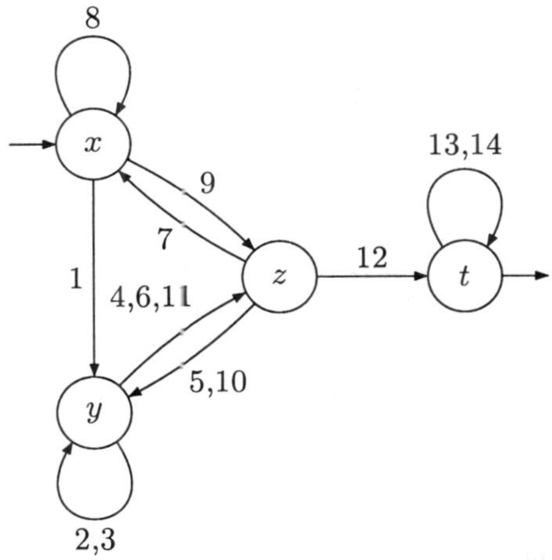

Figure 4. Another walk through the graph of Fig. 3

Proposition 2.1. *The semigroup A_2 satisfies the identity $u = v$ if and only if the graphs $G(u)$ and $G(v)$ are equal.*

A word w of length at least 2 is said to be *connected* if its directed graph $G(w)$ is strongly connected. In fact, this concept is known in the literature, although under different names. For instance, in [12] Mashevitsky calls a word w of length at least 2 *covered by cycles* if each its factor of length 2 occurs in a factor of w that starts and ends with the same letter. In the language of the graph $G(w)$, this property means that each arrow $x \to y$ of $G(w)$ belongs to a directed cycle (namely, to the walk induced by a factor of w that starts and ends with the same letter and contains xy). It is one of the basic facts of the theory of directed graphs (cf. [15], Theorem 8.1.5) that such a graph is strongly connected if and only if each its arrow belongs to a directed cycle. Yet another name for an obviously equivalent concept can be found in Pollák's posthumously published paper [22], where a word w of length at least 2 is said to be *prime* if it cannot be decomposed as $w \equiv w'w''$ with $\mathrm{alph}(w') \cap \mathrm{alph}(w'') = \varnothing$.

Our next proposition reveals the semigroup meaning of the concept of a connected word. Recall that an element s of a semigroup S is *regular in* S if there exists $s' \in S$ such that $ss's = s$.

Proposition 2.2. *A word $w \in \Sigma^+$ is connected if and only if for every* *morphism* $\varphi : \Sigma^+ \to S$ *with* $S \in \mathbf{A_2}$ *the element* $w\varphi$ *is regular in* S.

Proof. First suppose that w is a connected word, and let x and y be respectively the first and the last letters of w (we do not assume that $x \not\equiv y$). Since the graph $G(w)$ is strongly connected, there is a walk

$$y \equiv x_0 \to x_1 \to \cdots \to x_n \equiv x.$$

Let $u \equiv x_1 \cdots x_{n-1} w x_1 \cdots x_{n-1}$, then the graphs of the words w and wuw are equal. Here if $n = 0$ (which means that $y \equiv x$) or $n = 1$, the product $x_1 \cdots x_{n-1}$ interprets as the empty word. By Proposition 2.1 we have $w\psi = (wuw)\psi = w\psi \cdot u\psi \cdot w\psi$ where ψ denotes the canonical morphism from Σ^+ to $F_\Sigma(\mathbf{A_2})$, the free semigroup of the variety $\mathbf{A_2}$ over Σ. Thus, $w\psi$ is regular in $F_\Sigma(\mathbf{A_2})$ and, since every morphism $\varphi : \Sigma^+ \to S$ factors through ψ, the element $w\varphi$ is regular in S.

Now suppose that w is not connected. This means that the graph $G(w)$ contains a bridge whence the walk w splits into the part preceding the bridge, the bridge, and the part following the bridge. (The reader may see such a situation in Fig. 3 or 4 where the arrow $z \to t$ forms a bridge.) Accordingly, w decomposes as $w \equiv w'w''$ where the words w' and w'' correspond to the parts of the walk respectively before and after the bridge. Clearly, $\mathrm{alph}(w') \cap \mathrm{alph}(w'') = \varnothing$. Consider the subsemigroup $A_0 = \{b, ab, ba, 0\}$ of A_2 and let the morphism $\varphi : \Sigma^+ \to A_0$ be defined by:

$$x\varphi = \begin{cases} ba & \text{if } x \in \mathrm{alph}(w'), \\ ab & \text{otherwise.} \end{cases}$$

Then using the defining relations of A_2, one readily calculates that

$$w\varphi = w'\varphi \cdot w''\varphi = (ba)^{|w'|} \cdot (ab)^{|w''|} = ba \cdot ab = bab = b.$$

However, it is easy to check that b is not regular in A_0. □

Remark 2.1. The fact that, under the canonical morphism from Σ^+ onto $F_\Sigma(\mathbf{A_2})$, every word covered by cycles maps onto a regular element of $F_\Sigma(\mathbf{A_2})$ is a partial case of a similar result claimed by Mashevitsky in [12], Lemma 6, see also [13], Lemma 7. Since then, this result has been used (with reference to [12]) in several important papers including, for instance, [5] and [14]. However, its proof in [12] contains a fatal flaw (and so does the translation of the proof into English published in [13]). Namely, in [12] Lemma 6 is deduced from Lemma 5 which claims that every word u covered by cycles can be transformed modulo certain identities to a word of the form $z_1u_1z_1 \cdots z_ku_kz_k$ where $z_1, \ldots, z_k$ are letters and $z_{i+1} \in \mathrm{alph}(u_i)$ for all $i = 1, \ldots, k-1$ provided that $k > 1$. In order to justify the latter claim, Mashevitsky uses induction on $|\mathrm{alph}(u)|$ but in the course of the proof he illegitimately applies the induction assumption to a factor that in general is not covered by its cycles. The word $u \equiv xyxzy$ can be used as a concrete counter example showing that the argument from [12] does not work: here the induction assumption should have been applied to the factor zy which is certainly not covered by its cycles.

However, a correct proof of the described intermediate claim can be achieved by simple graph-theoretic means, and moreover, the claim can be avoided because we can prove Lemma 6 of [12] by a suitable modification of reasoning applied in the above proof of Proposition 2.2. Thus, results of [5] and [14] that rely on the lemma are correct.

For every word w, the walk w through the graph $G(w)$ naturally induces a linear order among the strongly connected components of $G(w)$, namely, the order in which the components are traversed. Let $G_1, \ldots, G_m$ be the strongly connected components of $G(w)$ listed in this order so that each bridge in $G(w)$ leads from G_i to G_{i+1}, $i = 1, \ldots, m-1$. Suppose we "open the bridges", that is, we remove each bridge arrow $x \to y$ but instead we equip its beginning x and its end y with an outgoing and an ingoing mark respectively. (Figure 5 demonstrates this procedure for the graph of the word $x^2yzxzy^2zt^2$ shown in Fig. 3.) Then the walk w splits into the walks $w_1, \ldots, w_m$ such that $G(w_i) = G_i$ for all $i = 1, \ldots, m$. Observe that

$$w \equiv w_1 \cdots w_m, \tag{2}$$

$\mathrm{alph}(w_i) \cap \mathrm{alph}(w_j) = \varnothing$ whenever $i \neq j$, and each factor w_i either is connected or has length 1. We call (2) the *canonical decomposition* of w. For instance, the canonical decomposition of the word $w \equiv x^2yzxzy^2zt^2$ is $w \equiv w_1 \cdot w_2 \equiv x^2yzxzy^2z \cdot t^2$, see Fig. 5. In the theory of semigroup varieties, canonical decompositions were studied and intensely used by Pollák who called them *closed decompositions* or IC-*decompositions* or *irreducible decompositions*, see [19, 20, 21] respectively.

We call a word w *semiconnected* if each factor in its canonical decomposition is connected. It is easy to see that this is equivalent to w being a product of connected words. This notion also has appeared in the literature under various names. Pollák [22] called semiconnected words *semiprime.* Trakhtman in [30] calls a word w *repeated* if each letter from $\mathrm{alph}(w)$ appears in a factor of w of length at least 2 that starts and ends with the same letter. In the language of the graph $G(w)$, this property means that each

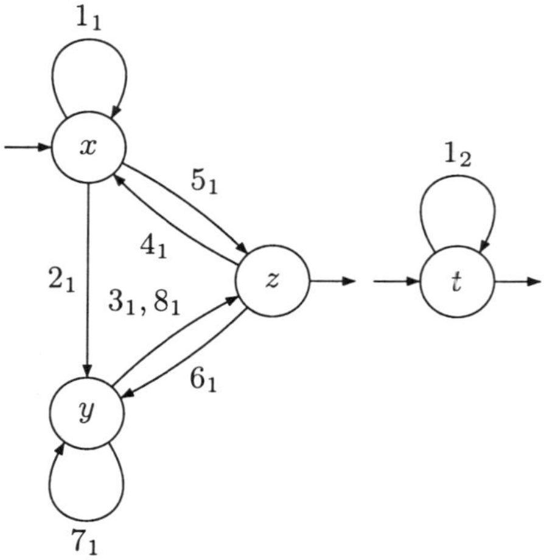

Figure 5. "Opening the bridge" in the graph of the word $x^2yzxzy^2z \cdot t^2$

vertex of $G(w)$ belongs to a directed cycle (that may reduce to a loop). Clearly, this is equivalent to saying that none of the strongly connected components of $G(w)$ are trivial which in turn amounts to saying that w is semiconnected.

We shall need the following corollary to Proposition 2.2.

Proposition 2.3. *A word $w \in \Sigma^+$ is semiconnected if and only if for every morphism $\varphi : \Sigma^+ \to S$ with $S \in \mathbf{A_2}$ the element $w\varphi$ is a product of regular elements of S.*

Proof. If w is semiconnected and (2) is its canonical decomposition, then $w\varphi = w_1\varphi \cdots w_m\varphi$ is a product of regular elements by Proposition 2.2.

Now suppose that w is not semiconnected. This means that w can be decomposed as $w \equiv w'yw''$ where y is a letter, $\mathrm{alph}(w') \cap \mathrm{alph}(w'') = \varnothing$ and $y \notin \mathrm{alph}(w'w'')$. Consider the subsemigroup $B_0 = \{d, cd, dc, 0\}$ of B_2 and let the morphism $\varphi : \Sigma^+ \to B_0$ be defined as follows:

$$x\varphi = \begin{cases} dc & \text{if } x \in \mathrm{alph}(w'), \\ d & \text{if } x \equiv y, \\ cd & \text{otherwise.} \end{cases}$$

Then using the defining relations of B_2, one readily calculates that

$$w\varphi = w'\varphi \cdot y\varphi \cdot w''\varphi = (dc)^{|w'|} \cdot d \cdot (cd)^{|w''|} = dc \cdot d \cdot cd = d.$$

However, since the regular elements of B_0 are cd, dc, and 0 and they form a subsemigroup, d is not a product of regular elements. □

3. Kublanovskii's lemma and the scheme of its usage

Many of our results arise as applications of the following partial case of an important lemma due to Kublanovskii, see [5], Lemma 3.2:

Proposition 3.1. *For any semigroup $S \in \mathbf{A_2}$ and distinct regular elements $s, s' \in S$ there exists a [0]-simple semigroup K and a surjective morphism $\varphi : S \to K$ such that $s\varphi \neq s'\varphi$.*

We use Proposition 3.1 to provide finite generating semigroups for the top varieties in the basic intervals. We shall apply it to semigroups with a certain reductivity property that we are going to introduce now.

As usual, for a semigroup S we denote by $E(S)$ the set of all its idempotents. We say that S is *e-separable on the right* (*on the left*) if for every pair p, q of distinct elements in S there exists an $e \in E(S)$ such that $pe \neq qe$ (respectively $ep \neq eq$). A semigroup that is e-separable on both the right and the left is called *e-separable*.

Any monoid is obviously an example of an e-separable semigroup. Another "mass" example (which is more relevant for the present paper) is any inverse semigroup. Indeed, suppose S is inverse and $p, q \in S$ are such that $pe = qe$ for every $e \in E(S)$. Taking the idempotent $p^{-1}p$ for e, we conclude that $p = pp^{-1}p = qp^{-1}p$ and, similarly, $q = pq^{-1}q$. Hence

$$p = qp^{-1}p = pq^{-1}qp^{-1}p = pp^{-1}pq^{-1}q = pq^{-1}q = q.$$

Two further easy examples that are quite important for us are the subsemigroup $A_0 = \{b, ab, ba, 0\}$ of A_2 and the subsemigroup $B_0 = \{d, cd, dc, 0\}$ of B_2 introduced in the proofs of Propositions 2.2 and 2.3. Observe that the direct product of e-separable semigroups is again an e-separable semigroup.

The next proposition relates the notion of e-separability to certain decompositions of words and also explains the importance of A_0 and B_0.

Proposition 3.2. (i) *Suppose S is e-separable on the right and $A_0 \in \operatorname{var} S$. Further, suppose S satisfies an identity $u = v$ such that the word u can be decomposed as $u \equiv u'u''$ with $\operatorname{alph}(u') \cap \operatorname{alph}(u'') = \varnothing$. Then the word v can be decomposed as $v \equiv v'v''$ with $\operatorname{alph}(v') = \operatorname{alph}(u')$, $\operatorname{alph}(v'') = \operatorname{alph}(u'')$, and S satisfies the identity $u' = v'$.*

(ii) *Suppose S is e-separable on the right and $B_0 \in \operatorname{var} S$. Further, suppose S satisfies an identity $u = v$ such that the word u can be decomposed as $u \equiv u'yu''$ where $\operatorname{alph}(u') \cap \operatorname{alph}(u'') = \varnothing$ and y is a letter with $y \notin \operatorname{alph}(u'u'')$. Then the word v can be decomposed as $v \equiv v'yv''$ with $\operatorname{alph}(v') = \operatorname{alph}(u')$, $\operatorname{alph}(v'') = \operatorname{alph}(u'')$, and S satisfies $u' = v'$.*

Proof. (i) Since the semigroup A_0 belongs to the variety $\operatorname{var} S$, it must satisfy the identity $u = v$. This implies that $\operatorname{alph}(u) = \operatorname{alph}(v)$—otherwise the identity $u = v$ would fail in the 2-element semilattice and could not be satisfied by A_0. Let the morphism $\zeta : \Sigma^+ \to A_0$ be defined as follows:

$$x\zeta = \begin{cases} ba & \text{if } x \in \operatorname{alph}(u'), \\ ab & \text{otherwise.} \end{cases} \tag{3}$$

As in the proof of Proposition 2.2, one readily obtains that $u\zeta = b$. Hence also $v\zeta = b$. By (3) the element $v\zeta$ is a product of the idempotents ab and ba in some order. Since $ab \cdot ba = 0$, if such a product is not equal to 0, then no occurrence of ab precedes an occurrence of ba. This implies that in the word v no occurrence of a letter from $\operatorname{alph}(u'')$ precedes an occurrence of a letter from $\operatorname{alph}(u')$. Therefore v decomposes as $v \equiv v'v''$ where $\operatorname{alph}(v') = \operatorname{alph}(u')$ and $\operatorname{alph}(v'') = \operatorname{alph}(u'')$.

It remains to show that the identity $u' = v'$ holds in S. Arguing by contradiction, consider a morphism $\varphi : \Sigma^+ \to S$ such that $p = u'\varphi \neq q = v'\varphi$. Since S is e-separable on the right, there exists an $e \in E(S)$ such that $pe \neq qe$. Now we define a "modification" ξ of the morphism φ by letting

$$x\xi = \begin{cases} x\varphi & \text{if } x \in \operatorname{alph}(u'), \\ e & \text{otherwise.} \end{cases}$$

Taking into account the equalities $\operatorname{alph}(v') = \operatorname{alph}(u')$, $\operatorname{alph}(v'') = \operatorname{alph}(u'')$ and $\operatorname{alph}(u') \cap \operatorname{alph}(u'') = \varnothing$, we obtain

$$u\xi = u'\xi \cdot u''\xi = u'\varphi \cdot e = pe \neq qe = v'\varphi \cdot e = v'\xi \cdot v''\xi = v\xi.$$

This contradicts the assumption that the identity $u = v$ holds in S. Thus, we conclude that $u'\varphi = v'\varphi$ under every morphism $\varphi : \Sigma^+ \to S$, that is, S satisfies the identity $u' = v'$.

We omit the proof of (ii) as it is completely analogous. In fact, the conclusion of (ii) can be deduced from a weaker version of e-separability.□

We are ready to present the main technical tool of our paper.

Proposition 3.3. *Let* $\mathbf{V}$ *be a subvariety of the variety* $\mathbf{A_2}$ *and* $A_0 \in \mathbf{V}$. *Suppose* S *is an e-separable semigroup in* $\mathbf{V}$ *such that the varieties* $\mathbf{V}$ *and* $\operatorname{var}(S \times A_0)$ *are* θ*-related. Then* $\mathbf{V} = \operatorname{var}(S \times A_0)$.

Proof. From the condition, we immediately see that $\mathbf{V} \supseteq \operatorname{var}(S \times A_0)$. Arguing by contradiction, assume that the inclusion is strict. Then there exists an identity that holds in the semigroup $S \times A_0$ but fails in the variety $\mathbf{V}$. We choose an identity $u = v$ with this property and with the least value of $|\operatorname{alph}(u)|$. We first check that the words u and v are connected.

Assume for a moment that, say, u is not connected. This means that it can be decomposed as $u \equiv u'u''$ with $\operatorname{alph}(u') \cap \operatorname{alph}(u'') = \varnothing$. Since the semigroup $S \times A_0$ is e-separable as the direct product of two e-separable semigroups, we are in a position to employ Proposition 3.2(i) and its dual. We thus conclude that v decomposes as $v \equiv v'v''$ where $\operatorname{alph}(v') = \operatorname{alph}(u')$, $\operatorname{alph}(v'') = \operatorname{alph}(u'')$ and both the identities $u' = v'$ and $u'' = v''$ hold in $S \times A_0$. Since $|\operatorname{alph}(u')|, |\operatorname{alph}(u'')| < |\operatorname{alph}(u)|$, our choice of the identity $u = v$ ensures that the identities $u' = v'$ and $u'' = v''$ hold also in the variety $\mathbf{V}$. However, together they obviously imply the identity $u = v$ that cannot hold in $\mathbf{V}$, a contradiction.

Now consider $F_\Sigma(\mathbf{V})$, the free semigroup of the variety $\mathbf{V}$ over the alphabet Σ, and let $\chi : \Sigma^+ \to F_\Sigma(\mathbf{V})$ be the canonical morphism. By the above claim and Proposition 2.2, $u\chi$ and $v\chi$ are distinct regular elements of $F_\Sigma(\mathbf{V})$. We are in a position to apply Kublanovskii's lemma (Proposition 3.1) according to which there exists a [0]-simple semigroup K and a surjective morphism $\varphi : F_\Sigma(\mathbf{V}) \to K$ such that $(u\chi)\varphi \neq (v\chi)\varphi$. However, the varieties $\mathbf{V}$ and $\operatorname{var}(S \times A_0)$ are supposed to be θ-related. By the definition this means that the two varieties contain the same simple semigroups and 0-simple semigroups with zero divisors but since they both contain the 2-element semilattice we can simply state that $\mathbf{V}$ and $\operatorname{var}(S \times A_0)$ contain the same [0]-simple semigroups. In particular, the [0]-simple semigroup K must belong to $\operatorname{var}(S \times A_0)$ whence it must satisfy the identity $u = v$. This implies that $u(\chi\varphi) = v(\chi\varphi)$, a contradiction. □

For the "B_0-version" of Proposition 3.3, we need an auxiliary result. Most probably, it belongs to semigroup folklore because the semigroup A_0 (under various names) frequently arises in the literature in results of a similar flavor, see, e. g., [28, 29, 40, 42].

Lemma 3.1. *If no subsemigroup of a given semigroup S has A_0 as a quotient then the regular elements of S form a subsemigroup.*

Proof. Let $p, q \in S$ be two arbitrary regular elements, and let p', q' be such that $pp'p = p$, $qq'q = q$. If $e = p'p$, then $pe = p$ whence $p \mathrel{\mathcal{L}} e$. Similarly, if $f = qq'$, then $q = fq$ whence $q \mathrel{\mathcal{R}} f$. Using the fact that the Green relations $\mathcal{L}$ and $\mathcal{R}$ are respectively a right congruence and a left congruence, we observe that $pq \mathrel{\mathcal{L}} eq \mathrel{\mathcal{R}} ef$. Thus, the products pq and ef are in the same $\mathcal{D}$-class.

Let T stand for the subsemigroup in S generated by the idempotents e and f. It is easy to see that $T = \{e, f, ef\} \cup feT \cup efeT$ and $I = feS \cup efeS$ is an ideal in T. There are two cases: $ef \notin I$ and $ef \in I$.

If $ef \notin I$, then we also have $e, f \notin I$ and $e \neq f$. Thus, the Rees quotient S/I consists of four elements: e, f, ef and 0, and the relations $e^2 = e$, $f^2 = f$ and $fe = 0$ hold. Then S/I is isomorphic to A_0, a contradiction.

If $ef \in I$, then $ef = fer$ or $ef = efer$ for some $r \in T$. Multiplying each of these equalities by e on the left and by f on the right yields the equality $ef = eferf$. Taking into account that r is an alternating product of e and f, we see that $ef = (ef)^{n+1}$ for some positive integer n, in particular, ef is a regular element. It is known (cf. [3], Chapter 2) that then the $\mathcal{D}$-class of ef consists of regular elements whence the product pq is regular. □

Proposition 3.4. *Let* $\mathbf{V}$ *be a subvariety of the variety* $\mathbf{A_2}$ *and* $B_0 \in \mathbf{V}$ *while* $A_0 \notin \mathbf{V}$. *Suppose* S *is an e-separable semigroup in* $\mathbf{V}$ *such that the varieties* $\mathbf{V}$ *and* $\mathrm{var}(S \times B_0)$ *are* θ*-related. Then* $\mathbf{V} = \mathrm{var}(S \times B_0)$.

Proof. We repeat mutatis mutandis the arguments in the proof of Proposition 3.3. Namely, assuming that $\mathbf{V} \supsetneqq \mathrm{var}(S \times B_0)$, we choose an identity $u = v$ with the minimum possible value of $|\,\mathrm{alph}(u)|$ such that $u = v$ holds in $\mathrm{var}(S \times B_0)$ but fails in $\mathbf{V}$. Using Proposition 3.2(ii) and its dual, we then conclude that the words u and v have to be semiconnected. By Proposition 2.3, the images of u and v under the canonical morphism $\chi : \Sigma^+ \to F_\Sigma(\mathbf{V})$ are distinct product of regular elements of $F_\Sigma(\mathbf{V})$. Now we employ the condition $A_0 \notin \mathbf{V}$ which implies that no subsemigroup of $F_\Sigma(\mathbf{V})$ can have A_0 as a quotient. By Lemma 3.1 the elements $u\chi$ and $v\chi$ are in fact regular, and we are in a position to apply Kublanovskii's lemma exactly as in the last paragraph of the proof of Proposition 3.3. □

As an illustration of the power of the above techniques, we recover an identity basis for the variety $\mathbf{B_2} = \mathrm{var}\, B_2$:

Proposition 3.5. *The variety* $\mathbf{B_2}$ *is defined by the identities*

$$x^2 = x^3, \ xyx = xyxyx, \ x^2y^2 = y^2x^2. \tag{4}$$

Remark 3.1. The identity basis (4) for $\mathbf{B_2}$ was found by Trakhtman, see [30], and was frequently cited and used in numerous applications, including quite important ones such as the positive solution to the finite basis problem for 5-element semigroups [32, 33] or a classification of finite inverse semigroups whose plain semigroup identities are finitely based [38]. However, Reilly [24] has recently observed that there is a serious lacuna in Trakhtman's argument in [30]. Reilly [24] has mastered a correct proof of Proposition 3.5 using a (fairly nontrivial) solution to the word problem for $\mathbf{B_2}$. Our machinery yields another, rather straightforward proof.

Proof. Let $\mathbf{V}$ stand for the variety defined by the identities (4). We just check the conditions of Proposition 3.4 for $\mathbf{V}$. It is easy to see that $\mathbf{V} \subseteq \mathbf{A_2}$. Indeed, by Proposition 2.1, this amounts to deducing the identity $xyxzx = xzxyx$ from the identities (4), and here is the deduction:

$$\begin{aligned} xyxzx &= (xy)^2(xz)^2x && \text{in view of } xyx = xyxyx \\ &= (xz)^2(xy)^2x && \text{in view of } x^2y^2 = y^2x^2 \\ &= xzxyx && \text{in view of } xyxyx = xyx. \end{aligned}$$

Further, the semigroup B_2 satisfies (4); the identities $x^2 = x^3$ and $xyx = xyxyx$ hold in any combinatorial completely 0-simple semigroup and, in the presence of $x^2 = x^3$, the identity $x^2y^2 = y^2x^2$ means precisely that idempotents commute which is true in B_2. Thus, $B_2 \in \mathbf{V}$ and also $B_0 \in \mathbf{V}$ as B_0 is a subsemigroup in B_2. In contrast, $A_0 \notin \mathbf{V}$ since the idempotents ab and ba of A_0 do not commute. Similarly, $\mathbf{V}$ contains no nontrivial left or right zero semigroup, and using the classification of subvarieties of $\mathbf{A_2}$ generated by simple semigroups and/or 0-simple semigroups with zero divisors (Proposition 1.4), we conclude that $\operatorname{var} \mathcal{J}(\mathbf{V}) = \mathbf{B_2}$. Thus, the varieties $\mathbf{V}$ and $\mathbf{B_2} = \operatorname{var}(B_2 \times B_0)$ are θ-related. Finally, the semigroup B_2 is e-separative because it is inverse. By Proposition 3.4 we then have $\mathbf{V} = \operatorname{var}(B_2 \times B_0) = \mathbf{B_2}$. $\square$

4. Main results

We are ready to start implementing the program announced at the end of Sec. 1. First we provide a finite generating semigroup and a finite identity basis for the top element of each of the nine basic intervals shown in Fig. 2. Some of these results are known but we give a uniform treatment basing on Proposition 3.3.

We begin with the bottom interval $[\mathbf{T}, \mathbf{T}^{\mathcal{J}}]$.

Theorem 4.1. *The variety $\mathbf{T}^{\mathcal{J}}$ is generated by the semigroup A_0 and is defined by the identities*

$$x^2 = x^3, \quad xyx = yxy = (xy)^2. \tag{5}$$

Proof. Let $\mathbf{V}$ stand for the variety defined by the identities (5). First we check that $\mathbf{V} \subseteq \mathbf{A_2}$. By Proposition 2.1, this amounts to deducing each of the identities $xyx = xyxyx$ and $xyxzx = xzxyx$ from (5). For the first, we just apply the identity $xyx = xyxy$ twice:

$$xyx = xyxy = x \cdot yxy = x \cdot yxyx = xyxyx.$$

Observe that combining $xyx = xyxy$ with $xyx = xyxyx$ yields the identity $xyx^2 = xyx$. Indeed,

$$xyx^2 = xyx \cdot x = xyxy \cdot x = xyxyx = xyx.$$

Now we are ready to deduce $xyxzx = xzxyx$:

$$\begin{aligned}
xyxzx &= x \cdot yxz \cdot x = yxz \cdot x \cdot yxz && \text{in view of } xyx = yxy \\
&= y \cdot xzxyx \cdot z = y \cdot zxzxyxy \cdot z && \text{in view of } yxy = xyxy \\
& && \text{and } xyx = xyxy \\
&= yz \cdot xzxyx \cdot yz = xzxyx \cdot yz \cdot xzxyx && \text{in view of } yxy = xyx \\
&= xzxyxy \cdot zxzxyx = xzxyx \cdot xzxyx && \text{in view of } xyxy = xyx \\
& && \text{and } xyxy = yxy \\
&= xzxyx^2 \cdot zxyx = xzxyx \cdot zxyx && \text{in view of } xyx^2 = xyx \\
&= x \cdot zxy \cdot x \cdot zxy \cdot x = xzxyx && \text{in view of } xyxyx = xyx.
\end{aligned}$$

Since nontrivial left or right zero semigroup and the Brandt semigroup B_2 do not satisfy the identity $xyx = yxy$, the variety $\mathbf{V}$ contains neither nontrivial simple semigroups nor 0-simple semigroups with zero divisors. This means that $\mathbf{V} \subseteq \mathbf{T}^{\mathcal{J}}$. On the other hand, it is easy to check that A_0 satisfies the identities (5) whence $\operatorname{var} A_0 \subseteq \mathbf{V}$. Now, applying Proposition 3.3 to the variety $\mathbf{T}^{\mathcal{J}}$ and its e-separative member A_0, we conclude that A_0 generates $\mathbf{T}^{\mathcal{J}}$ whence $\operatorname{var} A_0 = \mathbf{V} = \mathbf{T}^{\mathcal{J}}$, as required. □

Remark 4.1. Edmunds [4] proved that the identities

$$x^3 = x^2, \ xyx = x^2yx = xy^2x = xyx^2 = xyxy = yxy$$

constitute an identity basis for A_0 (denoted by `S(4,22)` in the catalogue of 4-element semigroups in [4]). Now we see that many of these identities are redundant. It is fair to say that Edmunds's goal was to find some finite basis rather than a minimal one—otherwise the complete analysis of identities of 4-element semigroups would become far too cumbersome.

Yet another, fairly straightforward finite identity basis for the variety $\operatorname{var} A_0 = \mathbf{T}^{\mathcal{J}}$ can be extracted from the fact that this variety is nothing but the class of all $\mathcal{J}$-trivial semigroups from $\mathbf{A_2}$. It is known (and easy to verify) that a semigroup satisfying $x^2 = x^3$ is $\mathcal{J}$-trivial if and only if it satisfies also the identity $(xy)^2 = (yx)^2$. Thus, one gets an identity basis for var $A_0 = \mathbf{T}^{\mathcal{J}}$ merely by adding the latter identity to the identities (1).

Further results in this section involve three oversemigroups of A_0 which we are going to introduce now. Let $\mathbb{B} = \langle\{0,1\}; +, \cdot\rangle$ be the Boolean semiring (that is, $0 + x = x + 0 = x$, $1 + x = x + 1 = 1$, $0 \cdot x = x \cdot 0 = 0$, $1 \cdot x = x \cdot 1 = x$ for all $x \in \{0,1\}$; in particular, $1 + 1 = 1$). Consider the following 2×2-matrices over $\mathbb{B}$:

$$0 = \begin{pmatrix} 0 & 0 \\ 0 & 0 \end{pmatrix}, \ e_1 = \begin{pmatrix} 1 & 0 \\ 0 & 0 \end{pmatrix}, \ e_2 = \begin{pmatrix} 1 & 1 \\ 0 & 0 \end{pmatrix}, \ f_1 = \begin{pmatrix} 0 & 0 \\ 0 & 1 \end{pmatrix}, \ f_2 = \begin{pmatrix} 0 & 1 \\ 0 & 1 \end{pmatrix}, \ b = \begin{pmatrix} 0 & 1 \\ 0 & 0 \end{pmatrix}.$$

It is routine to show that under usual matrix multiplication these six matrices form a semigroup; we denote this semigroup by C_0. The semigroup C_0 can be thought of as an amalgam of two copies of the semigroup A_0, namely, $\{e_1, f_2, b, 0\}$ and $\{e_2, f_1, b, 0\}$. One readily sees that both $C_0 \setminus \{e_2\}$ and $C_0 \setminus \{f_2\}$ are subsemigroups in C_0; we denote them by LC_0 and RC_0 respectively. Observe that $e_1 f_1 = 0$ while $e_1 f_2 = e_2 f_1 = e_2 f_2 = b$. This observation is the only ingredient which is needed to establish the following lemma and is not completely trivial.

Lemma 4.1. *The semigroups C_0, LC_0 and RC_0 are e-separable.*

Now we are well prepared to analyze all the remaining intervals. Our next theorem deals with the top elements of the intervals $[\mathbf{L}, \mathbf{L}^{\mathcal{J}}]$, $[\mathbf{R}, \mathbf{R}^{\mathcal{J}}]$ and $[\mathbf{RB}, \mathbf{RB}^{\mathcal{J}}]$.

Theorem 4.2. (i) *The variety $\mathbf{L}^{\mathcal{J}}$ is generated by the semigroup LC_0 and is defined by the identities*

$$x^2 = x^3, \ xyx = (xy)^2, \ xyxzx = xzxyx. \tag{6}$$

(ii) *The variety $\mathbf{R}^{\mathcal{J}}$ is generated by the semigroup RC_0 and is defined by the identities*

$$x^2 = x^3, \ yxy = (xy)^2, \ xyxzx = xzxyx. \tag{7}$$

(iii) *The variety $\mathbf{RB}^{\mathcal{J}}$ is generated by each of the semigroups $LC_0 \times RC_0$ or C_0 and is defined within $\mathbf{A_2}$ by the identity*

$$xyx = xy^2x. \tag{8}$$

The proof of this result follows exactly the same scheme as the proof of Theorem 4.1 as Lemma 4.1 allows us to apply Proposition 3.3 in each of the cases (i)–(iii).

Remark 4.2. The variety $\mathbf{RB}^{\mathcal{J}}$ was considered by the authors in [9, 41] under the name $\overline{\mathbf{B}}_\mathbf{2}$. The fact that the identities (1),(8) define this variety was first discovered in [9], Theorem 3.6, via direct manipulation with identities. In [41], Proposition 2(iv), it was shown that this can be easily extracted from an old result by Sapir and Sukhanov, see [26], Theorem 1. The fact that the variety $\mathbf{RB}^{\mathcal{J}}$ is finitely generated appears to be new.

The varieties $\mathbf{L}^{\mathcal{J}}$ and $\mathbf{R}^{\mathcal{J}}$ can be easily identified with the classes of respectively $\mathcal{R}$-trivial and $\mathcal{L}$-trivial semigroups in $\mathbf{A_2}$. This allows one to easily deduce that they are defined by the identities (6) and respectively (7). Again, the fact that these varieties are finitely generated seems to be new.

Our next theorem is also a direct application of Proposition 3.3 with the help of Lemma 4.1.

Theorem 4.3. (i) *The variety* $\mathbf{B_2}^{\mathcal{J}}$ *is generated by the semigroup* $A_0 \times B_2$ *and is defined within* $\mathbf{A_2}$ *by the identity*

$$x^2y^2x^2 = y^2x^2y^2. \tag{9}$$

(ii) *The variety* $\mathbf{LB_2}^{\mathcal{J}}$ *is generated by the semigroup* $LC_0 \times B_2$ *and is defined within* $\mathbf{A_2}$ *by the identity*

$$x^2y^2x^2 = (x^2y^2)^2. \tag{10}$$

(iii) *The variety* $\mathbf{RB_2}^{\mathcal{J}}$ *is generated by the semigroup* $RC_0 \times B_2$ *and is defined within* $\mathbf{A_2}$ *by the identity*

$$y^2x^2y^2 = (x^2y^2)^2. \tag{11}$$

(iv) *The variety* $\mathbf{NB_2}^{\mathcal{J}}$ *is generated by each of the semigroups* $LC_0 \times RC_0 \times B_2$ *or* $C_0 \times B_2$ *and is defined within* $\mathbf{A_2}$ *by the identity*

$$x^2y^2x^2 = x^2yx^2. \tag{12}$$

Remark 4.3. The variety $\mathbf{NB_2}^{\mathcal{J}}$ was considered in [9] under the name $\overline{\mathbf{A}}_\mathbf{2}$ where it was shown that the identity (12) defines this variety within $\mathbf{A_2}$ see [9], Theorem 2.7. The question of whether or not $\overline{\mathbf{A}}_\mathbf{2} = \overline{\mathbf{B}}_\mathbf{2} \vee \mathbf{B_2}$, the lattice join of the varieties $\overline{\mathbf{B}}_\mathbf{2} = \mathbf{RB}^{\mathcal{J}}$ and $\mathbf{B_2}$ was posed in [9] and then solved in the affirmative in [41], but the solution was rather involved. Now combining Theorem 4.2(iii) with Theorem 4.3(iv), we obtain this equality as an immediate corollary. The fact that the varieties $\mathbf{NB_2}^{\mathcal{J}}$, $\mathbf{LB_2}^{\mathcal{J}}$, $\mathbf{RB_2}^{\mathcal{J}}$ are finitely generated is new to the best of our knowledge.

To complete the picture, we mention that by the definition $\mathbf{A_2}^{\mathcal{J}} = \mathbf{A_2}$. Thus, all top varieties in the basic intervals are finitely based and finitely generated.

If ϑ is a complete congruence on a complete lattice $\langle L, \vee, \wedge\rangle$, then it is easy to see that the top elements of the ϑ-classes form a $\wedge$-subsemilattice in $\langle L, \vee, \wedge\rangle$ but they need not constitute a sublattice in general. However, Theorems 4.2 and 4.3 immediately imply

Corollary 4.1. *The top elements of the basic intervals form a sublattice in* $L(\mathbf{A_2})$.

Of course, this sublattice is isomorphic to the quotient $L(\mathbf{A_2})/\theta$.

It can be easily verified that the bottom elements of the basic intervals also form a sublattice in $L(\mathbf{A_2})$. If we "combine" these two sublattices, that is generate a sublattice in $L(\mathbf{A_2})$ by all the extreme elements of the basic intervals, we obtain a 31-element distributive sublattice shown in Fig 6. This sublattice may thought of as sort of a skeleton for $L(\mathbf{A_2})$. Besides the varieties introduced in Sec. 1, the following varieties are present in Fig. 6:

$\mathbf{A_0} = \operatorname{var} A_0 = \mathbf{T}^{\mathcal{J}}$;	$\mathbf{LA_0} = \operatorname{var}\{A_0, L\}$;
$\mathbf{RA_0} = \operatorname{var}\{A_0, R\}$;	$\mathbf{NA_0} = \operatorname{var}\{A_0, L, R\}$;
$\mathbf{B_0} = \operatorname{var} B_0$;	$\mathbf{LB_0} = \operatorname{var}\{L, B_0\}$;
$\mathbf{RB_0} = \operatorname{var}\{R, B_0\}$;	$\mathbf{NB_0} = \operatorname{var}\{L, R, B_0\}$;
$\mathbf{LC_0} = \operatorname{var} LC_0 = \mathbf{L}^{\mathcal{J}}$;	$\mathbf{RC_0} = \operatorname{var} RC_0 = \mathbf{R}^{\mathcal{J}}$;
$\mathbf{LC_0R} = \operatorname{var}\{LC_0, R\}$;	$\mathbf{RC_0L} = \operatorname{var}\{RC_0, L\}$;
$\mathbf{C_0} = \operatorname{var} C_0 = \mathbf{RB}^{\mathcal{J}}$;	$\mathbf{AB_2} = \operatorname{var}\{A_0, B_2\} = \mathbf{B_2}^{\mathcal{J}}$;
$\mathbf{LAB_2} = \operatorname{var}\{A_0, B_2, L\}$;	$\mathbf{RAB_2} = \operatorname{var}\{A_0, B_2, R\}$;
$\mathbf{NAB_2} = \operatorname{var}\{A_0, B_2, L, R\}$;	$\mathbf{LCB_2} = \operatorname{var}\{LC_0, B_2\} = \mathbf{LB_2}^{\mathcal{J}}$;
$\mathbf{RCB_2} = \operatorname{var}\{RC_0, B_2\} = \mathbf{RB_2}^{\mathcal{J}}$;	$\mathbf{LCB_2R} = \operatorname{var}\{LC_0, B_2, R\}$;
$\mathbf{RCB_2L} = \operatorname{var}\{RC_0, B_2, L\}$.	$\mathbf{CB_2} = \operatorname{var}\{C_0, B_2\} = \mathbf{NB_2}^{\mathcal{J}}$.

Verifying the relations shown in Fig. 6 modulo Theorems 4.2 and 4.3 and the results in [9], Section 5, amounts to routine calculations which are left to the reader.

Recall that an element z of a lattice L is said to *cover* an element $x \in L$ if $x < z$ and there is no $y \in L$ such that $x < y < z$. Of course, in general, if L' is a sublattice of L and $z \in L'$ covers $x \in L'$ within L', z may not cover x in the lattice L. We conclude this paper with a result stating that

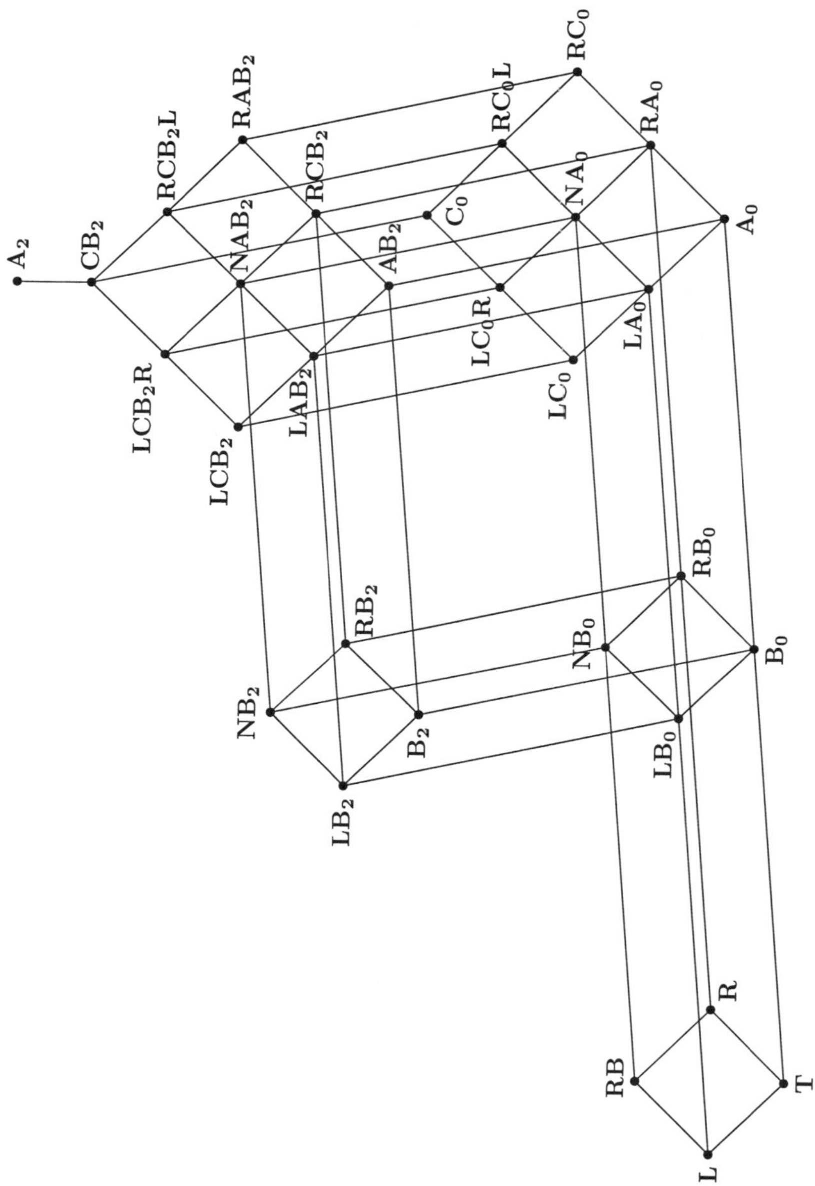

Figure 6. The sublattice generated by the extreme elements of the basic intervals

some coverings shown in Fig. 6 are in fact coverings in the whole lattice $L(\mathbf{A_2})$. This result has been published by the first author [9] but here we want to demonstrate that our new technique proves it in a transparent and calculation-free way.

Theorem 4.4. *The following coverings hold in the lattice* $L(\mathbf{A_2})$*:*

(i) $\mathbf{A_2}$ *covers* $\mathbf{CB_2}$*;*
(ii) $\mathbf{CB_2}$ *covers* $\mathbf{C_0}$*;*
(iii) $\mathbf{AB_2}$ *covers* $\mathbf{A_0}$ *and* $\mathbf{B_2}$*;*
(iv) $\mathbf{A_0}$ *and* $\mathbf{B_2}$ *cover* $\mathbf{B_0}$.

Proof. We shall prove (iii) as a typical item; all other statements are completely analogous.

Let $\mathbf{V}$ be an arbitrary variety such that $\mathbf{AB_2} \supsetneqq \mathbf{V} \supseteq \mathbf{B_2}$. Then $A_0 \notin \mathbf{V}$, $B_0 \in \mathbf{V}$ and $\mathbf{V}\ \theta\ \mathbf{B_2}$. By Proposition 3.4, $\mathbf{V}$ is generated by the e-separable semigroup B_2, that is $\mathbf{V} = \operatorname{var} B_2 = \mathbf{B_2}$.

Now let $\mathbf{V}$ be such that $\mathbf{AB_2} \supsetneqq \mathbf{V} \supseteq \mathbf{A_0}$. Then $B_2 \notin \mathbf{V}$, and therefore, $\mathbf{V}$ drops into the least θ-class whence $\mathbf{V}\ \theta\ \mathbf{A_0}$. Since $A_0 \in \mathbf{V}$, this implies $\mathbf{V} = \mathbf{A_0}$. □

Acknowledgments. The authors are very much indebted to Norman R. Reilly for providing his manuscripts [23, 24, 25] and for a number of extremely stimulating discussions. The second author also thanks the organizers of the McAlister conference for the honorable invitation to speak at that truly remarkable meeting.

References

1. K. Auinger, A method for the construction of complete congruences on lattices of pseudovarieties, *J. Pure Applied Algebra* **126**, 1–17 (1998).
2. S. Burris and H. P. Sankappanavar, *A Course in Universal Algebra*, Springer-Verlag, Berlin–Heidelberg–New York (1981).
3. A. H. Clifford and G. B. Preston, *The Algebraic Theory of Semigroups. Vol. I*, Amer. Math. Soc., Providence, RI (1961).
4. C. C. Edmunds, Varieties generated by semigroups of order four, *Semigroup Forum* **21**, 67–81 (1980).
5. T. E. Hall, S. I. Kublanovskii, S. Margolis, M. V. Sapir, and P. G. Trotter, Algorithmic problems for finite groups and finite 0-simple semigroups, *J. Pure Appl. Algebra* **119**, 75–96 (1997).
6. J. M. Howie, *Fundamentals of Semigroup Theory*, 2nd edition, Clarendon Press, Oxford (1995).

7. G. Lallement, *Semigroups and Combinatorial Applications*, John Wiley & Sons, New York–Chichester–Brisbane–Toronto (1979).
8. E. W. H. Lee, On the lattice of Rees–Sushkevich varieties, Ph. D. thesis, Simon Fraser Univ. (2002).
9. E. W. H. Lee, Identity bases for some non-exact varieties, *Semigroup Forum* **68**, 445–457 (2004).
10. E. W. H. Lee, Subvarieties of the variety generated by the five-element Brandt semigroup, *Int. J. Algebra and Computation* **16**, 417–441 (2006).
11. G. I. Mashevitsky, On identities in varieties of completely simple semigroups over abelian groups, in E. S. Lyapin (ed.), *Sovremennaya Algebra* [*Contemporary Algebra*], Leningrad State Pedagogical Institute, Leningrad, 81–89 (1978) [Russian].
12. G. I. Mashevitsky, Varieties generated by completely 0-simple semigroups, in E. S. Lyapin (ed.), *Polugruppy i ikh Gomomorfizmy* [*Semigroups and their Homomorphisms*], Leningrad State Pedagogical Institute, Leningrad, 53–62 (1991) [Russian].
13. G. Mashevitzky, Matrix rank 1 semigroup identities, *Comm. Algebra* **22**, 3553–3562 (1994).
14. G. Mashevitzky, The pseudovariety generated by completely 0-simple semigroups, *Semigroup Forum* **54**, 83–91 (1997).
15. O. Ore, *Theory of Graphs*, Amer. Math. Soc., Providence, RI (1962).
16. L. Polák, On varieties of completely regular semigroups. I, *Semigroup Forum* **32**, 97–123 (1985).
17. L. Polák, On varieties of completely regular semigroups. II, *Semigroup Forum* **36**, 253–284 (1987).
18. L. Polák, On varieties of completely regular semigroups. III, *Semigroup Forum* **37**, 1–30 (1988).
19. Gy. Pollák, On hereditarily finitely based varieties of semigroups, *Acta Sci. Math. Szeged* **37**, 339–348 (1975).
20. Gy. Pollák, A class of hereditarily finitely based varieties of semigroups, in Gy. Pollák (ed.), *Algebraic theory of semigroups* (*Colloq. Math. Soc. János Bolyai* **20**), North Holland, Amsterdam, 433–445 (1979).
21. Gy. Pollák, Some sufficient conditions for hereditarily finitely based varieties of semigroups, *Acta Sci. Math. Szeged* **50**, 299–330 (1986).
22. Gy. Pollák, Arithmetics in free semigroups, *Acta Sci. Math. Szeged* **68**, 107–115 (2002).
23. N. R. Reilly, Complete congruences on the lattice of Rees–Sushkevich varieties, preprint.
24. N. R. Reilly, The word problem for a five element Brandt semigroup, preprint.
25. N. R. Reilly, Varieties generated by completely 0-simple semigroups, preprint.
26. M. V. Sapir and E. V. Sukhanov, On varieties of periodic semigroups, *Izvestiya Vysshikh Uchebnykh Zavedenii. Matematika*, no.4, 48–55 (1981) [Russian; Engl. translation *Soviet Math. Izv. VUZ* **25**, no.4, 53–63 (1981)].
27. L. N. Shevrin and M. V. Volkov, Identities of semigroups, *Izvestiya Vysshikh Uchebnykh Zavedenii. Matematika*, no.11, 3–47 (1985) [Russian; Engl. translation *Soviet Math. Izv. VUZ* **29**, no.11, 1–64 (1985)].

28. A. V. Tishchenko and M. V. Volkov, A characterization of semigroup varieties of finite index in the "forbidden divisors" language, *Izvestiya Vysshikh Uchebnykh Zavedenii. Matematika*, no.1, 91–99 (1995) [Russian; Engl. translation *Russ. Math. Izv. VUZ* **39**, no.1, 84–92 (1995)].
29. N. G. Torlopova, Varieties of quasiorthodox semigroups, *Acta Sci. Math. Szeged* **47**, 297–301 (1984) [Russian].
30. A. N. Trakhtman, An identity basis of the five-element Brandt semigroup, *Mat. Zapiski UrGU* **12**, no.3, 147–149 (1981) [Russian].
31. A. N. Trakhtman, Graphs of identities of a completely 0-simple five-element semigroup, Ural Polytechnic Institute, Sverdlovsk, 1981, 6pp. [Russian]. (Deposited at VINITI [All-Union Institute for Scientific and Technical Information] on 07.12.81, Moscow, no.5558-81.)
32. A. N. Trakhtman, The finite basis problem for semigroups of order less than six, *Semigroup Forum* **27**, 387–389 (1983).
33. A. N. Trakhtman, Finiteness of identity bases of 5-element semigroups, in E. S. Lyapin (ed.), *Polugruppy i ikh Gomomorfizmy* [*Semigroups and their Homomorphisms*], Leningrad State Pedagogical Institute, Leningrad, 76–97 (1991) [Russian].
34. A. N. Trakhtman, Identities of a five-element 0-simple semigroup, *Semigroup Forum* **48**, 385–387 (1994).
35. B. M. Vernikov and M. V. Volkov, Lattices of nilpotent semigroup varieties, *Mat. Zapiski UrGU* **14**, no.3, 53–65 (1988) [Russian].
36. B. M. Vernikov and M. V. Volkov, Lattices of nilpotent semigroup varieties. II, *Izvestiya Ural'skogo Universiteta*, no.10 (*Matematika i Mekhanika*, no.1), 13–33 (1998) [Russian].
37. B. M. Vernikov and M. V. Volkov, The structure of lattices of nilsemigroup varieties, *Izvestiya Ural'skogo Universiteta*, no.18 (*Matematika i Mekhanika*, no.3), 34–52 (2000) [Russian].
38. M. V. Volkov, On the identity bases of Brandt semigroups, *Mat. Zapiski UrGU* **14**, no.1, 38–42 (1985) [Russian].
39. M. V. Volkov, Young diagrams and the structure of the lattice of overcommutative semigroup varieties, in P. M. Higgins (ed.), *Transformation Semigroups. Proc. Int. Conf. held at the Univ. of Essex*, Univ. of Essex, Colchester, 99–110 (1994).
40. M. V. Volkov, "Forbidden divisor" characterizations of epigroups with certain properties of group elements, in M. Ito (ed.), *Algebraic Systems, Formal Languages and Computations* (*Surikaisekikenkyusho Kokyuroku* **1166**), Research Institute for Math. Sci., Kyoto, 226–234 (2000).
41. M. V. Volkov, On a question by Edmond W. H. Lee, *Izvestija Ural'skogo Universiteta* **36** (*Matematika i Mekhanika* **7**), 167–178 (2005).
42. M. V. Volkov, M. V. Sapir, HFB property and structure of semigroups, *Contrib. Gen. Algebra* **6**, 303–310 (1988).

ON TRIVIALIZERS AND SUBSEMIGROUPS*

ANTÓNIO MALHEIRO

Centro de Álgebra da Universidade de Lisboa,
Av. Prof. Gama Pinto 2, 1649-003 Lisboa, Portugal
E-mail: malheiro@cii.fc.ul.pt
and
Departamento de Matemática,
Faculdade de Ciências e Tecnologia da Universidade Nova de Lisboa,
Quinta da Torre, 2829-516 Monte de Caparica, Portugal

The aim of this paper is to develop the calculus of trivializers for subsemigroups. Given a finite presentation $\mathcal{P}$ defining a semigroup S and a trivializer of the Squier complex of $\mathcal{P}$, we obtain an infinite trivializer of the Squier complex of a finite presentation defining a subsemigroup of S. Also, we give a method to find finite trivializers for special subsemigroups and hence to show that those subsemigroups have *finite derivation type* (FDT). An application of this method is given: we prove that if $S = \mathcal{B}[Y, S_\alpha]$ is a band of monoids having FDT, then so does S_α, for any $\alpha \in Y$.

1. Introduction

In this paper we develop the research program proposed by Pride in [13]. Pride suggests a general program for developing a calculus of trivializers (spherical pictures) for monoids. In particular, the development of the calculus of trivializers for various standard monoid constructions. Following this idea we obtain, in this paper, a general trivializer for special subsemigroups. Related to this general program there are various other results; see, for example, [8], [9], [10], [15], [19], [20], [21] and [22].

We are particularly interested in finite trivializers: we say that a finite monoid presentation $\mathcal{P}$ has *finite derivation type* (FDT) if there exists a finite trivializer of the Squier complex $\mathcal{D}(\mathcal{P})$. This homotopical property was introduced by Squier in [18]. (See also [13].) He also showed that this property is an invariant property of finite presentations [18] and so we can

*This work was developed within the project POCTI/0143/2003 of the Centro de Álgebra da Universidade de Lisboa, financed by FCT and FEDER.

talk of FDT monoids. It is straightforward to extend this property to the class of semigroups [11]. It is important to have in mind that FDT is not in general inherited by finitely presented submonoids (private communication of V. Diekert mentioned in [12]).

In our main theorem, Theorem 4.1, given a semigroup presentation $\mathcal{P}$ defining a semigroup S, and a semigroup presentation $\mathcal{Q}$ obtained from $\mathcal{P}$ under a particular pair of mappings, that defines a subsemigroup of S, we deduce an infinite trivializer $\mathbf{Z}$ of the Squier complex $\mathcal{D}(\mathcal{Q})$. Although we have an infinite trivializer it yields a method for finding a finite trivializer for some special kinds of subsemigroups, and hence to show that certain subsemigroups have FDT. This method has two steps:

(1) obtain an infinite trivializer $\mathbf{Z}$ using Theorem 4.1;
(2) seek for a finite trivializer that turns the closed paths in $\mathbf{Z}$ into null-homotopic paths.

We can apply this technique to certain kinds of semigroups, namely those where finite presentability was obtained using the methodology explained in [3].

In the paper [3] the preservation of finite presentability relatively to subsemigroups is studied. In general this property is, also, not inherited by subsemigroups. However, the main theorem ([3], Theorem 2.1) gives an infinite presentation for a subsemigroup T of a given semigroup S and it yields a constructive method to search for a finite presentation defining T. Their method consists of:

(1) find a generating set for the subsemigroup in question;
(2) define a specific rewriting mapping;
(3) seek for a finite number of relations which imply all the relations obtained in Theorem 2.1 of [3].

Using this method, Campbell *et al.* [3] proved that a large ideal of a finitely presented semigroup is also finitely presented. Later, the same authors proved that a large right (left) ideal of a finitely presented semigroup is also finitely presented [4]. In 1998, again the same authors, proved that if S is a free product of finitely many finite semigroups, then any finitely generated right ideal of S and any finitely generated large subsemigroup of S is finitely presented [5]. Making use of the same method, Ruškuc proved that a large subsemigroup of a finitely presented subsemigroup is also finitely presented [17]. Another successful result, which will be explored later in this paper, was obtained for submonoids of bands of monoids [1].

It says that if a band of monoids $B[Y, S_\beta]$ is finitely presented then every monoid S_α $(\alpha \in Y)$ is finitely presented.

Once the finite presentability is guaranteed using the methodology of [3] we can apply our main theorem. Doing so we have been able to prove in Theorem 5.1 that if $B[Y, S_\beta]$ is a band of monoids having FDT then so does S_β, for any $\beta \in Y$. With the same technique we show in a forthcoming paper that large ideals of semigroups having FDT, also have the same property [9].

We start this paper with some basic concepts on rewriting systems, 2-complexes and Squier complexes. Later, we introduce the general theory of presentations for subsemigroups given in [3]. Then follows a section dedicated to our main theorem where, given a finite presentation $\mathcal{P}$ defining a semigroup S and a trivializer of $\mathcal{D}(\mathcal{P})$, we obtain a trivializer for the Squier complex of a finite presentation defining a subsemigroup of S. In the last section an application of the main theorem is given: we prove that if a band of monoids $B[Y, S_\beta]$ has FDT then so does S_β, for any $\beta \in Y$.

2. Preliminaries

We give a brief description of the main concepts used along the paper. For further information the reader is referred to [2] and [16] for presentations and rewriting systems, to [7] for 2-complexes, and to [13] and [14] for Squier complexes.

2.1. *Presentations and rewriting systems*

A *semigroup presentation* $\mathcal{P}$ is a pair $\langle A \mid R\rangle$, where R is a *rewriting system* on the *alphabet* A, in other words, R is a binary relation on the free semigroup, denoted by A^+, over A. The elements of A^+ are referred to as *words* and the elements of R as *rewriting rules.* The set $A^+ \cup \{1\}$, where 1 denotes the *empty word*, is denoted by A^*. Each rewriting rule $(r_{+1}, r_{-1}) \in R$ is usually written in the form $r_{+1} = r_{-1}$. To avoid confusion, for any two words $w_1, w_2 \in A^+$, we write $w_1 \equiv w_2$ if they are identical words.

On the free semigroup over A we define a binary relation $\to_R$ called a *single-step reduction*, in the following way: for all $u, v \in A^+$, we have $u \to_R v$ if and only if $u \equiv w_1 r_{+1} w_2$ and $v \equiv w_1 r_{-1} w_2$, for some $(r_{+1}, r_{-1}) \in R$ and $w_1, w_2 \in A^*$. By $\overset{*}{\leftrightarrow}_R$ we denote the *Thue congruence* generated by R, which is the reflexive, symmetric and transitive closure of $\to_R$. The quotient of the free semigroup A^+ by $\overset{*}{\leftrightarrow}_R$ is said to be the *semigroup defined by* $\mathcal{P}$, usually denoted by $S(\mathcal{P})$. More generally, a semigroup S is said to be defined by a semigroup presentation $\mathcal{P}$ if $S \cong S(\mathcal{P})$.

The correspondent notions of *monoid presentation* and of *monoid defined by* $\mathcal{P}$ are obtained replacing the free semigroup A^+ by the free monoid A^*, in the above definitions.

2.2. *About 2-complexes*

For our convenience, we introduce a combinatorial model of 2-complexes.

A *graph* Γ is a quintuple $(V, E, \iota, \tau, ^{-1})$, where V and E are non-empty sets of *vertices* and *edges*, respectively. Each edge e has *extremities* ιe and τe, respectively the *initial vertex* and the *terminal vertex*. The orientation on the graph is given by the *inverse map* $^{-1} : E \longrightarrow E$ that satisfies $e \neq e^{-1}$, $\iota e^{-1} = \tau e$ and $(e^{-1})^{-1} = e$, for all $e \in E$.

A *non-empty path* p is a finite sequence of edges $e_1 \cdots e_n$ such that $\tau e_i = \iota e_{i+1}$, for $1 \leq i \leq n-1$. The notions of initial and terminal vertices are naturally extended to paths by $\iota p = \iota e_1$ and $\tau p = \tau e_n$, and p^{-1} represents the *inverse path* $e_n^{-1} \cdots e_1^{-1}$. A path p is said to be *closed* if $\iota p = \tau p = v$. For each vertex v, it is convention to introduce an *empty path* 1_v with no edges, satisfying $\iota 1_v = \tau 1_v = v$ and $1_v^{-1} = 1_v$. Given paths p and q such that $\tau p = \iota q$, we define by juxtaposition the product of p by q, the path pq consisting of the edges of p followed by the edges of q.

A 2-complex $\mathcal{D}$ is a pair (Γ, F), where Γ is a graph and F is a set of closed paths on the graph Γ, called *defining paths.* Based on these defining paths a *homotopy relation* $\sim$ on $\mathcal{D}$ can be introduced in the following way: two paths p and q are said to be homotopic, and we write $p \sim q$, if one can be obtained from the other by a finite sequence of the following operations:

(I) Insert or delete a subpath ee^{-1}, for any edge e;
(II) Insert or delete a subpath p, for any $p \in F \cup F^{-1}$.

A closed path which is homotopic to an empty path is said to be *null-homotopic.*

The set of all homotopy classes $[p]$, of closed paths p with initial vertex v, together with the multiplication $[p][q] = [pq]$, form a group known as the *fundamental group* $\pi_1(\mathcal{D}, v)$ of $\mathcal{D}$ at v.

Let Γ_1 and Γ_2 be graphs. A *mapping of graphs* $\phi : \Gamma_1 \longrightarrow \Gamma_2$ is a mapping from the set of vertices and edges of Γ_1 such that to vertices of Γ_1 correspond vertices in Γ_2, and to edges in Γ_1 correspond paths in Γ_2, satisfying $\phi(\iota e) = \iota\phi(e)$, $\phi(\tau e) = \tau\phi(e)$ and $\phi(e)^{-1} = \phi(e^{-1})$, for any edge e in Γ_1. Naturally, such a mapping can be extended to paths defining $\phi(1_v) = 1_{\phi(v)}$, for any vertex v in Γ_1, and $\phi(p) = \phi(e_1)\cdots\phi(e_n)$, for a non-empty path $p = e_1 \cdots e_n$ in Γ_1.

Given mappings of graphs $\phi : \Gamma_1 \longrightarrow \Gamma_2$ and $\psi : \Gamma_2 \longrightarrow \Gamma_1$ we say that $(\phi, \psi) \in \Omega_0$ if for any vertex $v \in \Gamma_1$ there exists a path from v to $\psi \circ \phi(v)$.

Let $\mathcal{D}_1 = (\Gamma_1, F_1)$ and $\mathcal{D}_2 = (\Gamma_2, F_2)$ be 2-complexes. A *mapping of 2-complexes* from $\mathcal{D}_1$ to $\mathcal{D}_2$ is a mapping of graphs $\phi : \Gamma_1 \longrightarrow \Gamma_2$ such that $\phi(p)$ is null-homotopic in $\mathcal{D}_2$, for any path p in F_1.

Let $\phi : \mathcal{D}_1 \longrightarrow \mathcal{D}_2$ and $\psi : \mathcal{D}_2 \longrightarrow \mathcal{D}_1$ be mappings of 2-complexes such that $(\phi, \psi) \in \Omega_0$. Let Λ_v be a fixed path from v to $\psi \circ \phi(v)$, for each vertex $v \in V$. We say that $(\phi, \psi) \in \Omega_1$ if for any edge e in $\mathcal{D}_1$ the closed path $e\Lambda_{\tau e}(\psi \circ \phi(e))^{-1}\Lambda_{\iota e}^{-1}$ is null-homotopic in $\mathcal{D}_1$.

The following proposition can be easily proved.

Proposition 2.1. *Let $\phi : \mathcal{D}_1 \longrightarrow \mathcal{D}_2$ and $\psi : \mathcal{D}_2 \longrightarrow \mathcal{D}_1$ be mappings of 2-complexes such that $(\phi, \psi) \in \Omega_1$. If $\mathcal{D}_2$ has trivial fundamental groups, then so does $\mathcal{D}_1$.*

Proof. Let v be a vertex in $\mathcal{D}_1$ and p be a closed path in $\mathcal{D}_1$ at v. Since $(\phi, \psi) \in \Omega_1$, we have $p \sim \Lambda_v \psi \circ \phi(p) \Lambda_v^{-1}$. But $\phi(p) \sim 1_{\phi(v)}$ because $\pi_1(\mathcal{D}_2, \phi(v))$ is a trivial group. Hence, $\psi \circ \phi(p) \sim 1_{\psi\circ\phi(v)}$ which means that $p \sim 1_v$. Therefore $\pi_1(\mathcal{D}_1, v)$ is also trivial. □

2.3. *The Squier complex*

Let $\mathcal{P} = \langle A \mid R \rangle$ be a semigroup presentation. The *Squier complex* $\mathcal{D}(\mathcal{P})$ associated to $\mathcal{P}$ has underlying graph $\Gamma(\mathcal{P})$ with vertices A^+ and edges of the form

$$\mathbb{E} = (w_1, r, \varepsilon, w_2), \qquad \text{with } w_1, w_2 \in A^*,\ r \in R \text{ and } \varepsilon = \pm 1,$$

where $\iota\mathbb{E} = w_1 r_\varepsilon w_2$, $\tau\mathbb{E} = w_1 r_{-\varepsilon} w_2$ and $\mathbb{E}^{-1} = (w_1, r, -\varepsilon, w_2)$. The concatenation product in A^*, induces natural left and right actions of A^* on this graph: for any $x, y \in A^*$ and any vertex $v \in A^+$, we define $x \cdot v = xv$ and $v \cdot y = vy$, and for any edge $\mathbb{E} = (w_1, r, \varepsilon, w_2)$, we let $x \cdot \mathbb{E} = (xw_1, r, \varepsilon, w_2)$ and $\mathbb{E} \cdot y = (w_1, r, \varepsilon, w_2 y)$. These actions are compatible, hence we say that A^* acts on the graph. This action can be naturally extended to paths. Clearly, from the definition, each connected component of the graph represents a congruence class of the Thue congruence $\overset{*}{\leftrightarrow}_R$, and hence there is a bijection between the set of all the connected components and the semigroup defined by $\mathcal{P}$.

Given any paths $\mathbb{P}_1$ and $\mathbb{P}_2$ in the above graph, we can consider two new paths $(\mathbb{P}_1 \cdot \iota\mathbb{P}_2)(\tau\mathbb{P}_1 \cdot \mathbb{P}_2)$ and $(\iota\mathbb{P}_1 \cdot \mathbb{P}_2)(\mathbb{P}_1 \cdot \tau\mathbb{P}_2)$, with the same initial vertex and the same terminal vertex. It is convenient to consider both paths as 'essentially the same'. With that purpose, let us first denote by $[\mathbb{P}_1, \mathbb{P}_2]$ the closed path $(\mathbb{P}_1 \cdot \iota\mathbb{P}_2)(\tau\mathbb{P}_1 \cdot \mathbb{P}_2)(\mathbb{P}_1 \cdot \tau\mathbb{P}_2)^{-1}(\iota\mathbb{P}_1 \cdot \mathbb{P}_2)^{-1}$. Now, by definition, the Squier complex $\mathcal{D}(\mathcal{P})$ has defining paths of the form $[\mathbb{E}_1, \mathbb{E}_2]$, for any edges $\mathbb{E}_1, \mathbb{E}_2$. It follows by induction that, for any paths $\mathbb{P}_1$ and $\mathbb{P}_2$, we get

$$(\mathbb{P}_1 \cdot \iota\mathbb{P}_2)(\tau\mathbb{P}_1 \cdot \mathbb{P}_2) \sim (\iota\mathbb{P}_1 \cdot \mathbb{P}_2)(\mathbb{P}_1 \cdot \tau\mathbb{P}_2).$$

This property will be referred to as *pull-down and push-up*, in the sequel.

Let $\mathbf{X}$ be a set of closed paths on $\mathcal{D} = \mathcal{D}(\mathcal{P})$. We can consider an extended 2-complex $\mathcal{D}^{\mathbf{X}}$ obtained from the 2-complex $\mathcal{D}$ adjoining new defining paths of the form $x \cdot \mathbb{P} \cdot y$, for any $\mathbb{P}$ in $\mathbf{X}$ and any $x, y \in A^*$. We get a new homotopy relation which will be denoted by $\sim_{\mathbf{X}}$. The set $\mathbf{X}$ is said to be a *trivializer* of $\mathcal{D}$ if the 2-complex $\mathcal{D}^{\mathbf{X}}$ has trivial fundamental groups. A finite presentation $\mathcal{P}$ is said to be of *finite derivation type* (FDT) if there is a finite trivializer of $\mathcal{D}$. This property is in fact invariant under finite presentations defining the same semigroup [10]. Hence we can talk about FDT semigroups.

An 'algebraic' definition of FDT was first introduced for monoids by Squier in [18]. The equivalent geometric definition above was first given by Pride [13] (for monoids rather than semigroups).

Let $\mathcal{Q} = \langle B \mid U \rangle$ and $\mathcal{P} = \langle A \mid R \rangle$ be semigroup presentations. A *mapping of presentations* $\psi : \mathcal{Q} \longrightarrow \mathcal{P}$ is an homomorphism ψ from B^+ to A^+ such that for any rewriting rule $r \in U$ there is a path $\mathbb{P}_r$ from $\psi(r_{+1})$ to $\psi(r_{-1})$. Choosing $\mathbb{P}_r$ to be as short as possible we can extend ψ to a mapping of graphs defining $\psi(\mathbb{E}) = \psi(w_1) \cdot \mathbb{P}_r^{\varepsilon} \cdot \psi(w_2)$, for any edge $\mathbb{E} = (w_1, r, \varepsilon, w_2)$, with $w_1, w_2 \in A^*$, $r \in U$ and $\varepsilon = \pm 1$. It follows easily that, for any given edges $\mathbb{E}_1, \mathbb{E}_2$ of $\mathcal{D}(\mathcal{Q})$, we have

$$\psi([\mathbb{E}_1, \mathbb{E}_2]) = [\psi(\mathbb{E}_1), \psi(\mathbb{E}_2)],$$

which is a null-homotopic path in $\mathcal{D}(\mathcal{P})$. Hence, a mapping of presentations can be extended to a mapping of 2-complexes $\psi : \mathcal{D}(\mathcal{Q}) \longrightarrow \mathcal{D}(\mathcal{P})$.

3. A general presentation for a subsemigroup

Let S be a semigroup defined by a semigroup presentation $\mathcal{P} = \langle A \mid R \rangle$ and let T be a subsemigroup of S. Let $\mathcal{L}(A, T)$ denote the set of all words in A^+ representing an element of the subsemigroup T. Let $\Gamma(\mathcal{P}, T)$ be the full subgraph of $\Gamma(\mathcal{P})$ induced by the set of vertices $\mathcal{L}(A,T)$. Clearly, each connected component of $\Gamma(\mathcal{P})$, where some vertex represents an element of the semigroup T, is also a connected component of $\Gamma(\mathcal{P}, T)$.

Our goal is to find a presentation $\mathcal{Q}$ defining T. First, let $X = \{\xi_i \mid i \in J\} \subseteq A^+$ be a generating set of the semigroup T. Then, we consider an alphabet $B = \{b_i \mid i \in J\}$ in one-to-one correspondence with X. Each word in B^+ will represent an element of the semigroup T. Hence, let

$$\psi : B^+ \longrightarrow A^+$$

be the homomorphism induced by that correspondence

$$\psi(b_i) \equiv \xi_i, \qquad \text{with } i \in J. \tag{1}$$

This homomorphism is called the *interpretation mapping* since it interprets each word of B^+ as an element of T. Clearly, the homomorphic image of ψ is contained in $\mathcal{L}(A,T)$.

Let $\phi : \mathcal{L}(A,T) \longrightarrow B^+$ be a mapping that satisfies $\psi(\phi(w)) \overset{*}{\leftrightarrow}_R w$, for all $w \in \mathcal{L}(A,T)$. Such a mapping is called a *rewriting mapping.*

In [3] the authors, with the above elements, were able to exhibit a presentation for T.

Theorem 3.1. *Let S be the semigroup defined by a presentation $\langle A \mid R \rangle$, and let T be the subsemigroup of S generated by $X = \{\xi_i \mid i \in J\}$. Let $B = \{b_i \mid i \in J\}$ be a new alphabet in one-to-one correspondence with X, let ψ be the interpretation mapping defined by (1) and let ϕ be a rewriting mapping. Then the semigroup T is defined by the semigroup presentation $\mathcal{Q}$ with generators B and rewriting rules of the form*

$$b_i = \phi(\xi_i), \text{ with } i \in J, \tag{2}$$

$$\phi(w_1 w_2) = \phi(w_1)\phi(w_2), \tag{3}$$

$$\phi(w_3 u w_4) = \phi(w_3 v w_4), \tag{4}$$

where $w_1, w_2 \in \mathcal{L}(A,T)$, $(u,v) \in R$ and $w_3, w_4 \in A^*$ are any words such that $w_3 u w_4 \in \mathcal{L}(A,T)$.

Campbell *et al.* [3] mentioned the disadvantages of the presentation obtained in the previous theorem. It depends on the rewriting mapping ϕ, which has not been defined constructively and it is always infinite. However, the theorem delineates a method to search for a finite presentation as explained in the introduction.

In this paper our goal is to obtain a trivializer of the Squier complex associated to a finite presentation defining the subsemigroup T, when it is given a trivializer of the Squier complex associated to a finite presentation defining S. This proposal will be achieved using adequate mappings ψ and ϕ as above. In fact, those mappings ψ and ϕ that were defined to obtain a finite presentation for T, will be kept allowing us to use the results on finite presentability obtained using the above method.

The next proposition is essential for our future work.

Proposition 3.1. *Let S be the semigroup defined by a presentation $\mathcal{P} = \langle A \mid R \rangle$, and let T be a subsemigroup of S generated by $X = \{\xi_i \mid i \in J\} \subseteq A^+$. Let $B = \{b_i \mid i \in J\}$ be a new alphabet in one-to-one correspondence with X. Let ψ be the interpretation mapping defined by (1) and let $\phi : \mathcal{L}(A,T) \longrightarrow B^+$ be a rewriting mapping. Given a presentation $\mathcal{Q} = \langle B \mid U \rangle$, where U is a rewriting system on B equivalent to the one given by (2), (3) and (4), it follows that*

(i) $\psi : \mathcal{Q} \longrightarrow \mathcal{P}$ *is a mapping of presentations,*
(ii) ϕ *can be extended to a mapping of graphs from* $\Gamma(\mathcal{P},T)$ *to* $\Gamma(\mathcal{Q})$ *and*
(iii) $w \overset{*}{\leftrightarrow}_U \phi \circ \psi(w)$, *for all* $w \in B^+$.

Proof. First, since ψ is an homomorphism and ϕ is a rewriting mapping, there exists a path in $\Gamma(\mathcal{P},T)$ from $\psi(s_{+1})$ to $\psi(s_{-1})$, for each rewriting rule (s_{+1}, s_{-1}) of the form (2), (3) or (4). Now, since U is a rewriting system equivalent to the one given by (2), (3) and (4), we conclude that ψ is a mapping of presentations.

With the rewriting rules of the form (4), we can easily extend the mapping ϕ to a mapping of graphs from $\Gamma(\mathcal{P},T)$ to $\Gamma(\mathcal{Q})$. Since U is a rewriting system equivalent to the one given in Theorem 3.1, for any $(u,v) \in R$ and $w_3, w_4 \in A^*$ such that $w_3uw_4 \in \mathcal{L}(A,T)$, we can fix a path $\mathbb{P}_{w_3,(u,v),w_4}$ in $\Gamma(\mathcal{Q})$ from $\phi(w_3uw_4)$ to $\phi(w_3vw_4)$. Hence, for each edge $\mathbb{E} = (w_3, u = v, \varepsilon, w_4)$ in $\Gamma(\mathcal{P},T)$, with $(u,v) \in R$ and $w_3, w_4 \in A^*$ such that $w_3uw_4 \in \mathcal{L}(A,T)$, we define $\phi(\mathbb{E})$ to be the path $\mathbb{P}^{\varepsilon}_{w_3,(u,v),w_4}$ in $\Gamma(\mathcal{Q})$.

Next, we show that the rewriting rules of the form (2) and (3), assure that, for any $w \in B^+$, we get $w \overset{*}{\leftrightarrow}_U \phi \circ \psi(w)$. Let w be a word in the alphabet B, and $b_1, \ldots, b_n \in B$ be such that $w \equiv b_1 \cdots b_n$. Each $\psi(b_i)$ is a generator in X. Consequently, by the rewriting rules of the form (2) and

(3), we get

$$\begin{aligned}\phi(\psi(w)) &\equiv \phi(\psi(b_1)\cdots\psi(b_n)) \\ &\overset{*}{\leftrightarrow}_U \phi(\psi(b_1))\cdots\phi(\psi(b_n))) \\ &\overset{*}{\leftrightarrow}_U b_1\cdots b_n \equiv w\end{aligned}$$

as required. □

The last proposition tells us that the mappings $\psi : B^+ \longrightarrow A^+$ and $\phi : \mathcal{L}(A,T) \longrightarrow B^+$, used in the above method to obtain a finite presentation that defines the subsemigroup in question, can be extended to mappings of graphs $\psi : \Gamma(\mathcal{Q}) \longrightarrow \Gamma(\mathcal{P},T)$ and $\phi : \Gamma(\mathcal{P},T) \longrightarrow \Gamma(\mathcal{Q})$ satisfying $(\psi,\phi) \in \Omega_0$ and $(\phi,\psi) \in \Omega_0$.

Given a semigroup S defined by a presentation $\mathcal{P} = \langle A \mid R\rangle$ and a subsemigroup T of S generated by $X = \{\xi_i \mid i \in J\} \subseteq A^+$, we say that T is *defined by the presentation* $\mathcal{Q} = \langle B \mid U\rangle$ *under* (ψ,ϕ) if $B = \{b_i \mid i \in J\}$ is a new alphabet in one-to-one correspondence with X, ψ is the interpretation mapping defined by the rule (1), $\phi : \mathcal{L}(A,T) \longrightarrow B^+$ is a rewriting mapping and U is a rewriting system on B equivalent to the one given by the rewriting rules (2), (3) and (4).

4. A general trivializer for a subsemigroup

Let S be a semigroup defined by a presentation $\mathcal{P} = \langle A \mid R\rangle$. Let T be a subsemigroup of S generated by a subset $X = \{\xi_i \mid i \in J\}$ of A^+. Suppose that $\mathcal{Q} = \langle B \mid U\rangle$ is a presentation defining T under (ψ,ϕ).

By Proposition 3.1 the homomorphism ψ can be extended to a mapping of presentations from $\mathcal{Q}$ to $\mathcal{P}$. As we have seen in Subsection 2.3, this mapping can be extended to a mapping between the associated Squier complexes $\mathcal{D}(\mathcal{Q})$ and $\mathcal{D}(\mathcal{P})$. In fact, ψ is a mapping of 2-complexes from the Squier complex $\mathcal{D}(\mathcal{Q})$ to the induced subcomplex $\mathcal{D}(\mathcal{P},T)$ of $\mathcal{D}(\mathcal{P})$ determined by the set of vertices $\mathcal{L}(A,T)$.

It follows, also from Proposition 3.1, that the rewriting mapping ϕ can be extended to a mapping of graphs. Furthermore, we have $(\psi,\phi) \in \Omega_0$ and $(\phi,\psi) \in \Omega_0$. Hence, for each vertex $w \in B^+$, we can fix a path Λ_w from w to $\phi\circ\psi(w)$.

Given a trivializer of $\mathcal{D}(\mathcal{P})$, our propose is to obtain a trivializer of the Squier complex $\mathcal{D}(\mathcal{Q})$. Let $\mathbf{X}$ be a trivializer of $\mathcal{D}(\mathcal{P})$, and let $\mathcal{D}(\mathcal{P},T)^{\mathbf{X}}$ denote the induced subcomplex of $\mathcal{D}(\mathcal{P})^{\mathbf{X}}$ with respect to $\mathcal{L}(A,T)$. Notice also that $\mathcal{D}(\mathcal{P},T)$ is a subcomplex of $\mathcal{D}(\mathcal{P},T)^{\mathbf{X}}$.

Now we can state our main theorem.

Theorem 4.1. *Let S be the semigroup defined by a presentation $\mathcal{P} = \langle A \mid R\rangle$. Let T be the subsemigroup of S defined by $\mathcal{Q} = \langle B \mid U\rangle$ under (ψ, ϕ), for a given generating set $X = \{\xi_i \mid i \in J\} \subseteq A^+$. If $\mathbf{X}$ is a trivializer of $\mathcal{D}(\mathcal{P})$ and $\mathbf{Z}$ is a set of closed paths in $\mathcal{D}(\mathcal{Q})$ such that the paths of the form*

- $(Z1)$ $\mathbb{E}\ \Lambda_{\tau\mathbb{E}}\ (\phi \circ \psi(\mathbb{E}))^{-1}\ \Lambda_{\iota\mathbb{E}}^{-1}$, *for any edge $\mathbb{E}$ in $\Gamma(\mathcal{Q})$;*
- $(Z2)$ $\phi([\mathbb{E}_1, \mathbb{E}_2])$, *for any edges $\mathbb{E}_1, \mathbb{E}_2$ in $\Gamma(\mathcal{P})$ such that $\iota\mathbb{E}_1\iota\mathbb{E}_2 \in \mathcal{L}(A, T)$;*
- $(Z3)$ $\phi(w_1 \cdot \mathbb{P} \cdot w_2)$, *for any $\mathbb{P} \in \mathbf{X}$ and $w_1, w_2 \in A^*$ such that $w_1(\iota\mathbb{P})w_2 \in \mathcal{L}(A, T)$*

are all null-homotopic in $\mathcal{D}(\mathcal{Q})^{\mathbf{Z}}$, then the 2-complex $\mathcal{D}(\mathcal{Q})^{\mathbf{Z}}$ has trivial fundamental groups.

Proof. Above, we mentioned that ψ is a mapping of 2-complexes from $\mathcal{D}(\mathcal{Q})$ to $\mathcal{D}(\mathcal{P})$. Now, extending ψ to $\mathcal{D}(\mathcal{Q})^{\mathbf{Z}}$ and $\mathcal{D}(\mathcal{P})^{\mathbf{X}}$ we still have a mapping of 2-complexes, since $\mathcal{D}(\mathcal{P})^{\mathbf{X}}$ has trivial fundamental groups. In fact, ψ is a mapping of 2-complexes from $\mathcal{D}(\mathcal{Q})^{\mathbf{Z}}$ to $\mathcal{D}(\mathcal{P}, T)^{\mathbf{X}}$

By Proposition 3.1, the mapping ϕ can be extended to a mapping of graphs from $\Gamma(\mathcal{P}, T)$ to $\Gamma(\mathcal{Q})$. Notice that the defining paths of $\mathcal{D}(\mathcal{P}, T)^{\mathbf{X}}$ have the form $[\mathbb{E}_1, \mathbb{E}_2]$, for any edges $\mathbb{E}_1, \mathbb{E}_2$ in $\Gamma(\mathcal{P})$ such that $\iota\mathbb{E}_1\iota\mathbb{E}_2 \in \mathcal{L}(A, T)$ and $w_1 \cdot \mathbb{P} \cdot w_2$, with $\mathbb{P} \in \mathbf{X}$ and $w_1, w_2 \in A^*$ such that $w_1(\iota\mathbb{P})w_2 \in \mathcal{L}(A, T)$. Now, observe that the images of those paths under ϕ, that is the paths of the form $(Z2)$ and $(Z3)$, are null-homotopic in $\mathcal{D}(\mathcal{Q})^{\mathbf{Z}}$ by assumption. Therefore, the mapping of graphs ϕ can be extended to a mapping of 2-complexes from $\mathcal{D}(\mathcal{P}, T)^{\mathbf{X}}$ to $\mathcal{D}(\mathcal{Q})^{\mathbf{Z}}$.

Regarding the paths of the form $(Z1)$, it follows that $(\psi, \phi) \in \Omega_1$. Therefore, the conditions of Proposition 2.1 are fulfilled. Since $\mathbf{X}$ is trivializer of $\mathcal{D}(\mathcal{P})$, that is, the 2-complex $\mathcal{D}(\mathcal{P})^{\mathbf{X}}$ has trivial fundamental groups, we conclude that $\mathcal{D}(\mathcal{Q})^{\mathbf{Z}}$ as also trivial fundamental groups. □

5. An application to bands of monoids

A semigroup S is said to be a *union of semigroups* if it is the disjoint union of some of its subsemigroups. If we can replace subsemigroups by subgroups then S is said to be a *union of groups*. It is proved [6] that unions of groups are precisely the *completely regular semigroups*, which are those semigroups S where for every element $a \in S$ there exists $x \in S$ such that $a = axa$ and $ax = xa$.

We will consider a particular case of unions of semigroups. First, recall that a semigroup Y where every element is an idempotent is called a band. Generally, a semigroup S is said to be a *band of semigroups* if there is a band Y and a collection $(S_\alpha)_{\alpha\in Y}$ of disjoint subsemigroups of S such that $S = \bigcup_{\alpha\in Y} S_\alpha$ and $S_\alpha S_\beta \subseteq S_{\alpha\beta}$. In this case the semigroup S is denoted by $\mathcal{B}[Y, S_\alpha]$.

We now consider the following question: If $\mathcal{B}[Y, S_\alpha]$ has FDT, have all S_α $(\alpha \in Y)$ FDT? We give a partial answer to this general problem, namely for the case where all S_α $(\alpha \in Y)$ are monoids.

Theorem 5.1. *Let $S = \mathcal{B}[Y, S_\alpha]$ be a band of monoids. If S has FDT, then so does S_α, for any $\alpha \in Y$.*

We will prove the above theorem using the methodology explained in the introduction. We first obtain an infinite trivializer $\mathbf{Z}$ using Theorem 4.1, and then we give a finite trivializer that turns the closed paths in $\mathbf{Z}$ into null-homotopic paths.

First, we would like to notice that in [1] the authors proved that if a band of monoids $\mathcal{B}[Y, S_\alpha]$ is finitely presented, then Y is finite and all S_α are finitely presented. In their proof the authors used the general method of finding a (finite) presentation for subsemigroups introduced in [3]. Thus we can use the presentation and the mappings defined in [1].

Let $S = \mathcal{B}[Y, S_\alpha]$ be a band of monoids and let $\mathcal{P} = \langle A \mid R\rangle$ be a presentation defining S. For each $\alpha \in Y$, let e_α denote a fixed word in A^+ that represents the identity of the monoid S_α.

Let us fix an element ζ of the band Y. The Theorem 5.1 in [1] states that the monoid S_ζ is generated by the subset $X = \{e_\zeta a e_\beta : a \in S_\alpha \cap A, \alpha, \beta \in Y, \zeta\alpha\beta = \zeta\}$ of A^+.

We introduce a left action and a right action of the monoid S on the band Y. Given elements $s \in S$ and $\alpha \in Y$ the left action $s\alpha$ (respectively, right action αs), of s on α, is the element $\beta\alpha$ (respectively, $\alpha\beta$), where $\beta \in Y$ is such that $s \in S_\beta$. Notice that the given actions are compatible in the sense that $(\alpha s)\gamma = \alpha(s\gamma)$, for all $\alpha, \gamma \in Y$ and $s \in S$.

Consider the new set B in one-to-one correspondence with X given by

$$\{[a, \beta] \mid a \in A, \beta \in Y, \zeta a\beta = \zeta\}.$$

Let $\psi : B^+ \longrightarrow A^+$ be the homomorphism induced by the correspondence

$$[a, \beta] \mapsto e_\zeta a e_\beta, \tag{5}$$

for all $[a, \beta] \in B$.

Our next step is to define a rewriting mapping ϕ from $\mathcal{L}(A, S_\zeta)$ to B^+. Following the ideas on [1], let $\mu : \{(w, \beta) \in A^+ \times Y : \zeta w\beta = \zeta\} \longrightarrow B^+$ be the mapping defined inductively by

$$\begin{cases} \mu(a,\beta) = [a,\beta], & \text{if } a \in A,\ \beta \in Y \text{ and } \zeta a\beta = \zeta, \text{ and} \\ \mu(wa,\beta) = \mu(w, a\beta)[a,\beta], & \text{if } w \in A^+,\ a \in A,\ \beta \in Y \text{ and } \zeta wa\beta = \zeta. \end{cases}$$

Notice that the element $\mu(wa, \beta)$ is well defined since $a\beta \in Y$ is an idempotent, and hence $\zeta = \zeta wa\beta = \zeta wa\beta a\beta = \zeta a\beta$.

The next lemma follows from the definition [1].

Lemma 5.1. *With the above notation, for all $w_1, w_2 \in A^+$ and $\beta \in Y$ such that $\zeta w_1 w_2 \beta = \zeta$, we have*

$$\mu(w_1 w_2, \beta) \equiv \mu(w_1, w_2\beta)\mu(w_2, \beta).$$

The mapping ϕ is now defined using the mapping μ: given $w \in \mathcal{L}(A, S_\zeta)$ we set $\phi(w) \equiv \mu(w, \zeta)$. From the above lemma it becomes clear that, for all $w_1, w_2 \in \mathcal{L}(A, S_\zeta)$, we have

$$\phi(w_1 w_2) \equiv \phi(w_1)\phi(w_2). \tag{6}$$

Also, from [1] we have

Lemma 5.2. *With the above notation,*

$$\psi(\mu(w, \beta)) \overset{*}{\leftrightarrow}_R e_\zeta w e_\beta$$

for all $w \in A^+$ and $\beta \in Y$ such that $\zeta w\beta = \zeta$. In particular, ϕ is a rewriting mapping.

Now, consider the rewriting system U on the alphabet B given by the rewriting rules

$$\begin{aligned} \phi \circ \psi([a,\beta]) &= [a,\beta] \quad ([a,\beta] \in B) \\ \mu(u, \gamma\zeta) &= \mu(v, \gamma\zeta) \quad ((u,v) \in R, \gamma \in Y, \zeta u\gamma\zeta = \zeta). \end{aligned}$$

Notice that, if A and Y are finite then also B and U are finite.

The Theorem 5.5 in [1] gives a finite presentation for S_ζ.

Theorem 5.2. *Let $S = \mathcal{B}[Y, S_\alpha]$ be a band of monoids, let $\langle A \mid R\rangle$ be a semigroup presentation defining S and let ζ be an arbitrary element of Y. With the above notation, the monoid S_ζ is defined by the semigroup presentation $\mathcal{Q} = \langle B \mid U\rangle$.*

Using our terminology introduced in Section 3, we say that the monoid S_ζ is defined by the semigroup presentation $\mathcal{Q} = \langle B \mid U \rangle$ under (ψ, ϕ). Hence, by Proposition 3.1, $\psi : \mathcal{Q} \longrightarrow \mathcal{P}$ is a mapping of presentations and we can extend ϕ to a mapping of graphs from $\Gamma(\mathcal{P}, S_\zeta)$ to $\Gamma(\mathcal{Q})$. With that proposal, we first define a mapping

$$\mu : \{(\mathbb{E}, \beta) \in \Gamma(\mathcal{P}) \times Y : \zeta\iota\mathbb{E}\beta = \zeta\} \longrightarrow \Gamma(\mathcal{Q})$$

in the following way: given an edge $\mathbb{E} = (w_1, u = v, \varepsilon, w_2)$ in $\Gamma(\mathcal{P})$ and $\beta \in Y$ such that $\zeta\iota\mathbb{E}\beta = \zeta$ we set

$$\mu(\mathbb{E}, \beta) = (\mu(w_1, uw_2\beta), \mu(u, w_2\beta) = \mu(v, w_2\beta), \varepsilon, \mu(w_2, \beta)).$$

This definition can be naturally extended to paths and we easily get the following

Lemma 5.3. *For all $\beta \in Y$, $w_1, w_2 \in A^+$ and any path $\mathbb{P}$ in $\Gamma(\mathcal{P})$ such that $\zeta w_1 \iota\mathbb{P} w_2 \beta = \zeta$, we have*

$$\mu(w_1 \cdot \mathbb{P} \cdot w_2, \beta) = \mu(w_1, \iota\mathbb{P}w_2\beta) \cdot \mu(\mathbb{P}, w_2\beta) \cdot \mu(w_2, \beta).$$

Proof. It is clear having in mind the above definition and the Lemma 5.1. □

We may now extend ϕ to a mapping of graphs defining

$$\phi(\mathbb{E}) = \mu(\mathbb{E}, \zeta), \tag{7}$$

for any edge $\mathbb{E}$ in $\Gamma(\mathcal{P}, S_\zeta)$.

As in Section 4 we will fix some paths in the graph $\Gamma(\mathcal{Q})$. For each letter $[a, \beta] \in B$, we fix a path

$$\Lambda_{[a,\beta]} = (1, [a, \beta] = \phi \circ \psi([a, \beta]), +1, 1)$$

from $[a, \beta]$ to $\phi \circ \psi([a, \beta])$. Inductively, we extend this definition to any word $[a, \beta]u$, with $u \in B^+$, by letting

$$\Lambda_{[a,\beta]u} = (\Lambda_{[a,\beta]} \cdot u)(\phi \circ \psi([a, \beta]) \cdot \Lambda_u).$$

Observe that, since ψ is a homomorphism, and by equality (6), the path $\Lambda_{[a,\beta]u}$ ends at the vertex $\phi \circ \psi([a, \beta]u)$.

Also, it is clear from the definition that

$$\Lambda_{u_1 u_2} = (\Lambda_{u_1} \cdot u_2)(\phi \circ \psi(u_1) \cdot \Lambda_{u_2}) \tag{8}$$

and that by pull-down and push-up we have

$$\Lambda_{u_1 u_2} \sim (u_1 \cdot \Lambda_{u_2})(\Lambda_{u_1} \cdot \phi \circ \psi(u_2)), \tag{9}$$

for any words $u_1, u_2 \in B^+$.

Lemma 5.4. *With the above notation, let $\mathbb{E}_1, \mathbb{E}_2$ be edges in $\Gamma(\mathcal{P})$ such that $\iota\mathbb{E}_1\iota\mathbb{E}_2 \in \mathcal{L}(A, S_\zeta)$. Then we have*

$$\phi([\mathbb{E}_1, \mathbb{E}_2]) = [\mu(\mathbb{E}_1, \iota\mathbb{E}_2\zeta), \mu(\mathbb{E}_2, \zeta)].$$

Proof. By definition we have

$$\begin{aligned}\phi([\mathbb{E}_1, \mathbb{E}_2]) &= \phi\left((\mathbb{E}_1 \cdot \iota\mathbb{E}_2)(\tau\mathbb{E}_1 \cdot \mathbb{E}_2)(\mathbb{E}_1 \cdot \tau\mathbb{E}_2)^{-1}(\iota\mathbb{E}_1 \cdot \mathbb{E}_2)^{-1}\right)\\ &= \phi(\mathbb{E}_1 \cdot \iota\mathbb{E}_2)\phi(\tau\mathbb{E}_1 \cdot \mathbb{E}_2)\phi(\mathbb{E}_1 \cdot \tau\mathbb{E}_2)^{-1}\phi(\iota\mathbb{E}_1 \cdot \mathbb{E}_2)^{-1}.\end{aligned}$$

Now, regarding Lemma 5.3, we get

$$\begin{aligned}\phi([\mathbb{E}_1, \mathbb{E}_2]) = (\mu(\mathbb{E}_1, \iota\mathbb{E}_2\zeta) \cdot \mu(\iota\mathbb{E}_2, \zeta))(\mu(\tau\mathbb{E}_1, \iota\mathbb{E}_2\zeta) \cdot \mu(\mathbb{E}_2, \zeta))\\ (\mu(\mathbb{E}_1, \tau\mathbb{E}_2\zeta) \cdot \mu(\tau\mathbb{E}_2, \zeta))^{-1}(\mu(\iota\mathbb{E}_1, \iota\mathbb{E}_2\zeta) \cdot \mu(\mathbb{E}_2, \zeta))^{-1}.\end{aligned}$$

The elements $\iota\mathbb{E}_2\zeta$ and $\tau\mathbb{E}_2\zeta$ are equal, and hence we get the required result. □

We now obtain a trivializer of $\mathcal{D}(\mathcal{Q})$.

Theorem 5.3. *Let $S = \mathcal{B}[Y, S_\alpha]$ be a band of monoids, let $\mathcal{P} = \langle A \mid R\rangle$ be a semigroup presentation defining S, and let $\zeta \in Y$ be arbitrary. Let $\mathcal{Q} = \langle B \mid U\rangle$ be a semigroup presentation defining S_ζ under (ψ, ϕ), as introduced in (5) and (7), in terms of the generating set $\{e_\zeta a e_\beta : a \in A, \beta \in Y, \zeta a\beta = \zeta\}$. If $\mathbf{X}$ is a trivializer of the Squier complex $\mathcal{D}(\mathcal{P})$, then the set $\mathbf{B}$ of closed paths in $\Gamma(\mathcal{Q})$ of the form*

(*B*1) $\mathbb{E}\Lambda_{\tau\mathbb{E}}(\psi \circ \phi(\mathbb{E}))^{-1}\Lambda_{\iota\mathbb{E}}^{-1}$, *with* $\mathbb{E} = (1, r, \varepsilon, 1)$, $r \in U$ *and* $\varepsilon = \pm 1$;
(*B*2) $\mu(\mathbb{P}, \gamma\zeta)$, *with* $\mathbb{P} \in \mathbf{X}$ *and* $\gamma \in Y$ *such that* $\zeta\iota\mathbb{P}\gamma\zeta = \zeta$

is a trivializer of the Squier complex $\mathcal{D}(\mathcal{Q})$.

Proof. Regarding Theorem 4.1 it is sufficient to show that the paths of the form $(Z1)$, $(Z2)$ and $(Z3)$ are null-homotopic in $\mathcal{D}(\mathcal{Q})^{\mathbf{B}}$.

To show that the paths of the form $(Z1)$ are null-homotopic, it is equivalent to show that, for any edge $\mathbb{E}$ in $\Gamma(\mathcal{Q})$, we have

$$\mathbb{E}\Lambda_{\tau\mathbb{E}} \sim_{\mathbf{B}} \Lambda_{\iota\mathbb{E}}\phi \circ \psi(\mathbb{E}).$$

Let $\mathbb{E}$ be an edge in $\Gamma(\mathcal{Q})$ having the form $(1, r, \varepsilon, 1)$, with $r \in U$ and $\varepsilon = \pm 1$. Any edge in $\Gamma(\mathcal{Q})$ has the form $u_1 \cdot \mathbb{E} \cdot u_2$, with $u_1, u_2 \in B^*$. Suppose that

u_1 and u_2 are non-empty. The other cases where u_1 or u_2 are empty can be dealt with similarly. From (8) and (9) we obtain

$$(u_1 \cdot \mathbb{E} \cdot u_2)\Lambda_{u_1\tau\mathbb{E}u_2} \sim (u_1 \cdot \mathbb{E} \cdot u_2)(u_1 \cdot \Lambda_{\tau\mathbb{E}} \cdot u_2)(\Lambda_{u_1} \cdot \phi \circ \psi(\tau\mathbb{E})u_2)$$
$$(\phi \circ \psi(u_1\tau\mathbb{E}) \cdot \Lambda_{u_2}).$$

Thus by $(B1)$ we get

$$(u_1 \cdot \mathbb{E} \cdot u_2)\Lambda_{u_1\tau\mathbb{E}u_2} \sim_{\mathbf{B}} (u_1 \cdot \Lambda_{\iota\mathbb{E}} \cdot u_2)(u_1 \cdot \phi \circ \psi(\mathbb{E}) \cdot u_2)(\Lambda_{u_1} \cdot \phi \circ \psi(\tau\mathbb{E})u_2)$$
$$(\phi \circ \psi(u_1\tau\mathbb{E}) \cdot \Lambda_{u_2}).$$

Now, by pull-down and push-up and by (8) we have

$$\begin{aligned}(u_1 \cdot \mathbb{E} \cdot u_2)\Lambda_{u_1\tau\mathbb{E}u_2} &\sim_{\mathbf{B}} (\Lambda_{u_1} \cdot \tau\mathbb{E}u_2)(\phi \circ \psi(u_1) \cdot \Lambda_{\iota\mathbb{E}} \cdot u_2)(\phi \circ \psi(u_1\iota\mathbb{E}) \cdot \Lambda_{u_2})\\ &\qquad (\phi \circ \psi(u_1) \cdot \phi \circ \psi(\mathbb{E}) \cdot \phi \circ \psi(u_2))\\ &= \Lambda_{u_1\iota\mathbb{E}u_2}(\phi \circ \psi(u_1) \cdot \phi \circ \psi(\mathbb{E}) \cdot \phi \circ \psi(u_2)).\end{aligned}$$

Therefore, it follows from Lemma 5.3 that

$$(u_1 \cdot \mathbb{E} \cdot u_2)\Lambda_{u_1\tau\mathbb{E}u_2} \sim_{\mathbf{B}} \Lambda_{u_1\iota\mathbb{E}u_2}\phi \circ \psi(u_1 \cdot \mathbb{E} \cdot u_2),$$

as required.

Now, by Lemma 5.4, for any edges $\mathbb{E}_1$ and $\mathbb{E}_2$, with $\iota\mathbb{E}_1\iota\mathbb{E}_2 \in \mathcal{L}(A, S_\zeta)$, we have $\phi([\mathbb{E}_1, \mathbb{E}_2]) = [\mu(\mathbb{E}_1, \iota\mathbb{E}_2\zeta), \mu(\mathbb{E}_2, \zeta)]$, where $\mu(\mathbb{E}_1, \iota\mathbb{E}_2\zeta)$ and $\mu(\mathbb{E}_2, \zeta)$ are in $\Gamma(\mathcal{Q})$. Hence, the paths of the form $(Z2)$ are null-homotopic in $\mathcal{D}(\mathcal{Q})^{\mathbf{B}}$.

Finally, it remains to prove that the paths of the form $(Z3)$ are null-homotopic. Let $\mathbb{P} \in \mathbf{X}$ and $w_1, w_2 \in A^*$ be such that $w_1(\iota\mathbb{P})w_2 \in \mathcal{L}(A, S_\zeta)$. Regarding Lemma 5.3, we have

$$\phi(w_1 \cdot \mathbb{P} \cdot w_2) = \mu(w_1, \iota\mathbb{P}w_2\zeta) \cdot \mu(\mathbb{P}, w_2\zeta) \cdot \mu(w_2, \zeta).$$

Now, observe that the path $\mu(\mathbb{P}, w_2\zeta)$ has the form $(B2)$. Hence, the path $\phi(w_1 \cdot \mathbb{P} \cdot w_2)$ is null-homotopic in $\mathcal{D}(\mathcal{Q})^{\mathbf{B}}$ as required. □

We now use the above theorem to prove Theorem 5.1. Starting with a finite presentation $\mathcal{P} = \langle A \mid R \rangle$ defining the band of monoids $S = \mathcal{B}[Y, S_\alpha]$, with Y finite, and a finite trivializer $\mathbf{X}$ of the Squier complex $\mathcal{D}(\mathcal{P})$, we obtain a finite presentation $\mathcal{Q} = \langle B \mid U \rangle$ defining S_ζ ([1]) and, from Theorem 5.3, a finite trivializer $\mathbf{B}$ of the Squier complex $\mathcal{D}(\mathcal{Q})$. Therefore, the monoid S_ζ has FDT.

Acknowledgments

I would like to express my deep gratitude to my supervisor Professor Gracinda Gomes for her comments and suggestions. Thanks are also due to the anonymous referee for his comments and suggestions which improved a previous version of this paper.

References

1. I.M. Araújo, M.J.J. Branco, V.H. Fernandes, G.M.S. Gomes and N. Ruškuc, On generators and relations for unions of semigroups, *Semigroup Forum* **63**, 49-62 (2001).
2. R.V. Book and F. Otto, *String-Rewriting Systems*, Springer–Verlag, New York (1993).
3. C.M. Campbell, E.F. Robertson, N. Ruškuc and R.M. Thomas, Reidemeister-Schreier type rewriting for semigroups, *Semigroup Forum* **51**, 47-62 (1995).
4. C.M. Campbell, E.F. Robertson, N. Ruškuc and R.M. Thomas, On subsemigroups of finitely presented semigroups, *J. Algebra* **180**, 1-21 (1996).
5. C.M. Campbell, E.F. Robertson, N. Ruškuc and R.M. Thomas, Presentations for subsemigroups - applications to ideals of semigroups, *J. Pure Applied Algebra* **124**, 47-64 (1998).
6. A.H. Clifford, Semigroups admitting relative inverses, *Ann. Math.* **42**, 1037-1049 (1941).
7. R.C. Lyndon and P.E. Schupp, *Combinatorial Group Theory*, Springer–Verlag, Berlin (1977).
8. A. Malheiro, *Finite derivation type for bands of monoids*, preprint (2006).
9. A. Malheiro, *Finite derivation type for Large ideals*, preprint (2006).
10. A. Malheiro, Finite derivation type for Rees matrix semigroups, *Theoret. Comput. Sci.* **355**, 274-290 (2006).
11. A. Malheiro, *Finiteness conditions in semigroup presentations*, Universidade de Lisboa, in preparation.
12. F. Otto, *On properties of monoids that are modular for free products and for certain free products with amalgamated submonoids*, Technical Report, Universität Kassel (1997).
13. S.J. Pride, Geometric methods in combinatorial semigroup theory, *Semigroups, Formal Languages and Groups*, J. Fountain (editor), Kluwer, 215-232 (1995).
14. S.J. Pride, Low-dimensional homotopy theory for monoids, *Int. J. Algebr. Comput.* **5**, 631-649 (1995).
15. S.J. Pride and X. Wang, Second Order Dehn functions of groups and monoids, *Int. J. Algebr. Comput.* **10**, 425-456 (2000).
16. N. Ruškuc, *Semigroup Presentations*, Ph.D. Thesis, University of St. Andrews (1995).
17. N. Ruškuc, Presentations for Subgroups of Monoids, *J. Algebra* **220**, 365-380 (1999).

18. C.C. Squier, F. Otto and Y. Kobayashi, A finiteness condition for rewriting systems, *Theoret. Comput. Sci.* **131**, 271-294 (1994).
19. J. Wang, Finite complete rewriting systems and finite derivation type for small extensions of monoids, *J. Algebra* **204**, 493-503 (1998).
20. J. Wang, Finite derivation type for semi-direct products of monoids, *Theoret. Comput. Sci.* **191**, 219-228 (1998).
21. J. Wang, *Rewriting systems, finiteness conditions and second order Dehn functions of monoids*, Ph.D. Thesis, University of Glasgow (1998).
22. X. Wang, *Second order Dehn functions of groups and monoids*, Ph.D. Thesis, University of Glasgow (1996).

SEMILATTICE ORDERED INVERSE SEMIGROUPS

DONALD B. MCALISTER*

Department of Mathematical Sciences,
Northern Illinois University,
DeKalb, Illinois 60115, U.S.A.
E-mail: don@math.niu.edu

Centro de Álgebra da Universidade de Lisboa,
Av. Prof. Gama Pinto, 2,
1649-003 Lisboa, Portugal

In this expository talk we give some recent results on the structure of inverse semigroups endowed with a compatible semilattice ordering. In addition we consider some open questions regarding these semigroups.

1. Introduction

A semigroup S is said to be ($\vee$-)semilatticed if it is a semilattice under $\vee$ and, for all a, $b \in S$, and c, $d \in S^1$,

$$c(a \vee b)d = cad \vee cbd$$

or, equivalently, $\vee$ is compatible with multiplication on both the left and the right. Such a semigoup is necessarily partially ordered; that is the partial order associated with $\vee$ is compatible with multiplication on both the left and the right. When S is a partially ordered group and a semilattice under the partial ordering, it is necessarily a semilatticed semigroup. Indeed it is lattice ordered in the sense that it is a lattice - in fact a distributive lattice - under the partial ordering and both lattice operations are compatible with multiplication. In this case we say that the group is an l-group. Thus,

*The author acknowledges support from the *Fundação Luso-Americana para o Desenvolvimento*, Proj. L-V-014/2005, and from the *Fundação para a Ciência e a Tecnologia* through the project POCTI/0143/2003 of the *Centro de Álgebra da Universidade de Lisboa* which is partially supported by FEDER.

for groups, there is a very strong relationship between the order and the structure of the group. For example, it is well known, that

Proposition 1.1. *Each non-trivial subgroup of an l-group is infinite. In particular, each non identity element has infinite order.*

We remark that the usual proof, for example that in [3], uses order properties within the l-group very strongly. But the result has a beautiful semigroup theoretic proof. Indeed, more generally, cf. [10].

Proposition 1.2. *Each non-trivial subgroup of a lattice ordered semigroup is infinite.*

The situation with semigroups is very different. It is possible for a partially ordered semigroup to be a semilattice, or even a lattice, under the partial ordering without being a semilatticed semigroup. For example, suppose that L is a lattice and consider it as a semigroup with $e.f = e \wedge f$. Then L is a partially ordered semigroup, and indeed a lattice, under the lattice ordering. However it is a $\vee$-semilatticed semigroup if and only if L is a distributive lattice. Thus the relationship between order and multiplication is more tenuous in the case of semigroups than in the case of groups.

Inverse semigroups form a class intermediate between semigroups in general and groups. So it is not surprising that there are similarities to the situation of l-groups and also differences. For example

Proposition 1.3. *[10] Let S be a lattice ordered inverse semigroup. Then every finite subsemigroup of S consists entirely of idempotents.*

However, the situation is different for semilattice orderings.

Proposition 1.4. *Let S be an E^*-unitary inverse semigroup; that is S is an inverse semigroup with zero in which $ea = e = e^2 \neq 0$ implies $a^2 = a$. Then S is a $\vee$-semilatticed semigroup under the reverse of the natural partial ordering.*

Since there are lots of finite E-unitary inverse semigroups and every Rees factor of such a semigroup is E^*-unitary, [2], it follows that finite $\vee$-semilatticed inverse semigroups are very common. [Bulman-Flemming, Fountain, and Gould [2] have also shown that not every E^*-unitary inverse semigroup is a Rees factor of an E-unitary inverse semigroup.] A class of interesting examples of such semigroups is the following.

Let G be a finite group. Then Schein's [20] coset semigroup $\mathcal{K}(G)$ consisting of all cosets Hx, $x \in G$, is a $\vee$-semilatticed inverse semigroup under inclusion. Here

$$Hx.Ky = \langle H, xKx^{-1}\rangle xy$$
$$Hx \vee Ky = \langle H, K, xy^{-1}\rangle y$$

where, for any subset X of G, $\langle X\rangle$ denotes the subgroup generated by X. $\mathcal{K}(G)$ has group of units isomorphic to G – the singleton cosets – and semilattice of idempotents anti-isomorphic to the lattice of subgroups of G under inclusion. In addition, $\mathcal{K}(G)$ is generated as a $\vee$-semilattice by the singleton cosets. Indeed

Theorem 1.1. *(Ana Paula Garrão [6]) Let G be a finite group. Then $\mathcal{K}(G)$ is the free $\vee$-semilatticed inverse semigroup over the group G. That is, more precisely, given any homomorphism of G into a $\vee$-semilatticed inverse semigroup S, there is a unique $\vee$-homomorphism $\phi : \mathcal{K}(G) \longrightarrow S$ such that the diagram*

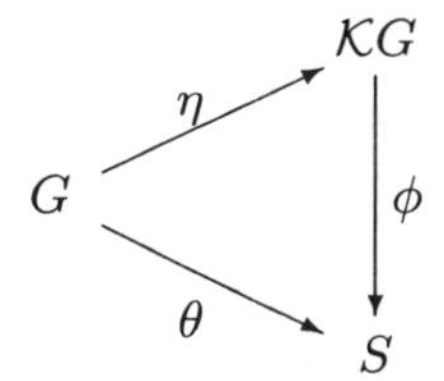

commutes, where, for each $g \in G$, $g\eta = \{g\}$.

This result shows that, although any finite group G can be embedded in a $\vee$-semilatticed inverse semigroup, the ways in which this can be done are fairly restricted. The ensuing semilatticed inverse subsemigroup is restricted to be isomorphic to a quotient of $\mathcal{K}(G)$ by a $\vee$-congruence.

Jonathan Leech [7] considers a class of inverse semigroups which he calls inverse algebras. These are complete $\wedge$-semilattice ordered semigroups under the natural partial ordering. He proves a dual analog to the result above for embeddings of groups into inverse algebras. Garrão's result applies to the more general situation when the partial ordering in not necessarily the natural partial order. It is worth noting, also, that the analog of her result is not valid for infinite groups.

2. The free monogenic semilatticed inverse semigroup

Given that $\vee$-semilatticed inverse semigroups form a variety of algebras, of type $(2, 2, 1)$, there are free $\vee$-semilatticed inverse semigroups on sets of arbitrary cardinality. Note that the theorem above does not assert that $\mathcal{K}(G)$ is the free $\vee$-semilattice inverse semigroup on the set G. Indeed, if $G = \{e\}$ has just one element then so does $\mathcal{K}(G)$ on the other hand, the infinite cyclic group generated by e is, naturally, a totally ordered inverse semigroup and therefore an image of the free $\vee$-semilatticed inverse semigroup on e. So this cannot be $\mathcal{K}(G)$.

It is therefore natural to ask what free $\vee$-semilatticed inverse semigroups look like. I don't know. However even free lattice ordered groups have a very complicated structure, [12], [13]; so this is not surprising and it would perhaps be better to ask a simpler question. What do free monogenic $\vee$-semilatticed inverse semigroups look like? Here we are on a little firmer ground given that free monogenic l-groups have a simple structure: The free monogenic l-group is isomorphic to $\mathbb{Z}\times\mathbb{Z}$ under the usual cartesian ordering. It is easy to see that the minimum group congruence σ on any semilatticed inverse semigrooup is a semilattice congruence; [9]. It follows from this that the free l-group on a set X is the maximum (l-)group homomorphic image of the free semilatticed inverse semigroup on X. Thus in particular

Proposition 2.1. *Let $S = FISL_1$ be the free monogenic semilatticed inverse semigroup. Then S has maximun (l-)group homomorphic image $\mathbb{Z}\times\mathbb{Z}$.*

From the remarks above, each of the semigroups $K_n = \mathcal{K}(\mathbb{Z}_n)$ is a monogenic semilatticed inverse semigroup and therefore a $\vee$-homomorphic image of $FISL_1$. Each of these semigroups and also $\mathbb{Z}\times\mathbb{Z}$ is commutative. However $FISL_1$ is not commutative. To see this we note that neither the free monogenic inverse semigroup FI_1 nor the bicyclic semigroup is commutative. Yet the results below show that each is a homomorphic image of $FISL_1$

Theorem 2.1. *[8] The bicyclic semigroup $B = \langle a, b : ab = 1\rangle$ admits two infinite families of compatible $\vee$-semilattice orderings. In one of these families, a, $ba > 1$ in the other $a < 1 < ba$.*

T. Saitô [19] showed that the idempotents of a totally ordered inverse semigroup form a binary tree under the natural partial ordering. In particular, the idempotents of any totally ordered inverse monoid form a chain under the natural partial ordering. It follows from this that no free inverse semigroup can be totally ordered; [16]. However

Theorem 2.2. *[5] While the free inverse semigroup $FI(X)$ on a set X cannot be totally ordered, it admits a $\vee$-semilattice ordering.*

In fact

Theorem 2.3. *(Giraldes, Gomes, McAlister [5]) The free monogenic inverse monoid admits 8 infinite families of distinct compatible lattice orderings.*

Each of the semigroups above is a $\vee$-homomorphic image of the free semilatticed inverse semigroup $FISL_1$. Consequently, there is a $\vee$-homomorphism of $FISL_1$ onto a subdirect product of all these semigroups. However I don't know if this homomorphism is an isomorphism. The varied quality of these examples does show that, whether or not the homomorphism is an isomorphism, $FISL_1$ is certainly a complicated object!

Question *Is $FISL_1$ an E-unitary semigroup?*

More generally,

Question *Is $FISL(X)$ an E-unitary semigroup?*

3. Totally ordered ω-regular semigroups

As I remarked above, T. Saitô showed that the idempotents of any totally ordered inverse monoid form a chain under the natural partial ordering. In view of this, and the fact that an ω-chain is the simplest example of a infinite chain, it is natural to ask if it is possible to give an explicit construction for the compatible total orderings on ω-regular semigroups. Saitô [17] has given a general construction of total orderings in terms of special subsemigroups – cones – with various properties But in the case the semigroup is ω-regular, it is possible to be quite explicit, modulo total orderings on groups. The results in this section are taken from the thesis of Paulo Medeiros (2006).

Theorem 3.1. *(Medeiros [14]) Let B_d denote the fundamental ω-simple semigroup with d $\mathcal{D}$-classes. Then B_d admits precisely 2^{d+1} compatible total orderings.*

Note that this implies the earlier result of McAlister [8], that the bicyclic semigroup admits precisely 4 distinct compatible total orderings. This was itself a consequence of a result of Saitô which shows that the total orderings on E-unitary inverse semigroups are completely described by total orderings on the idempotents and on the maximum group homomorphic image.

Theorem 3.2. *(Saitô [18]) Let S be an E-unitary inverse semigroup whose set E of idempotents forms a binary tree under the natural partial ordering.*

Let $\leq_E$ be a compatible ordering on E with the property that $e \leq_E f$ implies $a^{-1}ea \leq_E a^{-1}fa$ for each $a \in S$. Suppose further that $\leq_G$ is a total ordering on the maximum group homomorphic image $G = S/\sigma$ of S. Then $\leq_S$ defined by

$$a \leq_S b \Leftrightarrow \begin{cases} a\sigma <_G b\sigma \\ \text{or } a\sigma = b\sigma \text{ and } aa^{-1} \leq_E bb^{-1} \end{cases}$$

is the unique compatible total ordering on S which coincides with $\leq_E$ on the idempotents and such that $\sigma^\natural : S \to G$ is an isotone homomorphism onto the totally ordered group G. Every total ordering on S has this form for a unique $\leq_E$ and $\leq_G$.

Medeiros shows how this result can be extended to non E-unitary inverse semigroups whose idempotents form a chain under the natural partial ordering. Using this, he is able to describe explicitly the compatible total orderings on ω-regular semigroups. We quote the results for bisimple ω-semigroups $S = B(G,\theta) = \{b^r ga^s : g \in G\}$ for which the homomorphism $\phi : S \to \mathbb{Z}$ is isotone.

Theorem 3.3. *(Medeiros [14]) Let G be a totally ordered group, under the total order $\leq_G$ and let $\theta : G \to G$ an isotone endomorphism of G. Then $\leq_S$ definded by*

$$b^r ga^s \leq_S b^m ha^n \Longleftrightarrow \begin{cases} n - m > s - r \\ or \\ n - m = s - r \text{ and } \begin{cases} g\theta^{m-r} \leq h, \text{ if } m \geq r \\ g < h\theta^{r-m}, \text{ if } m < r \end{cases} \end{cases}$$

is a compatible total ordering on $S = B(G,\theta)$ which agrees with $\leq_G$ on G and is such that $ba > 1$ and ϕ is isotone. Conversely each such total ordering has this form for a unique total ordering on G.

We end this section with an interesting example of a non total ordering on $S = B(G,\theta)$ in the case when S is far from E-unitary. We shall say that θ is a nil homomorphism if $G = \bigcup\{\ker\theta^n : n \geq 1\}$; that is, for each $g \in G$, there exists n such that $g\theta^n = 1$.

Theorem 3.4. *(Medeiros [14]) Let G be a group and θ an endomorphism of G. Then $S = B(G,\theta)$ is a partially ordered inverse semigroup under the ordering $\leq$ defined by*

$$b^r ga^s \leq_S b^m ha^n \Longleftrightarrow \begin{cases} n - m > s - r \\ or \\ m - r = n - s \geq 0 \text{ and } g\theta^{m-r} = h \end{cases}$$

S is a $\vee$-semilatticed semigoup under this ordering if and only if θ is nil.

4. *E*-unitary inverse semigroups

Suppose that $S = P(G, \mathcal{X}, \mathcal{Y})$ is an E-unitary inverse semigroup. Then the set of idempotents of S, under the natural partial ordering, is isomorphic to $\mathcal{Y}$. Thus compatible partial orderings on the idempotents correspond to compatible partial orderings on $\mathcal{Y}$. Now any compatible partial ordering on S has the property that

$$e \leq f \Longrightarrow x^{-1}ex \leq x^{-1}fx$$

for any idempotents e, f and any $x \in S$. This condition translates to the following condition on $\mathcal{Y}$,

$$u \leq_{\mathcal{Y}} v \text{ with } gu,\, gv \in \mathcal{Y} \text{ implies } gu \leq_{\mathcal{Y}} gv.$$

Thus Saitô's theorem on totally ordered E-unitary inverse semigroups can be stated in the form

Theorem 4.1. [5] *Let $S = P(G, \mathcal{X}, \mathcal{Y})$ be an E-unitary inverse semigroup where G is a totally ordered group under $\leq_G$ and that $\mathcal{Y}$ is a totally ordered semilattice under a total order $\leq_{\mathcal{Y}}$ with the property that*

$$u \leq_{\mathcal{Y}} v \textit{ with } gu,\, gv \in \mathcal{Y} \textit{ implies } gu \leq_{\mathcal{Y}} gv.$$

Then $\leq$ defined by

$$(e, g) \leq (f, h) \Longleftrightarrow \begin{cases} g <_G h \\ \textit{or } g = h \textit{ and } e \leq_{\mathcal{Y}} f \end{cases}$$

is a compatible total order on S. Every compatible total order on S has this form for a unique total order $\leq_G$ on G and a unique compatible total order $\leq_{\mathcal{Y}}$ on $\mathcal{Y}$ which obeys the G-compatibility condition above.

The existence of a compatible total order on $\mathcal{Y}$ requires that $\mathcal{Y}$ is a binary tree and so, because G acts by order automorphisms on $\mathcal{X}$, that $\mathcal{X}$ is a tree.

The partial order $\leq$ in the theorem is a lexicographic ordering. Giraldes, Gomes and McAlister, [5], showed that every compatible lexicographic ordering on an E-unitary inverse semigroup has the form above where $\leq_{\mathcal{Y}}$ is a compatible semilattice ordering on $\mathcal{Y}$ which obeys the G-compatibility condition. This condition is automatically satisfied if $\leq_{\mathcal{Y}}$ is the natural partial order. Using this, they showed that every inverse semigroup is an idempotent separating homomorphic image of a semilattice ordered E-unitary inverse semigroup. As a consequence, being a semilatticed inverse semigroup imposes no restrictions on the ideal structure of the semigroup; [5]. The result also accentuates the naturalness of the question of constructing semilattice ordered E-unitary covers for semilatticed inverse semigroups. More precisely,

Definition 4.1. Let S be a semilatticed inverse semigroup. Then a semilattice ordered E-unitary inverse semigroup T is said to be an $E^{\vee}$-cover for S if there is an idempotent separating $\vee$-homomorphism of T onto S.

In this case the maximum group homomorphic image G of T is an l-group and we say that T is an $E^{\vee}$-unitary cover of S over G. Reilly and McAlister, [11], showed that E-unitary covers of an inverse semigroup S over a group G correspond to idempotent separating pre-homomorphisms of S into the semigroup $\mathcal{K}(G)$ of cosets of G. As we saw, this is a semilatticed inverse semigroup under inclusion. Paula Garrão in her thesis gives analogous necessary and sufficient conditions for constructing $E^{\vee}$-unitary covers.

Theorem 4.2. *(Garrão [6]) Let S be a semilatticed inverse semigroup and G an l-group. Then there exists an $E^{\vee}$-unitary cover of S over G if and only if there is an idempotent separating and idempotent pure pre-homomorphism ϕ of S into $\mathcal{K}(G)$ such that $G = \bigcup\{s\phi : s \in S\}$ – these are the conditions for an E-unitary cover – such that*

$$a\phi \vee b\phi \subseteq (a \vee b)\phi$$

for all a, $b \in S$.

Given the pre-homomorphism, one can construct the covering but this still leaves the question – does every semilattice ordered inverse semigroup have an $E^{\vee}$-unitary covering? This, in turn leads us back to the question with which we began.

Theorem 4.3. *Every semilattice ordered inverse semigroup has an $E^{\vee}$-unitary covering if and only if every free semilattice ordered inverse semigroup is E-unitary.*

When it comes to totally ordered $E^{\vee}$-unitary covers, the situation is rather different.

Theorem 4.4. *(Garrão [6]) Let S be a semilattice ordered inverse semigroup. Then S has a totally ordered $E^{\vee}$-unitary covering if and only if S is E-unitary – and totally ordered.*

Corollary 4.1. *(Garrão [6]) Let S be a totally ordered E-unitary inverse semigroup. Then every isotone homomorphic image of S is E-unitary.*

5. Semidirect product embeddings

There is a natural way of embedding any partially ordered set X in a semilattice. Namely, take the semilattice $\hat{X}$ of all (down) order ideals of X. The the map $a \to a \downarrow = \{x \in X : x \leq a\}$ gives such an embedding. If G is a group which acts on X by order automorphisms then G also acts, in a natural way, on $\hat{X}$ so that one can form the semidirect product $\hat{X} * G$ of $\hat{X}$ by G. The map $(a, g) \to (a \downarrow, g)$ then gives an embedding of $P(G, X, Y)$ into $\hat{X} * G$ for any subemilattice and order ideal Y of X. Thus

Theorem 5.1. *(O'Carroll [15]) Every E-unitary inverse semigroup can be embedded in the semidirect product of a semilattice by a group.*

In general, this embedding is rather sparse. However a more dense one is obtained if we take, instead of $\hat{X}$, the semilattice X^f of finite intersections $\bigcap\{a_1 \downarrow, \cdots, a_n \downarrow\}$ of principal ideals of X. Then the same function $a \to a \downarrow$ gives an embedding of $P(G, X, Y)$ in $X^f * G$.

This last section, which represents joint work with Gracinda Gomes, considers the question of whether this result of O'Carroll can be extended to totally ordered E-unitary inverse semigroups. Namely, can every totally ordered E-unitary inverse semigroup be embedded in a totally ordered semidirect product of a totally ordered semilattice by a totally ordered group? [Note that the order on the semidirect product will be the lexicographic one.]

We begin with a simple example.

Let $G = \langle g \rangle$ be an infinite cyclic group and $E = \{0, 1\}$. Then $E \times G$ is totally ordered by

$$(a, g^m) \leq (b, g^n) \Leftrightarrow \begin{cases} m < n \\ or\ m = n \text{ and } a \leq b \end{cases}.$$

This totally ordered inverse semigroup contains, as an inverse subsemigroup, the semigroup

$$S = \{(1, g^{2n}) : n \in \mathbb{Z}\} \cup \{(0, g^n) : n \in \mathbb{Z}\}.$$

Thus S is a totally ordered E-unitary inverse semigroup which, while not a semidirect product of a semilattice by a group can be embedded in the (semi)direct product of a semilattice by a group.

In fact, $S \approx P(G, X, Y)$ where $Y = E$ and X has the diagram

$$\begin{array}{ccc} v & & u = 1 \\ & \nwarrow \quad \nearrow & \\ & 0 & \end{array} .$$

The group G acts by fixing 0 and g interchanges u and v. In this case X is a semilattice, in fact a binary tree, while $\hat{X}$ consists of the principal ideals $x \downarrow$, $x \in X$, together with the ideals X and $\varnothing$. Note that $\hat{X}$ cannot be totally ordered because it is not a tree.

[Note that, in general, $\hat{X}$ can be totally ordered if and only if it is a chain – which is equivalent to X being a chain – under the natural partial order.]

X^f consists of the ideals $x \downarrow$, $x \in X$ and is isomorphic to X itself. Suppose that there is a compatible total order on $X^f * G$. Then there are two mutually exclusive possibilities, $u \downarrow < v \downarrow$ or $v \downarrow < u \downarrow$. In the first case, we get $g(u \downarrow) \leq g(v \downarrow)$ that is $v \downarrow \leq u \downarrow$ which is a contradiction. The second situation is similarly impossible. Hence $X^f * G$ cannot be totally ordered. This shows that we cannot expect the procedure which worked so well in the un-ordered case to work in this case also.

The construction of an embedding in the example is prototypical of the situation of E-unitary Clifford semigroups; that is, for those whose idempotents are central.

Proposition 5.1. *Let S be a totally ordered E-unitary Clifford semigroup. Then S can be embedded in the (semi)-direct product of a totally ordered semilattice by a totally ordered group.*

We have already seen that the idempotents of a totally ordered inverse semigroup S form a binary tree under the natural partial ordering. In particular, the idempotents of each local submonoid of S form a chain under the natural partial ordering. It is therfore natural to consider inverse semigoups S whose idempotents form a chain under the natural partial ordering when we want to study totally ordered inverse semigroups. Then each $\mathcal{D}$-class and each $\mathcal{J}$-class is an inverse subsemigroup of S. Let e be an idempotent of S and set $L = eSe \cap L_e$, where L_e is the $\mathcal{L}$-class of e.

Proposition 5.2. *Let S be a totally ordered inverse semigroup, under a total order $\leq$, whose idempotents form a chain under the natural partial order, and let e be an idempotent of S. If the intersection of non-empty left ideals of L is non-empty, then either $\leq$ is the natural partial order on the idempotents of D_e or is the inverse of the natural partial order.*

Corollary 5.1. *Let S be a totally ordered E-unitary inverse semigroup whose idempotents form a chain under the natural partial order. If S/σ is commutative then in each $\mathcal{D}$-class of S the imposed total order on the idempotents is either the natural partial order or the inverse of the natural partial order.*

Let us say that an inverse monoid M is *reflexive* if the intersection of principal left ideals of $L = \{x \in M : x^{-1}x = 1\}$ is non-empty.

Theorem 5.2. *Let $S = P(G, X, Y)$ be a bisimple E-bisimple inverse monoid where Y is a chain (and X is a tree). Then the following are equivalent:*

(1) X is up-directed;
(2) S is reflexive;
(3) the principal left ideals of L form a chain under inclusion;
(4) X is a chain.

Corollary 5.2. *Let S be a totally ordered reflexive E-unitary bisimple inverse monoid. Then S can be embedded in the totally ordered semidirect product of a totally ordered chain by a totally ordered group.*

We end this section and the talk with another example. This time, an example of a totally ordered inverse semigroup in which not all the $\mathcal{D}$-classes are subsemigroups. Thus the idempotents do not form a chain under the natural partial order. Such a semigroup S must contain an element x such that

$$xx^{-1}.x^{-1}x \notin \{xx^{-1}, x^{-1}x\}.$$

Because S is totally ordered, one can show that either $x^{-3}x^3 = x^{-2}x^2$ or $x^3x^{-3} = x^2x^{-2}$.

Proposition 5.3. *Let $T = \langle x, x^{-1}\rangle$ be the inverse subsemigroup of S generated by x. Then T is isomorphic to the inverse semigroup with presentation $\langle x : x^{-3}x^3 = x^{-2}x^2\rangle$ and so is unique, up to isomorphism.*

Corollary 5.3. *Let S be a totally ordered inverse semigroup then the idempotents of each D-class form a chain under the natural partial order if and only if S does not contain a copy of T. In this case, each D-class, and each J-class of S, is a subsemigroup.*

In light of this corollary, it is natural to investigate the structure of T.

Theorem 5.3. *T is E-unitary; with two D-classes. It is an ideal extension of the bicyclic semigroup by a 2×2 Brandt semigroup. Indeed $T \approx P(\mathbb{Z}, X, Y)$ where Y has the diagram below*

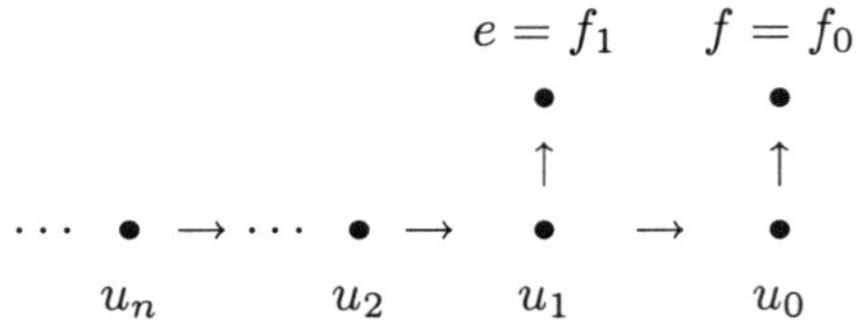

and X has the diagram

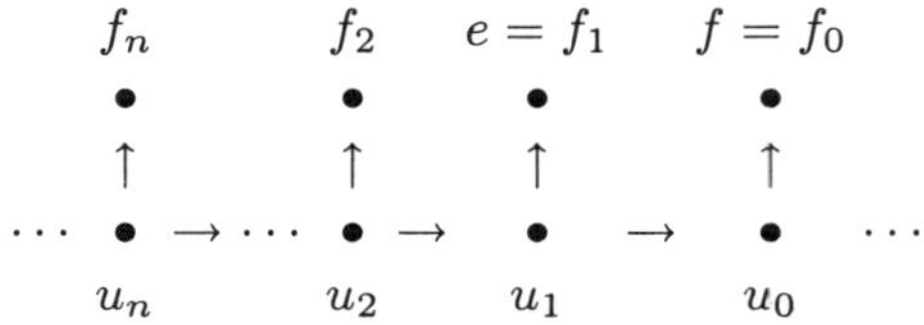

The action of $\mathbb{Z}$ is the obvious one.

Because T is E-unitary, any compatible total order on it is given by a total order on $T/\sigma \approx \mathbb{Z}$ and a total order $\leq_E$ on the set E of idempotents which has the property that $x \leq_E y$ implies $t^{-1}xt \leq_E t^{-1}yt$ for each t in T.

Theorem 5.4. *T admits precisely four total orders, depending on whether*

(1) the order on $T/\sigma \approx \mathbb{Z}$ is the usual one or its inverse;
(2) $e < f$ or $f < e$.

In the case that $e < f$ the order on the idempotents translates to the order with diagram

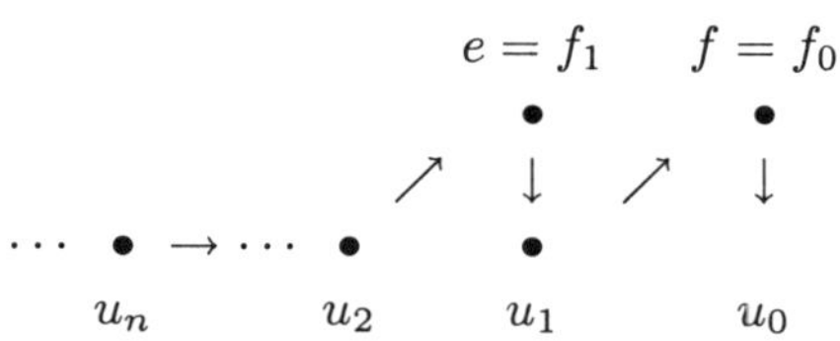

on Y and by the group action to the total order

$$\begin{array}{ccccccccccc} & & f_n & & & f_2 & & e=f_1 & & f=f_0 & \\ & & \bullet & & & \bullet & & \bullet & & \bullet & \\ & \nearrow & \downarrow & \nearrow & & \downarrow & \nearrow & \downarrow & \nearrow & \downarrow & \nearrow \\ \cdots & & \bullet & & \cdots & \bullet & & \bullet & & \bullet & \cdots \\ & & u_n & & & u_2 & & u_1 & & u_0 & \end{array}$$

on X.

Note that, in each of the four cases, the order extends to $P(G, X, X)$ so that in each case T can be embedded into the semidirect product of a totally ordered chain by a totally ordered group. However the question: can every totally orderd E-unitary inverse semigroup be embedded in a totally ordered semidirect product of a totally ordered semigroup by a totally ordered group still remains to be solved.

References

1. G. Birkhoff, Lattice Theory, *American Mathematical Society Colloquium Publications* **25**, Providence, R.I. (1948).
2. S. Bullman-Fleming, J. Fountain and V. Gould, Inverse semigroups with zero: covers and their structure, *J. Australian Math. Soc. (A)* **67**, 15-30 (1999).
3. M.R. Darnel, *Theory of Lattice Ordered Groups*, Dekker, New York (1995).
4. L. Fuchs, *Partially Ordered Algebraic Systems*, Pergamon Press, Oxford (1963)
5. E. Giraldes, G.M.S. Gomes and D.B. McAlister, On a class of lattice ordered inverse semigroups, *J. Algebra* **240**, 496-517 (2000).
6. A.P. Garrão, *∨-Congruências e coberturas de semigrupos ∨-semireticulados*, Ph.D. thesis, Universidade dos Açores (2005).
7. J. Leech, Inverse monoids with a natural semilattice ordering, *Proc. London Math. Soc.* **70 (3)**, 146-182 (1995).
8. D.B. McAlister, Compatible orders on the bicyclic semigroup, *Comm. Algebra* **27 (9)**, 4179-4208 (1999).
9. D.B. McAlister, *Lattice ordered groups and the structure of inverse semigroups*, (notes for a course) C.A.U.L., 1-53 (1998).
10. D.B. McAlister, Amenably ordered inverse semigroups, *J. Algebra* **65**, 118-146 (1980).
11. D.B. McAlister and N.R. Reilly, E-unitary covers for inverse semigroups, *Pacific J. Math.* **68**, 161-174 (1977).
12. S.H. McCleary, Free lattice groups represented as o-2 transitive l permutation groups, *Trans. Amer. Math. Soc.* **290**, 69-79 (1985).
13. S.H. McCleary, An even better representation for free lattice ordered groups, *Trans. Amer. Math. Soc.* **290**, 81-100 (1985).
14. P.J. Medeiros, *Ordens compatíveis em semigrupos ω-regulares*, Ph.D. thesis, Universidade dos Açores (2006).

15. L. O'Carroll, Embedding theorems for proper inverse semigroups, *J. Algebra* **42**, 26-40 (1976).
16. T. Saitô, No free inverse semigroup is orderable, *Proc. Japan Acad.* **50**, 837-838 (1974).
17. T. Saitô, Ordered inverse semigroups, *Trans. Amer. Math. Soc.* **153**, 99-138 (1971).
18. T. Saitô, Proper ordered inverse semigroups, *Pacific J. Math.* **15**, 649-666 (1965).
19. T. Saitô, Ordered idempotent semigroups, *J. Math. Soc. Japan* **14**, 150-169 (1962).
20. B.M. Schein, Semigroups of strong subsets, *Volzhskii Matematicheskiĭ Sbornik* **4**, 180-186 (1966).

ISOMORPHISM PROBLEMS FOR TRANSFORMATION SEMIGROUPS

SUZANA MENDES-GONÇALVES
Centro de Matemática,
Campus de Gualtar, Universidade do Minho,
4710-057 Braga, Portugal
E-mail: smendes@math.uminho.pt

Many people have studied the problem of describing all isomorphisms between transformation semigroups defined on sets and between linear transformation semigroups defined on vector spaces. In this paper, we summarise some of that work, as well as recent work showing that Baer-Levi semigroups defined on sets are never isomorphic to their linear analogue, despite all appearances to the contrary.

1. Introduction

Let X be an arbitrary set and let $P(X)$ denote the set of all *partial transformations* of X: that is, all transformations α whose *domain*, $\operatorname{dom}\alpha$, and *range*, $\operatorname{ran}\alpha$, are subsets of X. As usual, the *composition* $\alpha \circ \beta$ of $\alpha, \beta \in P(X)$ is the transformation with domain $Y = (\operatorname{ran}\alpha \cap \operatorname{dom}\beta)\alpha^{-1}$ such that, for all $x \in Y$, $x(\alpha \circ \beta) = (x\alpha)\beta$, and we write $\alpha \circ \beta$ more simply as $\alpha\beta$. It is well-known that $(P(X), \circ)$ is a semigroup. Let $T(X)$ denote the subsemigroup of $P(X)$ consisting of all $\alpha \in P(X)$ with domain X, and let $I(X)$ denote the *symmetric inverse semigroup* on X: that is, the set of all injective elements of $P(X)$.

If $\alpha \in T(X)$, we define the *rank* of α to be $r(\alpha) = |\operatorname{ran}\alpha|$ and we define another two cardinal numbers, called the *defect* and the *collapse* of α, respectively, as follows.

$$D(\alpha) = X \setminus X\alpha, \qquad d(\alpha) = |D(\alpha)|,$$
$$C(\alpha) = \bigcup\{y\alpha^{-1} : |y\alpha^{-1}| \geq 2\}, \quad c(\alpha) = |C(\alpha)|.$$

If X is an infinite set and $|X| = p \geq q \geq \aleph_0$, we write

$$BL(p, q) = \{\alpha \in T(X) : c(\alpha) = 0,\ d(\alpha) = q\}.$$

That is, $BL(p,q)$ consists of all one-to-one transformations α of X into itself such that $X \setminus X\alpha$ has cardinal q. Following Clifford and Preston, we call this the *Baer-Levi semigroup of type* (p,q) *on the set* X. It is well known that $BL(p,q)$ is a right simple, right cancellative semigroup without idempotents; and, according to Clifford and Preston [4] Vol. 2, p. 82, Teissier showed in 1953 that any semigroup with these properties can be embedded in some Baer-Levi semigroup.

Now let V be a vector space over a field F and let $P(V)$ denote the set of all *partial linear transformations* of V: that is, all linear transformations $\alpha : A \to B$ where A, B are subspaces of V. It is not difficult to see that $(P(V), \circ)$ is a semigroup, with composition of transformations defined as before. Let $T(V)$ denote the linear analogue of $T(X)$, and $I(V)$ that of $I(X)$. Note that we use the 'V' in place of 'X' to denote the fact that we are considering *linear* transformations.

In this paper we briefly survey some results on isomorphisms between transformation semigroups and between linear transformation semigroups. In section 2, we give a short account of work on the first problem, focusing on semigroups that are 'normal' in the usual group-theoretic sense (that is, which are invariant with respect to conjugation by permutations of the underlying set) and on semigroups that contain 'sufficient' constant maps. It is well-known that some important transformation semigroups have these properties. Likewise, in section 3, we illustrate the ongoing interest in the isomorphism problems between linear transformation semigroups, focusing on semigroups which are closed under conjugation by elements of the general linear group defined on the underlying vector space, as well as on semigroups which contain a 'big enough' number of linear transformations with rank 1. This motivates the definition of a linear version of the Baer-Levi semigroup: in section 4, we present some results which suggest that the two types of Baer-Levi semigroup are similar and finally, we show that the two versions – one defined on sets, the other on vector spaces – are never isomorphic. All the work in section 4 has appeared in [16].

2. Isomorphisms between transformation semigroups

In this section, we consider the problem of describing all isomorphisms between certain semigroups of transformations defined on *unstructured* sets. That is, we omit all mention of work on the corresponding problem for transformations which preserve some structure on the sets (for example, an order, graph, ultrafilter or topology). We begin with some background concerning automorphisms of transformation semigroups, as a preamble to our discussion of isomorphisms.

We denote the *symmetric group* on X by $G(X)$, and we let $\mathrm{Alt}(X)$ denote the *alternating group* on X: that is, the set of all $\alpha \in G(X)$ whose *shift* $S(\alpha) = \{x \in X : x\alpha \neq x\}$ is finite and for which $\alpha|S(\alpha)$ is an even permutation (see [21] p. 301). Given a subsemigroup S of $P(X)$, we denote the automorphism group of S by $\mathrm{Aut}(S)$. We say that $\phi \in \mathrm{Aut}(S)$ is *inner* if there exists some $g \in G(X)$ such that $\alpha\phi = g^{-1}\alpha g$ for every $\alpha \in S$.

In [22] Sullivan observed that many people had described the automorphism group of various transformation semigroups defined on a set X. For example, Lyapin in 1955, Magill in 1967, and Schreier in 1936, each described the automorphisms of the full transformation semigroup $T(X)$, and Mal'cev in 1952 determined the ideals of $T(X)$ and their automorphism groups. One year later, Liber characterised the automorphism group of each ideal of $I(X)$ and, several years later, Šutov studied the automorphisms of the subsemigroup of $T(X)$ consisting of all (total) transformations which shift at most a finite number of elements. Also, Magill in 1967 described all automorphisms of $P(X)$, and Šutov in 1961 determined all ideals of $P(X)$ and their automorphism groups. In all these cases, the automorphisms are inner and each semigroup has $G(X)$ as its automorphism group.

To unify these results, in [22] Sullivan introduced two simple concepts: a semigroup S of partial transformations on X is said to *cover* X if for every $x \in X$ there exists an idempotent map in S with range $\{x\}$; and S is $G(X)$-*normal* if $g^{-1}\alpha g \in S$ for every $\alpha \in S$ and every $g \in G(X)$. Using these ideas, he proved the following in [22] Theorems 1 and 2 (this generalises the result in [9] which was unknown to him at the time).

Theorem 2.1. *If S is a subsemigroup of $P(X)$ covering X then every automorphism of S is inner. Moreover, if S is also $G(X)$-normal, then its automorphism group is isomorphic to $G(X)$.*

Subsequently, in [28], Symons characterised all $G(X)$-normal subsemigroups of $T(X)$ for finite X and, for this case, he observed a partial converse of Theorem 2.1. That is, for comparison later with its linear analogue, we quote his remark in [28] p. 390 as a formal result.

Theorem 2.2. *Let X be a finite set and let S be a subsemigroup of $T(X)$ which is not contained in $G(X)$. Then the following are equivalent.*
(a) S is $G(X)$-normal;
(b) every automorphism of S is induced by a permutation of X.

It may be possible for some transformation semigroups without constant maps (and which are not permutation groups) to have all its automorphisms inner; and in fact, in [5], Fitzpatrick and Symons proved that if X is infinite and S is a subsemigroup of $T(X)$ containing $G(X)$, then every automorphism of S is inner. Later, this result was extended in [24] to include subsemigroups of $I(X)$ which contain $\mathrm{Alt}(X)$.

As noted by Sullivan in [26] p. 213, the Baer-Levi semigroups $BL(p,q)$ are $G(X)$-normal and they contain no constants or permutations, hence "they became a test case for Sullivan's Conjecture in the early 1980s" that every automorphism of a $G(X)$-normal transformation semigroup is inner. In fact, Levi, Schein, Sullivan and Wood together showed in [12] that the automorphisms of a Baer-Levi semigroup are inner; and a few years later, Levi proved Sullivan's Conjecture in [10] and [11].

Most of the above work was motivated by earlier results on infinite permutation groups. For example, in [21] Theorem 11.4.6, Scott shows that if $|X| > 3$ and $|X| \neq 6$, then every automorphism of a subgroup of $G(X)$ which contains $\mathrm{Alt}(X)$ is inner. Also, in [21] sections 10.8 and 11.3, he describes all normal subgroups of $G(X)$ for any set X with $|X| > 4$. In particular, such subgroups always contain $\mathrm{Alt}(X)$. Consequently, every automorphism of a normal subgroup N of $G(X)$ is inner if $|X| > 4$ and $|X| \neq 6$; and moreover, in this case, $\mathrm{Aut}(N)$ is isomorphic to $G(X)$ (see [21] Theorem 11.4.8).

Often, these results on automorphisms of transformation semigroups can be converted to results concerning isomorphisms. For example, it is well-known that $T(X)$ is isomorphic to $T(Y)$ if and only if $|X| = |Y|$ (this follows from [4] Vol. 1, Exercise 1.1.7). In fact, each isomorphism $\phi : T(X) \to T(Y)$ is *induced* by a bijection $g : X \to Y$ in the sense that $\alpha\phi = g^{-1}\alpha g$ for every $\alpha \in T(X)$. And it is not difficult to see that this result can be extended to $P(X)$ and $I(X)$, and their ideals. More generally, Mendes-Gonçalves has proven the following extension of Theorem 2.1 above.

Theorem 2.3. *Let S be a subsemigroup of $P(X)$ covering X and S' a subsemigroup of $P(Y)$ covering Y. If $\phi : S \to S'$ is an isomorphism from S onto S', then ϕ is induced by a bijection $g : X \to Y$.*

In [12], the authors proved that every automorphism of a Baer-Levi semigroup is inner. Our next Theorem extends their result to isomorphisms between Baer-Levi semigroups: its proof can be found in [16] Theorem 3.2. For clarity, we write $BL(X,p,q)$ instead of $BL(p,q)$.

Theorem 2.4. *Let X and Y be infinite sets with $|X| = p$ and $|Y| = m$ and let q and n be infinite cardinals such that $q \leq p$ and $n \leq m$. Then, the semigroups $BL(X,p,q)$ and $BL(Y,m,n)$ are isomorphic if and only if $p = m$ and $q = n$. Moreover, for each isomorphism $\theta : BL(X,p,q) \to BL(Y,m,n)$, there is a bijection $h : X \to Y$ such that $\alpha\theta = h^{-1}\alpha h$ for every $\alpha \in BL(X,p,q)$.*

In passing, we recall that in 1967 Magill considered a major generalisation of $T(X)$ and in [15] Theorem 3.1 he described the isomorphisms between such semigroups. Also, Schein in [19] described the homomorphisms between certain semigroups of endomorphisms of various algebraic systems; and, as he observed in [19] p. 31, in general "every homomorphism (excluding some trivial ones) is an isomorphism induced by an isomorphism or an anti-isomorphism" between the underlying sets. For a similar result concerning infinite permutation groups, see [21] Theorem 11.3.7.

3. Isomorphisms between linear transformation semigroups

There are significant results in the theory of transformation semigroups that have corresponding results in the context of linear algebra. For example, in [18] Reynolds and Sullivan showed that Howie's characterisation in 1966 of the elements of $E(X)$, the semigroup generated by the non-identity idempotents of $T(X)$, has an analogue for the linear case.

In [6], Fountain and Lewin found a way to unify these areas: they introduced the concept of a *strong independence algebra* $\mathcal{A}$, of which sets and vector spaces are prime examples; and in [6] and [7] they described the semigroup generated by the non-identity idempotents in the semigroup of endomorphisms of $\mathcal{A}$. Likewise, in [14], Lima extended the work by Howie and Marques-Smith in 1984 on the semigroup generated by all nilpotents of $P(X)$ of index 2 to strong independence algebras, and thus also to vector spaces.

To give a brief account of some results on isomorphisms between semigroups of linear transformations, we introduce some concepts. Let V and W be vector spaces over fields F and K, respectively. A *semilinear transformation* from V to W is a bijection $g : V \to W$ for which there is an isomorphism $\omega : F \to K$ such that, for every $u, v \in V$ and $k \in F$,

$$(u+v)g = ug + vg, \qquad (kv)g = (k\omega)(vg).$$

Let S and S' be subsemigroups of $P(V)$ and $P(W)$, respectively. As for the set case, we say that an isomorphism θ from S onto S' is *induced* by a semilinear transformation $g : V \to W$ if $\alpha\theta = g^{-1}\alpha g$ for every $\alpha \in S$. According to Baer [3] chapter 6, p. 201, several people studied the isomorphisms and automorphisms of certain groups of linear transformations. To state a major result in this regard, we let $G(V)$ denote the general linear group of V. Then, in [3] chapter 6, p. 229, Structure Theorem, it is shown that if neither F nor K has characteristic 2, and if V and W are vector spaces with dimension at least 3, then $G(V)$ and $G(W)$ are isomorphic if and only if $\dim V = \dim W$ and one of the following holds.

(a) F and K are isomorphic fields.

(b) $\dim V$ is finite and F and K are anti-isomorphic fields.

Moreover, a complete description of the isomorphisms from $G(V)$ onto $G(W)$ is given in [3] chapter 6, p. 231, Isomorphism Theorem.

Given a vector space V over a field F, we let $K(V)$ denote the subsemigroup of $T(V)$ which consists of zero and all linear transformations of V into itself with rank one. In [8] section 2, Gluskin studied the structure of $K(V)$ and some of its subsemigroups: in particular, he considered certain right ideals of $K(V)$ which we denote by $K^*(V)$ and of which $K(V)$ is a particular case (see [8] 2.7, pp. 113-114). In [8] Theorem 3.5, he proved the following.

Theorem 3.1. *Let V and W be vector spaces over the fields F and K, respectively, with $\dim V \geq 2$. If S and S' are subsemigroups of $T(V)$ and $T(W)$ containing $K^*(V)$ and $K^*(W)$, respectively, as two-sided ideals, then every isomorphism ϕ of S onto S' is induced by a semilinear transformation of V onto W. Moreover, $K^*(V)\phi = K^*(W)$.*

Subsequently, a weaker version of this result was proved in [20]. In fact, [20] Theorem 6.2 is an analogue of Theorem 2.3 for vector spaces. Recall our remark before Theorem 2.3 and note that, from Theorem 3.1, it follows that $T(V)$ and $T(W)$ are isomorphic if and only if there is a semilinear transformation from V onto W. Likewise, using Theorem 3.1, in [2] Araújo and Silva proved a linear version of Theorem 2.2 above. By analogy with the set case, we say that a subsemigroup S of $P(V)$ is *$G(V)$-normal* if $g^{-1}\alpha g \in S$ for every $\alpha \in S$ and every $g \in G(V)$.

Theorem 3.2. *Let V be a finite dimensional vector space over a field F and let S be a subsemigroup of $T(V) \setminus G(V)$. Then the following are equivalent.*

(a) S is $G(V)$-normal;

(b) S is an ideal of $T(V)$;

(c) every automorphism of S is induced by a semilinear transformation.

These results for semigroups of transformations and for semigroups of linear transformations are clearly connected. But, as observed by Sullivan in [25] p. 291, there are some 'puzzling' differences. For example, if X is an infinite set then certain total transformations of X can be written as a product of four idempotents in $T(X)$, and '4' is best possible. On the other hand, Reynolds and Sullivan showed in [18] that, if V is infinite-dimensional, then similarly-defined total linear transformations of V can be written as a product of three idempotents, and '3' is best possible. Consequently, in these cases, $E(X)$ and its linear analogue $E(V)$ can never be isomorphic.

As stated before, in [5] Fitzpatrick and Symons showed that, if X is infinite, then all automorphisms of a subsemigroup S of $T(X)$ which contains $G(X)$ are inner (note that if $G(X) \subseteq S$ then S is $G(X)$-normal). In fact, they first proved a weaker result: namely, that every automorphism of $G(X)$ can be extended to at most one automorphism of S (for contrast, we note that an automorphism of a normal subgroup of a group G is not necessarily 'extendible' to an automorphism of G: for example, see [21] exercise 9.2.28).

A simple analogue of their result for vector spaces would be as follows: if V is an infinite-dimensional vector space over a field F and S is a subsemigroup of $T(V)$ which contains $G(V)$, then every automorphism of $G(V)$ can be extended to at most one automorphism of S. However, as observed by Araújo in [1] pp. 57-58 this result does not hold: in [1] Lemmas 21 and 39, Araújo gives two simple examples of subsemigroups A_1 and A_2 of $T(V)$ which contain $G(V)$, but where the identity automorphism of $G(V)$ can be extended in infinitely many ways to automorphisms of A_1 and A_2, respectively. In order to produce a result close to the linear version of Fitzpatrick and Symons' result, Araújo in [1] Theorem 12 assumes that S is a subsemigroup of $P(V)$ containing $G(V) \cup E$, where E is the subset of $P(V)$ consisting of all identity transformations on one-dimensional subspaces of V, and he shows that every automorphism of $G(V)$ can be extended to at most one automorphism of S. In fact, [1] Theorem 12 is more general: the result is proved for strong independence algebras with at most one constant and rank at least 3.

4. Isomorphisms between Baer-Levi semigroups

In the remainder of this paper, we examine a semigroup related to the Baer-Levi semigroup $BL(p, q)$, which we define as follows. Let V be a vector space over a field F and suppose $\dim V = m \geq \aleph_0$. If $\alpha \in T(V)$, we write $\ker\alpha$ for the *kernel* of α, and put

$$n(\alpha) = \dim\ker\alpha,\ \ r(\alpha) = \dim\operatorname{ran}\alpha,\ \ d(\alpha) = \operatorname{codim}\operatorname{ran}\alpha.$$

As usual, these are called the *nullity, rank* and *defect* of α, respectively. For each cardinal n such that $\aleph_0 \leq n \leq m$, we write

$$GS(m,n) = \{\alpha \in T(V) : n(\alpha) = 0,\ d(\alpha) = n\}$$

and call this the *linear Baer-Levi semigroup* on V. It can be shown that this is indeed a semigroup with the same properties as $BL(p,q)$: that is, $GS(m,n)$ is a right cancellative, right simple semigroup without idempotents. This fact extends work by Lima [14] Proposition 4.1 on $GS(m,m)$. More importantly, however, these two types of Baer-Levi semigroup – one defined on sets, the other on vector spaces – are never isomorphic. Hence, it is natural to look for properties of $GS(m,n)$ which mimic those of $BL(p,q)$. Next, we consider two of these: namely, the left ideals of $GS(m,n)$ and some of its maximal subsemigroups (for details, see [16] sections 4 and 5).

First we transfer results of Šutov [27] and Sullivan [23] on the left ideals of $BL(p,q)$ to the linear Baer-Levi semigroup on V. By analogy with their work, the most natural way to do this is to show that the left ideals of $GS(m,n)$ are the subsets L of $GS(m,n)$ which satisfy the condition:

$$\big[\alpha \in L,\ \beta \in GS(m,n),\ \operatorname{ran}\beta \subseteq \operatorname{ran}\alpha,\\ \dim(\operatorname{ran}\alpha/\operatorname{ran}\beta) = n\big] \quad \text{implies} \quad \beta \in L.$$

Although this result is valid, to obtain more information about the left ideals of $GS(m,n)$ we proceed as follows.

If Y is a non-empty subset of $GS(m,n)$, we let $L_Y^+ = Y \cup L_Y$, where

$$L_Y = \{\beta \in GS(m,n) : \operatorname{ran}\beta \subseteq \operatorname{ran}\alpha,\\ \dim(\operatorname{ran}\alpha/\operatorname{ran}\beta) = n \ \text{ for some } \ \alpha \in Y\}.$$

To show L_Y is non-empty, we need some notation: namely, we write $\{e_i\}$ for a linearly independent subset of V, and take this to mean that the subscript i belongs to some (unmentioned) index set I. Let $\alpha \in Y$, suppose $\{e_i\}$ is a basis for V with $|I| = m$, and write $e_i\alpha = a_i$ for each $i \in I$. Since α is one-to-one, $\{a_i\}$ is linearly independent and so it can be expanded to a basis $\{a_i\} \cup \{b_j\}$ for V. Note that $|J| = d(\alpha) = n \leq m$, therefore we can write $\{a_i\} = \{c_i\} \cup \{d_j\}$. Now let $e_i\beta = c_i$ for every i and extend this by linearity to the whole of V. Clearly, β is in $GS(m,n)$ since it is one-to-one and $d(\beta) = \dim\langle d_j, b_j\rangle = n$. We have $\operatorname{ran}\beta \subseteq \operatorname{ran}\alpha$ and $\dim(\operatorname{ran}\alpha/\operatorname{ran}\beta) = \dim\langle d_j\rangle = n$. Hence $\beta \in L_Y$ and so L_Y is non-empty.

Theorem 4.1. *If Y is a non-empty subset of $GS(m,n)$, then L_Y^+ is a left ideal of $GS(m,n)$. Conversely, if I is a left ideal of $GS(m,n)$, then $I = L_I^+$.*

We can say a lot about the poset under $\subseteq$ of the (proper) left ideals in $GS(m,n)$. For example, it does not form a chain, and it has no minimal or maximal elements. In addition, it is not difficult to show that the principal left ideal generated by $\alpha \in GS(m,n)$ is $L^+_{\{\alpha\}}$.

In [13] Theorem 1, Levi and Wood described some maximal subsemigroups of $BL(p,q)$. To do this, they chose a non-empty subset A of X such that $|X \setminus A| \geq q$ and proved that the set

$$\begin{aligned} M_A = \{\alpha \in BL(p,q) : A \not\subseteq \operatorname{ran}\alpha \ \text{ or} \\ (A\alpha \subseteq A \ \text{ or } \ |\operatorname{ran}\alpha \setminus A| < q)\} \end{aligned}$$

is a maximal subsemigroup of $BL(p,q)$. By analogy with this, we let U be a non-zero subspace of V with $\operatorname{codim} U \geq n$ and define

$$\begin{aligned} M_U = \{\alpha \in GS(m,n) : U \not\subseteq \operatorname{ran}\alpha \ \text{ or} \\ (U\alpha \subseteq U \ \text{ or } \ \dim(\operatorname{ran}\alpha/U) < n)\}. \end{aligned}$$

Our next result determines some maximal subsemigroups of $GS(m,n)$.

Theorem 4.2. *For each non-zero subspace U of V with* $\operatorname{codim} U \geq n$, *$M_U$ is a maximal subsemigroup of $GS(m,n)$.*

We note that although the proofs of [13] Theorem 1 and of [16] Theorem 5.1 are similar in outline, they are quite different in detail. In fact, despite the similarity between some results for $P(X)$ and others for $P(V)$, the ideas and techniques in these two areas are quite different.

The main result of [16] can be stated as follows.

Theorem 4.3. *The semigroups $BL(p,q)$ and $GS(m,n)$ are not isomorphic for any infinite cardinals p, q, m, n with $q \leq p$ and $n \leq m$.*

Section 2 illustrates the ongoing interest in automorphisms and isomorphisms for semigroups of transformations of sets; and the corresponding results in section 3 illustrate the development of the work on automorphisms and isomorphisms for semigroups of linear transformations of vector spaces. Clearly, the above result overlaps both.

In passing, we note that in [17] an entirely different method from that used in the proof of Theorem 4.3 is used to show that $I(X)$ and $I(V)$ are almost never isomorphic, and that any inverse semigroup can be embedded in some $I(V)$. Likewise, since the semigroups $BL(p,q)$ and $GS(m,n)$ are never isomorphic, it is worth observing the following result (see [16] Theorem 3.12).

Theorem 4.4. *Any right simple, right cancellative semigroup S without idempotents can be embedded in some $GS(m,m)$.*

Acknowledgments

The author would like to thank Prof. R. P. Sullivan for the time he spent in writing an outline of this survey and in reading and commenting on its first draft and consequent versions. She would also like to express her gratitude for his ideas to emphasize the similarity and/or contrast between the set and linear versions of results and for his wise suggestions to improve the readability and the appearance of the paper.

The author acknowledges the support of the Portuguese Foundation for Science and Technology (FCT) through the research program POCTI.

References

1. J. Araújo, *Aspects of the endomorphism monoids of independence algebras*, PhD thesis, University of York, England, UK (2000).
2. J. Araújo and F. C. Silva, Semigroups of linear endomorphisms closed under conjugation, *Comm. Algebra* **28 (8)**, 3679-3689 (2000).
3. R. Baer, *Linear Algebra and Projective Geometry*, Academic Press, New York (1952).
4. A. H. Clifford and G. B. Preston, *The Algebraic Theory of Semigroups*, Mathematical Surveys, No. 7, vols. 1 and 2, American Mathematical Society, Providence, RI (1961 and 1967).
5. S. P. Fitzpatrick and J. S. V. Symons, Automorphisms of transformation semigroups, *Proc. Edinburgh Math. Soc.* **19A (4)**, 327-329 (1974/75).
6. J. Fountain and A. Lewin, Products of idempotent endomorphisms of an independence algebra of finite rank, *Proc. Edinburgh Math. Soc.* **35 (3)**, 493-500 (1992).
7. J. Fountain and A. Lewin, Products of idempotent endomorphisms of an independence algebra of infinite rank, *Math. Proc. Cambridge Philos. Soc.* **114 (2)**, 303-319 (1993).
8. L. M. Gluskin, Semigroups and rings of endomorphisms of linear spaces (Russian) *Izv. Akad. Nauk. SSSR* **23** (1959) 841-870; English translation in *Amer. Math. Soc. Translations* **45**, 105-137 (1965).

9. N. Iwahori and H. Nagao, On the automorphism group of the full transformation semigroups, *Proc. Japan Acad.* **48 (9)**, 639-640 (1972).
10. I. Levi, Automorphisms of normal transformation semigroups, *Proc. Edinburgh Math. Soc.* **28A (2)**, 185-205 (1985).
11. I. Levi, Automorphisms of normal partial tansformation semigroups, *Glasgow Math. J.* **29 (2)**, 149-157 (1987).
12. I. Levi, B. M. Schein, R. P. Sullivan and G. R. Wood, Automorphisms of Baer-Levi semigroups, *J. London Math. Soc.* **28 (3)**, 492-495 (1983).
13. I. Levi and G. R. Wood, On maximal subsemigroups of Baer-Levi semigroups, *Semigroup Forum* **30 (1)**, 99-102 (1984).
14. L. M. Lima, Nilpotent local automorphisms of an independence algebra, *Proc. Royal Soc. Edinburgh* **124A (3)**, 423-436 (1994).
15. K. D. Magill, Jr., Semigroup structures for families of functions, I: some homomorphism theorems, *J. Austral. Math. Soc.* **7**, 81-94 (1967).
16. S. Mendes-Gonçalves and R. P. Sullivan, Baer-Levi semigroups of linear transformations, *Proc. Royal Soc. Edinburgh* **134A**, 477-499 (2004).
17. S. Mendes-Gonçalves and R. P. Sullivan, Maximal inverse subsemigroups of the symmetric inverse semigroup on a finite-dimensional vector space, *Comm. Algebra* (to appear).
18. M. A. Reynolds and R. P. Sullivan, Products of idempotent linear transformations, *Proc. Royal Soc. Edinburgh*, **100A**, 123-138 (1985).
19. B. M. Schein, Ordered sets semilattices, distributive lattices and Boolean algebras with homomorphic endomorphism semigroups, *Funda. Math.* **68**, 31-50 (1970).
20. M. Schwachhöfer and M. Stroppel, Isomorphisms of linear semigroups, *Geom. Dedicata* **65 (3)**, 355-366 (1997).
21. W. R. Scott, *Group Theory*, Prentice Hall, Englewood Cliffs, New Jersey (1964).
22. R. P. Sullivan, Automorphisms of transformation semigroups, *J. Austral. Math. Soc.* **20 (1)**, 77-84 (1975).
23. R. P. Sullivan, Ideals in transformation semigroups, *Commen. Math. Univ. Carolinae* **19 (3)**, 431-446 (1978).
24. R. P. Sullivan, Automorphisms of injective transformation semigroups, *Studia Sci. Math. Hungar.* **15 (1-3)**, 1-4 (1980).
25. R. P. Sullivan, Transformation semigroups and linear algebra, pp 290-295 in *Monash Conference on Semigroup Theory (Melbourne, 1990)*, ed T. E. Hall, P. R. Jones and J. C. Meakin, World Scientific, Singapore (1991).
26. R. P. Sullivan, *Transformation semigroups: past, present and future*, pp 191-243 in *Proceedings of the International Conference on Semigroups (University of Minho, Portugal, June 1999)*, ed P. Smith, P. Mendes and E. Giraldes, World Scientific, Singapore (2000).
27. E. G. Šutov, On a certain semigroup of one-to-one transformations (Russian), *Uspehi Mat. Nauk.* **18** (1963), 231-235; English translation by Boris Schein in *Amer. Math. Soc. Translations* **139**, 191-196 (1988).
28. J. S. V. Symons, Normal transformation semigroups, *J. Austral. Math. Soc.* **22A (4)**, 385-390 (1976).

ON MCALISTER'S MONOID AND ITS CONTRACTED ALGEBRA

W. D. MUNN

Department of Mathematics, University of Glasgow,
University Gardens, Glasgow G12 8QW, U.K.
E-mail: wdm@maths.gla.ac.uk

The first part of this paper is devoted to an account of a certain inverse monoid with zero that is a Rees quotient of a free inverse monoid; the second part describes some recent results on its contracted semigroup algebra over a field.

1. The McAlister monoid M_X

Let X be a set with at least two elements and let S denote the free inverse monoid on X. We regard X as a subset of S; thus $\{x^{-1} : x \in X\}$ is a subset of S disjoint from X. By *the McAlister monoid* M_X *on* X we mean the Rees quotient S/SAS, where A is the nonempty subset of S defined by

$$A := \{x^{-1}y;\ x, y \in X, x \neq y\} \cup \{xy^{-1} :\ x, y \in X, x \neq y\}.$$

This was first studied by Lawson [3, Section 9.4], who gave it the title above because it is closely related to a monoid that appeared earlier in [4]. Since M_X is a Rees quotient of a free inverse monoid ($\equiv$ free inverse semigroup with identity adjoined), it inherits many of the properties of such monoids. In particular, it is a combinatorial completely semisimple inverse monoid; further, it is 0-E-unitary, since S is E-unitary.

We begin by giving a description of M_X which differs somewhat from that in [3]. For this purpose we introduce the following notation:

- G_X denotes the free group on X, with identity 1 (the empty word);
- P $(= P_X)$ denotes the submonoid of G_X consisting of all reduced words in X with no negative exponents (and so $P = X^*$, the free monoid on X);
- $(\forall a, b \in P)$ $\quad a$ *pref* b means 'a is a prefix of b';
- $(\forall a, b \in P)$ $\quad a$ *suff* b means 'a is a suffix of b';
- $P^{-1} := \{g^{-1} \in G_X :\ g \in P\}$.

Observe that $P \cap P^{-1} = \{1\}$. Note further that, for all $a, b \in P$,

$$a^{-1}b \in P \cup P^{-1} \quad \Leftrightarrow \quad (a \textit{ pref } b) \text{ or } (b \textit{ pref } a)$$

and

$$ab^{-1} \in P \cup P^{-1} \quad \Leftrightarrow \quad (a \textit{ suff } b) \text{ or } (b \textit{ suff } a).$$

The theorem below describes M_X in terms of P and the multiplication in G_X.

Theorem 1.1. *Let X be a set with at least two elements. With the notation above, define a set M'_X by*

$$M'_X := \{(a, r, b) \in P \times (P \cup P^{-1}) \times P : \; ar, rb \in P\} \cup \{0\}$$

and define a multiplication in M'_X by the following rules:

(i) $(a, r, b)(c, s, d) = (p, rs, q)$ *if* $arc^{-1}, b^{-1}sd \in P \cup P^{-1}$, *where*

$$p := \left\{ \begin{array}{cl} cr^{-1} & \textit{if } ar \textit{ suff } c \\ a & \textit{if } c \textit{ suff } ar \end{array} \right\} \quad \textit{and} \quad q := \left\{ \begin{array}{cl} s^{-1}b & \textit{if } sd \textit{ pref } b \\ d & \textit{if } b \textit{ pref } sd \end{array} \right\};$$

(ii) all other products $= 0$.

Then $M'_X \cong M_X$.

Proof. In [5], the elements of the free inverse semigroup on X are represented by birooted word trees with at least two vertices. If we adjoin the tree with a single vertex and no edges to represent the identity, we have a description of the free inverse monoid S on X. As before, let A denote the subset

$$\{x^{-1}y : \; x, y \in X, x \neq y\} \cup \{xy^{-1} : \; x, y \in X, x \neq y\}$$

of S. Clearly, the elements of $S \backslash SAS$ are characterised by the fact that their birooted word trees contain no vertices of the form $\xleftarrow{\;x\;}\circ\xrightarrow{\;y\;}$ or $\xrightarrow{\;x\;}\circ\xleftarrow{\;y\;}$ $(x, y \in X, x \neq y)$: for otherwise there would exist a spanning walk labelled by a word containing a syllable $x^{-1}y, xy^{-1}, y^{-1}x$ or yx^{-1}. In particular, no vertex corresponds to a fork in the tree. Thus, a typical nonzero element of M_X is uniquely represented by a birooted word tree of the form

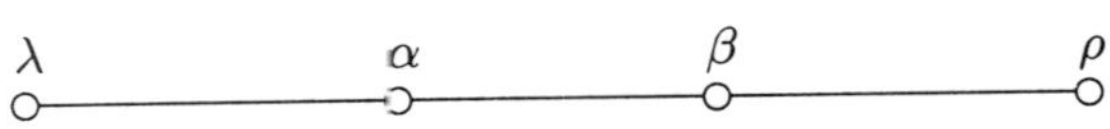

,

where all the arrows are from left to right. Conversely, every such tree represents a nonzero element of M_X. Such a tree is completely described by a triple (a, r, b), where a is the label on the path $\lambda \to \alpha$, r is the label on the path $\alpha \to \beta$ and b is the label on the path $\beta \to \rho$. Note that $a, b \in P$; also, for $r \neq 1$, $r \in P$ or $r \in P^{-1}$ according as α is to the left or to the right of β. Note, too, that $ar \in P$ and $rb \in P$. Conversely, every triple $(a, r, b) \in P \times (P \cup P^{-1}) \times P$ with $ar \in P$ and $rb \in P$ corresponds to a unique tree of the form described above.

The product of two nonzero elements of M_X is again nonzero if and only if the 'product' of the corresponding trees is another tree of the same type. There are essentially four cases, as illustrated below. In each case, to obtain a tree of the same type from those corresponding to the triples (a, r, b) and (c, s, d), it is necessary and sufficient that arc^{-1}, $b^{-1}sd \in P \cup P^{-1}$. The triples corresponding to the product $(a, r, b)(c, s, d)$ can then be read off the diagrams.

(i)

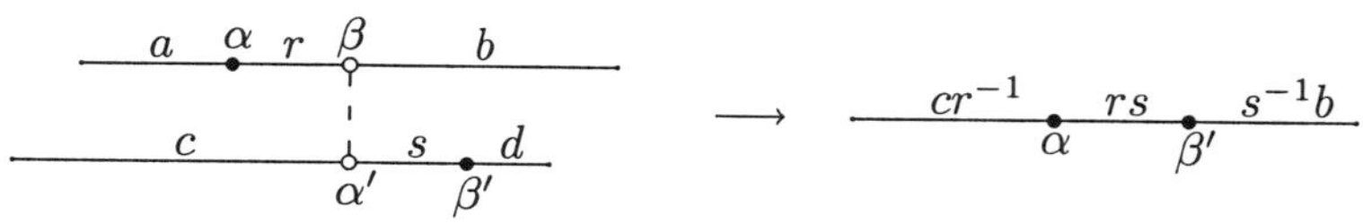

ar suff c, sd pref b

(ii)

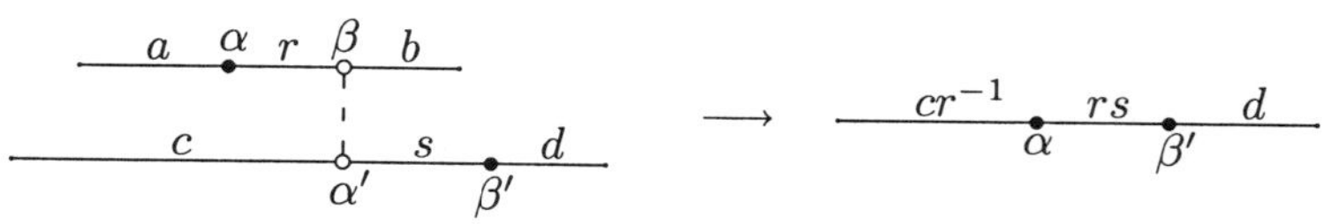

ar suff c, b pref sd

(iii)

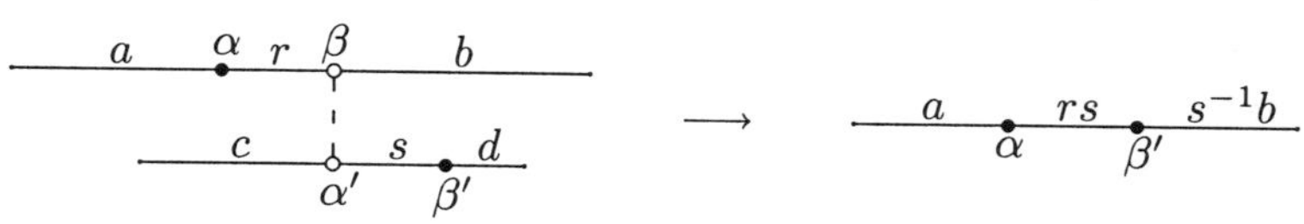

c suff ar, sd pref b

(iv) 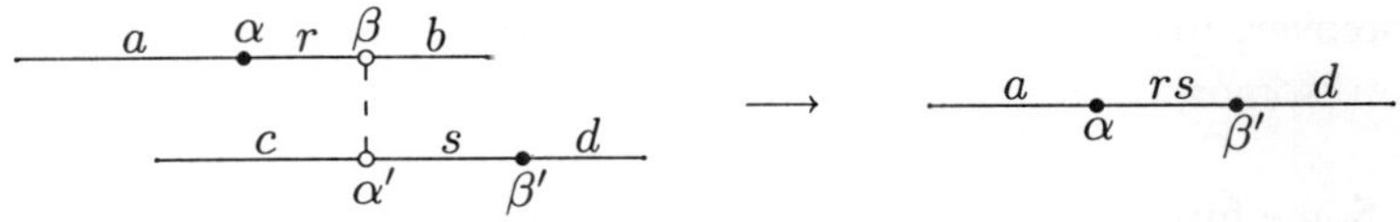

c *suff* ar, b *pref* sd

The stated multiplication in M'_X reflects this situation exactly. Hence $M'_X \cong M_X$. $\square$

Despite the apparent complexity of its multiplication rule, it is easy to establish properties of M'_X. Accordingly, for the remainder of the discussion, we take $M_X = M'_X$.

Denote the semilattice of M_X by E. For $e, f \in E$ we write $e \succ f$ ('e covers f') to mean that $e > f$ and that there is no $g \in E$ for which $e > g > f$. The main properties of E are listed below.

Corollary 1.1.

(i) $E = \{(a, 1, b) : a, b \in P\} \cup \{0\}$;

(ii) for all $a, b, c, d \in P$

$$(a, 1, b) \geq (c, 1, d) \quad \Leftrightarrow \quad (a \textit{ suff } c) \textit{ and } (b \textit{ pref } d);$$

(iii) for all $a, b, c, d \in P$

$$(a, 1, b) \succ (c, 1, d) \Leftrightarrow \left\{ \begin{array}{l} \textit{either } c = xa \textit{ for some } x \in X \textit{ and } d = b \\ \quad \textit{or} \quad\; d = bx \textit{ for some } x \in X \textit{ and } c = a \end{array} \right\};$$

(iv) E *is* 0*-uniform.*

Proof. Since birooted word trees represent idempotents if and only if the roots coincide, we obtain (i). Then (ii) is an immediate consequence of the multiplication rule (see case (ii) in the proof of the theorem). For (iii), we note that $(a, 1, b) \succ (c, 1, d)$ if and only if the tree for $(c, 1, d)$ can be obtained from that for $(a, 1, b)$ by adjoining one extra edge at either end.

For (iv), we have to show that any two nonzero principal ideals of E are isomorphic. Consider typical elements $(a, 1, b)$, $(c, 1, d)$ in $E \backslash \{0\}$. In view of (ii), and the fact that P is cancellative, we can define a bijection $\theta : E(a, 1, b) \to E(c, 1, d)$ by the rule that, for all $p, q \in P$,

$$(pa\; 1, bq)\theta = (pc, 1, dq).$$

Moreover, from (ii), both θ and θ^{-1} are strictly order-preserving. Thus θ is an isomorphism. □

Some further properties of M_X, including a description of Green's relations, are included in the next corollary.

Corollary 1.2.

(i) For all $(a, r, b) \in M_X \backslash \{0\}$

$$(a, r, b)^{-1} = (ar, r^{-1}, rb).$$

(ii) For all $(a, r, b), (c, s, d) \in M_X \backslash \{0\}$

$$\begin{aligned} (a, r, b)\mathcal{R}(c, s, d) &\Leftrightarrow a = c \text{ and } rb = sd, \\ (a, r, b)\mathcal{L}(c, s, d) &\Leftrightarrow b = d \text{ and } ar = cs, \\ (a, r, b)\mathcal{D}(c, s, d) &\Leftrightarrow arb = csd, \end{aligned}$$

where, in each case, the products on the right-hand side are computed in G_X. *Moreover,* $\mathcal{H}$ *is the identity relation on* M_X *and* $\mathcal{J} = \mathcal{D}$.

(iii) The mapping $\theta : P \to M_X$ *defined by* $a\theta = (1, a, 1)$ *is an injective homomorphism and im* θ *is a transversal of the nonzero* $\mathcal{D}$*-classes of* M_X.

Proof. For (i), we note that the birooted word tree corresponding to $(a, r, b)^{-1}$ is simply obtained from that for (a, r, b) by interchanging the two roots. Note, further, that

$$(a, r, b)(a, r, b)^{-1} = (a, 1, rb), \qquad (a, r, b)^{-1}(a, r, b) = (ar, 1, b).$$

From the first of these,

$$\begin{aligned} (a, r, b)\mathcal{R}(c, s, d) &\Leftrightarrow (a, 1, rb) = (c, 1, sd) \\ &\Leftrightarrow a = c \text{ and } rb = sd. \end{aligned}$$

Similarly, $(a, r, b)\mathcal{L}(c, s, d)$ if and only if $b = d$ and $ar = cs$. The stated characterisation of $\mathcal{D}$ follows from those of $\mathcal{R}$ and $\mathcal{L}$ — or, more directly, from the observation that two elements of a free inverse monoid are $\mathcal{D}$-equivalent if and only if their corresponding trees are identical when the roots are ignored. Since, as already noted, M_X is combinatorial and completely semisimple, we have that $\mathcal{H}$ is the identity relation and $\mathcal{J} = \mathcal{D}$ on M_X. Thus we have established (ii).

Finally, for (iii), we note that θ is clearly injective; moreover, it is easily checked that, for all $a, b \in P$, $(1, a, 1)(1, b, 1) = (1, ab, 1)$. That im θ is a transversal of the nonzero $\mathcal{D}$-classes follows from the characterisation of $\mathcal{D}$ in (ii). $\square$

We next discuss an embedding theorem for M_X. First, we recall that *the polycyclic monoid on* X is the monoid $Q_X := (P \times P) \cup \{0\}$ with multiplication defined by

$$(a,b)(c,d) = \begin{cases} (ap, d) & \text{if } c = bp \text{ for some } p \in P, \\ (c, dq) & \text{if } b = cq \text{ for some } q \in P, \\ 0 & \text{otherwise}, \end{cases}$$

$$0(a,b) = (a,b)0 = 0^2 = 0$$

[6; 3, §9.3]. This is a combinatorial 0-bisimple inverse monoid with semilattice $\{(a,a) : a \in P\} \cup \{0\}$ and such that, for all $(a,b) \in P \times P$, $(a,b)^{-1} = (b,a)$.

Now let Z denote the ideal $(Q_X \times \{0\}) \cup (\{0\} \times Q_X)$ of the direct square $Q_X \times Q_X$ and let R_X denote the Rees quotient $(Q_X \times Q_X)/Z$. With this notation we have the following theorem, which is another version of [3, Section 9.4, Theorem 16].

Theorem 1.2. *Let* X *be a set with at least two elements, let* R_X *be as above and let* * *denote word-reversal in the free monoid* P *on* X. *Then*

(i) R_X *is a combinatorial 0-bisimple inverse monoid;*

(ii) the mapping $\theta : M_X \to R_X$ *defined by*

$$(a,r,b)\theta = ((a^*, (ar)^*),\ (rb, b)), \quad 0\theta = 0$$

is an injective homomorphism and im θ *is a full inverse subsemigroup of* R_X.

Proof.

(i) It is a routine matter to verify that R_X is an inverse monoid, with semilattice $\{((a,a),\ (b,b)) : a, b \in P\} \cup \{0\}$. Let $a, b, c, d \in P$ and let $u := ((a,c),\ (b,d))$ Then

$$((a,a),\ (b,b)) = uu^{-1}, \quad u^{-1}u = ((c,c),\ (d,d))$$

and so R_X is 0-bisimple; further, if u and u^{-1} commute then $a = c$ and $b = d$, which shows that R_X is combinatorial.

(ii) Evidently θ is injective. Let $(a,r,b),\ (c,s,d) \in M_X\backslash\{0\}$. Then, with

$$p := \left\{ \begin{matrix} cr^{-1} & \text{if} & ar\ \mathit{suff}\ c \\ a & \text{if} & c\ \mathit{suff}\ ar \end{matrix} \right\} \text{ and } q := \left\{ \begin{matrix} s^{-1}b & \text{if} & sd\ \mathit{pref}\ b \\ d & \text{if} & b\ \mathit{pref}\, sd \end{matrix} \right\},$$

we have that

$$\begin{aligned}
&[(a,r,b)(c,s,d)]\theta = \\
&= \begin{cases} (p,rs,q)\theta & \text{if } arc^{-1},\ b^{-1}sd \in P \cup P^{-1}, \\ 0 & \text{otherwise}, \end{cases} \\
&= \begin{cases} ((p^*,(prs)^*),\ (rsq,\ q)) & \text{if } arc^{-1},\ b^{-1}sd \in P \cup P^{-1}, \\ 0 & \text{otherwise}, \end{cases} \\
&= \begin{cases} (((cr^{-1})^*,(cs)^*),\ (rb,s^{-1}b)) & \text{if } ar\ \mathit{suff}\, c \text{ and } sd\ \mathit{pref}\, b, \\ (((cr^{-1})^*,(cs)^*),\ (rsd,d)) & \text{if } ar\ \mathit{suff}\, c \text{ and } b\ \mathit{pref}\, sd, \\ ((a^*,(ars)^*),\ (rb,s^{-1}b)) & \text{if } c\ \mathit{suff}\, ar \text{ and } sd\ \mathit{pref}\, b, \\ ((a^*,(ars)^*),\ (rsd,d)) & \text{if } c\ \mathit{suff}\, ar \text{ and } b\ \mathit{pref}\, sd, \\ 0 & \text{otherwise}. \end{cases}
\end{aligned} \quad (1)$$

Now

$$\begin{aligned}
&(a^*,(ar)^*)(c^*,(cs)^*) = \\
&= \begin{cases} (a^*u,(cs)^*) & \text{if } c^* = (ar)^*u \text{ for some } u \in P, \\ (a^*,(cs)^*v) & \text{if } (ar)^* = c^*v \text{ for some } v \in P, \\ 0 & \text{otherwise}, \end{cases} \\
&= \begin{cases} ((cr^{-1})^*,(cs)^*) & \text{if } ar\ \mathit{suff}\, c, \\ (a^*,(ars)^*) & \text{if } c\ \mathit{suff}\, ar, \\ 0 & \text{otherwise} \end{cases}
\end{aligned} \quad (2)$$

and

$$\begin{aligned}
&(rb,b)(sd,d) = \\
&= \begin{cases} (rb,du) & \text{if } b = sdu \text{ for some } u \in P, \\ (rbv,d) & \text{if } sd = bv \text{ for some } v \in P, \\ 0 & \text{otherwise}, \end{cases} \\
&= \begin{cases} (rb,s^{-1}b) & \text{if } sd\ \mathit{pref}\, b, \\ (rsd,d) & \text{if } b\ \mathit{pref}\, sd, \\ 0 & \text{otherwise}. \end{cases}
\end{aligned} \quad (3)$$

Hence

$$\begin{aligned}
(a,r,b)\theta\ (c,s,d)\theta &= ((a^*,(ar)^*)(c^*,(cs)^*),\ (rb,b)(sd,d)) \\
&= [(a,r,b)(c,s,d)]\theta,
\end{aligned}$$

from (2),(3) and (1). Thus θ is a homomorphism. Consequently, im θ is an inverse subsemigroup of R_X; and it is full, since for all $a, b \in P$,

$$((a, a),\ (b, b)) = (a^*, 1, b)\theta. \qquad \square$$

Remark 1.1. As in [3, §5.2], denote the monoid consisting of all isomorphisms between principal ideals of E under composition of relations by T_E. This is an inverse monoid with semilattice isomorphic to E. By Corollary 1.1(iv), E is 0-uniform and so T_E is 0-bisimple. Further, by [3, §5.2, Theorem 9], since M_X is combinatorial, it can be embedded in T_E as a full inverse subsemigroup. However T_E is *not* combinatorial; for the automorphism group of E is nontrivial.

2. The contracted monoid algebra of M_X

In this section we discuss briefly some recent results on an algebra associated with M_X. First, we recall some definitions concerning an algebra A with a unity.

(i) A is *prime* if and only if, for all $a, b \in A$, $aAb = 0$ implies $a = 0$ or $b = 0$;

(ii) A is *primitive* if and only if A has a proper left ideal L such that $L + I = A$ for every nonzero two-sided ideal I of A.

It is well known that if A is primitive then it is prime, but not conversely.

For a nontrivial monoid S with zero z, we define *the contracted monoid algebra* $F_0[S]$ *of* S *over a field* F to be the quotient algebra $F[S]/Z$, where $F[S]$ is the monoid algebra of S over F and Z is the ideal Fz of $F[S]$. Thus $F_0[S]$ can be described as the algebra obtained from $F[S]$ by identifying z with the zero of $F[S]$.

The results below, which are established in [2], concern primeness and primitivity, respectively.

Theorem 2.1. *Let X be a set with at least two elements and let F be a field. Then $F_0[M_X]$ is prime if and only if X is infinite.*

Theorem 2.2. *Let X be a countably infinite set and let F be a field. Then $F_0[M_X]$ is primitive.*

To illustrate some of the techniques involved, we sketch a proof of Theorem 2.2. First, we state a lemma.

Lemma 2.1. *Let X be an infinite set, let F be a field and let I be a nonzero two-sided ideal of $F_0[M_X]$. Then there exists $e \in (E\backslash\{0\}) \cap I$.*

With this in mind, we proceed as follows. Since X is countably infinite, so also is P. Denote the elements of P by $a_1, a_2, \ldots, a_n, \ldots$ and, for each positive integer n, define $f_n \in M_X$ by

$$f_n := (1, 1, a_1 a_2 \cdots a_n).$$

Then, by Corollary 1.1(i) and (ii), $f_n \in E\backslash\{0\}$ and $f_1 > f_2 > \cdots > f_n > \cdots$.

Let L denote the left ideal of $F_0[M_X]$ generated by the elements $1 - f_n$ $(n = 1, 2, \ldots)$. Suppose that L is not proper. Then for some positive integer n there exist $u_1, u_2, \ldots, u_n \in F_0[M_X]$ such that

$$1 = u_1(1 - f_1) + u_2(1 - f_2) + \cdots + u_n(1 - f_n).$$

Now $(1 - f_i)f_n = 0$ $(i = 1, 2, \ldots, n)$. Hence, multiplying both sides above on the right by f_n, we get $f_n = 0$ — which is false. Thus L is proper.

Let I be a nonzero two-sided ideal of $F_0[M_X]$. By the lemma, there exists $e \in (E\backslash\{0\}) \cap I$. Now, by Corollary 1.1(i), $e = (b, 1, c)$ for some $b, c \in P$; and $bc = a_k$ for some positive integer k. A simple calculation gives

$$(1, b, c)e(b, c, 1) = (1, bc, 1) = (1, a_k, 1).$$

Write $a := a_1 a_2 \cdots a_k$. Then

$$\begin{aligned}(1, aa_k^{-1}, 1)(1, b, c)e(b, c, 1)(a, a^{-1}, a) &= (1, aa_k^{-1}, 1)(1, a_k, 1)(a, a^{-1}, a)\\ &= (1, a, 1)(a, a^{-1}, a)\\ &= (1, 1, a) = f_k\end{aligned}$$

and so $f_k \in I$. Then

$$1 = (1 - f_k) + f_k \in L + I,$$

which shows that $L + I = F_0[M_X]$. Hence $F_0[M_X]$ is primitive.

In [1], it was shown that, for the free inverse monoid S on X and a field F, $F[S]$ is primitive if and only if X is infinite. It is natural to conjecture that an analogous result holds for $F_0[M_X]$; but at present it is not known whether $F_0[M_X]$ is primitive for an uncountably infinite set X.

Acknowledgments

I wish to record my thanks to Dr Michael Crabb for making many helpful comments on the manuscript and to Dr Colin McGregor for providing the diagrams.

References

1. M.J. Crabb and W. D. Munn, On the algebra of a free inverse monoid, *J. Algebra* **184**, 297–303 (1996).
2. M.J. Crabb and W. D. Munn, Contracted semigroup algebras of McAlister monoids, University of Glasgow, Department of Mathematics Preprint.
3. M.V. Lawson, Inverse semigroups, Singapore, World Scientific (1999).
4. D.B. McAlister, Inverse semigroups separated over a subsemigroup, *Trans. Amer. Math. Soc.* **182**, 85–117 (1973).
5. W.D. Munn, Free inverse semigroups, *Proc. London Math. Soc.* (3) **29**, 385–404 (1974).
6. M. Nivat and J.-F. Perrot, Une généralisation du monoïde bicyclique, *Comptes Rendus de l'Acad. Sci. de Paris* **271**, 824–827 (1970).

RELATIVE MONOID PRESENTATIONS AND FINITE DERIVATION TYPE*

FRIEDRICH OTTO
Fachbereich Mathematik/Informatik,
Universität Kassel,
34109 Kassel, Germany
Email: otto@theory.informatik.uni-kassel.de

The notion of finite derivation type (FDT) is carried over to relative monoid presentations of the form $\mathcal{P} = \langle \mathcal{G}, T; R \rangle$, where $\mathcal{G}$ is a group, T is a finite alphabet, and R is a set of rules over the free product $\mathcal{G} * T^*$ satisfying $|\ell|_T, |r|_T > 0$ for each rule $(\ell \to r) \in R$. As an application we derive an algebraic characterization of the property FDT for monoids that are given through certain relative monoid presentations.

1. Introduction

In the theory of finite string-rewriting systems, two related geometric properties have been introduced. The first, a homotopical property, *finite derivation type* (FDT), was introduced by Squier [7]. The second, *finite homological type* (FHT), was introduced in [8]. Both these properties are invariants in the sense that if two finite rewriting systems represent the same monoid, then if one has the property so does the other. Hence, we can talk about FDT or FHT *monoids*. In general, FDT implies FHT, and for groups these two properties are equivalent [8], but it has been an open question whether this is also true for monoids in general.

Recently, this question has been answered in the negative in [5] by giving an example of a finitely presented monoid $\mathcal{M}$ that is FHT, but not FDT. This monoid is obtained from a finitely presented group $\mathcal{G}$ and a finitely generated subgroup $\mathcal{H}$ of $\mathcal{G}$ through a presentation of the form

$$[\mathbf{a}, \mathbf{a}^{-1}, t; Y = 1\,, a^{\varepsilon}a^{-\varepsilon} = 1,\, tb^{\varepsilon} = b^{\varepsilon}t\,(Y \in \mathbf{y}, a \in \mathbf{a}, b \in \mathbf{b}, \varepsilon = \pm 1)], \quad (1)$$

*This paper reports on joint work with Stephen J. Pride from the University of Glasgow, which was supported by a grant from the EPSRC under the MathFIT 2000 initiative (Grant GR/R29888).

where $\langle \mathbf{a}; \mathbf{y} \rangle$ is a finite group presentation of $\mathcal{G}$, and $\mathbf{b}$ is a subset of $\mathbf{a}$ generating the subgroup $\mathcal{H}$ of $\mathcal{G}$. The main result of [5] states that $\mathcal{M}$ is FDT if and only if $\mathcal{G}$ is FDT and $\mathcal{H}$ is finitely presented, and that $\mathcal{M}$ is FHT if and only if $\mathcal{G}$ is FHT and $\mathcal{H}$ is of type FP_2. The intended example monoid is then obtained by taking a group $\mathcal{G}$ from Bestvina and Brady [2] that is a right-angled Artin group (and therewith FDT), and that has a finitely generated subgroup $\mathcal{H}$ that is of type FP_2, but not finitely presented.

In [5] homotopy theory is used to establish the above result on the property FDT for the monoid $\mathcal{M}$. The aim of the current paper is to make the connections between the FDT property of $\mathcal{M}$ and the subgroup $\mathcal{H}$ more explicit. This will be done in three steps.

First we consider *relative monoid presentations* of the form

$$\mathcal{P} = \langle \mathcal{G}, T; R \rangle, \tag{2}$$

where $\mathcal{G}$ is a group, T is a finite alphabet, and R is a finite set of rules $\ell \to r$ over the free product $\mathcal{G} * T^*$ satisfying $|\ell|_T > 0$ and $|r|_T > 0$. Relative monoid presentations were introduced by Kilgour [3], who generalized Adjan's cancellativity test [1] to these relative presentations. Using them it is shown in [6] that the monoid $\mathcal{M}$ presented by $\mathcal{P}$ is FDT if and only if $\mathcal{G}$ is, provided the left graph or the right graph associated with (2) is cycle-free. In Section 3 we will generalize this result. We associate a 2-complex $\mathcal{D}(\mathcal{P})$ to the presentation $\mathcal{P}$, and we prove that this complex has a finite (homotopy) trivializer if and only if the monoid $\mathcal{M}$ presented by $\mathcal{P}$ has FDT, provided the group $\mathcal{G}$ has FDT (Corollary 3.1).

Next we associate a set of finite sequences Σ with two groups $\mathcal{F}$ and $\mathcal{G}$. On Σ we define a partial multiplication that makes Σ into a groupoid, another total operation, the concatenation of sequences, and we introduce the notion of *trivializer* for Σ. Then we associate a set Σ of sequences with a relative presentation of the form (2), and we prove that this groupoid Σ is isomorphic to the fundamental groupoid of the complex $\mathcal{D}(\mathcal{P})$ (Theorem 4.1). In particular, under this isomorphism the trivializers of the complex $\mathcal{D}(\mathcal{P})$ correspond to the trivializers of Σ.

Finally, we apply the above construction to a relative monoid presentation of the form

$$\mathcal{P} = \langle \mathcal{G}, t; tb_i = a_i t \, (i = 1, \ldots, n) \rangle, \tag{3}$$

where $\mathcal{G}$ is a finitely presented group, and the subgroups $\mathcal{A}$ and $\mathcal{B}$ of $\mathcal{G}$ that are generated by the sets $\{a_1, \ldots, a_n\}$ and $\{b_1, \ldots, b_n\}$, respectively, are isomorphic under the mapping induced by $a_i \mapsto b_i$ $(1 \le i \le n)$. It then turns out that the monoid $\mathcal{M}$ presented by $\mathcal{P}$ has property FDT if and only if the group $\mathcal{G}$ is FDT and the subgroup $\mathcal{A}$ (and therewith the subgroup $\mathcal{B}$ as well) is finitely presented (Corollary 5.1).

Because of the page limit we do not provide full proofs. They can be found in the technical report [4].

2. Relative monoid presentations

Let $\mathcal{P}$ be *relative monoid presentation* as in (2), where we require that $\ell \neq r$ for each rewrite rule $\ell \to r$ of R. The system R induces a *reduction relation* $\Rightarrow_R^*$ on the free product $F := \mathcal{G} * T^*$, which is the reflexive transitive closure of the *single-step reduction* relation $\Rightarrow_R$ that is defined as follows:

$$W = U_0 t_1 U_1 \dots U_{r-1} t_r U_r \Rightarrow_R W' \quad (U_j \in \mathcal{G}, t_j \in T)$$

if and only if there exist an index α and a rule

$$(\ell \to r) = (g_0 t_\alpha U_\alpha t_{\alpha+1} U_{\alpha+1} \dots t_{\alpha+k-1} g_k \to h_0 t'_1 h_1 \dots t'_m h_m) \in R,$$

where $g_i, h_i \in \mathcal{G}$ and $t_j, t'_j \in T$, such that

$$W' = U_0 t_1 \dots t_{\alpha-1} (U_{\alpha-1} g_0^{-1} h_0) t'_1 h_1 \dots t'_m (h_m g_k^{-1} U_{\alpha+k-1}) t_{\alpha+k} \dots t_r U_r,$$

that is, the factor $U_{\alpha-1} t_\alpha U_\alpha \dots t_{\alpha+k-1} U_{\alpha+k-1}$ of W is replaced by the factor $(U_{\alpha-1} g_0^{-1} h_0) t'_1 h_1 \dots t'_m (h_m g_k^{-1} U_{\alpha+k-1})$.

By $\Leftrightarrow_R^*$ we denote the smallest equivalence relation on F that contains the relation $\Rightarrow_R$. It is easily verified that $\Leftrightarrow_R^*$ is actually a congruence on F. By $\mathcal{M}$ we denote the monoid $F/\Leftrightarrow_R^* = (\mathcal{G} * T^*)/\Leftrightarrow_R^*$. We say that $\mathcal{M}$ is the *monoid* that is *presented by the relative presentation* $\mathcal{P}$.

From $\mathcal{P}$ we obtain a monoid presentation for $\mathcal{M}$ as follows. Let $\langle \Omega; S \rangle$ be a group presentation of $\mathcal{G}$, where Ω is a finite alphabet disjoint from T, and S is a subset of $\underline{\Omega}^*$. Here $\underline{\Omega} := \Omega \cup \overline{\Omega}$, where $\overline{\Omega} := \{ \bar{a} \mid a \in \Omega \}$ is the set of formal inverses of Ω. Thus, $\mathcal{G} = \underline{\Omega}^* / \leftrightarrow_{S_0}^*$, where $\leftrightarrow_{S_0}^*$ is the Thue congruence on $\underline{\Omega}^*$ that is induced by the string rewriting system

$$S_0 := \{ w \to 1 \mid w \in S \} \cup \{ a\bar{a} \to 1, \bar{a}a \to 1 \mid a \in \Omega \}.$$

Let $>'$ be a linear ordering of $\underline{\Omega}$, and let $>$ be the length-lexicographical extension of $>'$ to $\underline{\Omega}^*$. It is well-known that $>$ is a well-ordering on $\underline{\Omega}^*$. Thus, for each $w \in \underline{\Omega}^*$, the congruence class $[w]_{S_0}$ of w with respect to $\leftrightarrow_{S_0}^*$ contains a uniquely determined minimal element $\hat{w}$.

We define a mapping $\psi : F \to (\underline{\Omega} \cup T)^*$ by taking

$$\psi(U_0 t_1 U_1 \dots t_r U_r) := \hat{U}_0 t_1 \hat{U}_1 \dots t_r \hat{U}_r \quad (U_j \in \mathcal{G}, t_j \in T), \tag{4}$$

where $\hat{U}_j \in \underline{\Omega}^*$ denotes the minimal element from the congruence class mod S_0 that represents the group element U_j, and we define a mapping $\varphi : (\underline{\Omega} \cup T)^* \to F$ by setting

$$\varphi(u_0 t_1 u_1 \dots t_r u_r) := [u_0]_{S_0} t_1 [u_1]_{S_0} \dots t_r [u_r]_{S_0} \quad (u_j \in \underline{\Omega}^*, t_j \in T). \tag{5}$$

These mappings show that $(\underline{\Omega} \cup T; S_0)$ is a monoid presentation of the free product F. Finally we take the monoid presentation

$$\tilde{\mathcal{P}} := (\underline{\Omega} \cup T; S_0 \cup \{ \psi(\ell) \to \psi(r) \mid (\ell \to r) \in R \}). \tag{6}$$

The following result is straightforward.

Lemma 2.1. *$\tilde{\mathcal{P}}$ is a monoid presentation of the monoid $\mathcal{M}$.*

3. The Squier complex of a relative monoid presentation

With the relative monoid presentation $\mathcal{P} = \langle \mathcal{G}, T; R \rangle$ we associate a 2-complex $\mathcal{D} := \mathcal{D}(\mathcal{P})$ as follows. The underlying graph $\Gamma := \Gamma(\mathcal{P})$ has vertex set $F = \mathcal{G} * T^*$, and the edge set consists of all *atomic relative pictures*

$$\mathbb{E} = (U, \ell \to r, V, \varepsilon),$$

where $(\ell \to r) \in R$, $U, V \in F$, and $\varepsilon \in \{\pm 1\}$. The initial, terminal and inverse functions are respectively given by

$$\iota(\mathbb{E}) := \begin{cases} U\ell V \text{ if } \varepsilon = 1, \\ UrV \text{ if } \varepsilon = -1, \end{cases} \qquad \tau(\mathbb{E}) := \begin{cases} UrV \text{ if } \varepsilon = 1, \\ U\ell V \text{ if } \varepsilon = -1, \end{cases}$$

and $\mathbb{E}^{-1} := (U, \ell \to r, V, -\varepsilon)$. Thus, if

$$(\ell \to r) = (g_0 t_1 g_1 \dots t_k g_k \to h_0 t'_1 h_1 \dots t'_m h_m),$$

$$U = U_0 t_{i_1} U_1 \dots t_{i_\alpha} U_\alpha, \; V = V_0 t_{r_1} V_1 \dots t_{r_\beta} V_\beta, \text{ and } \varepsilon \in \{\pm 1\},$$

where $U_j, V_j, g_j, h_j \in \mathcal{G}$ and $t_\mu, t'_\nu \in T$, then $\iota(U, \ell \to r, V, 1) = U\ell V = U_0 t_{i_1} U_1 \dots t_{i_\alpha} (U_\alpha g_0) t_1 g_1 \dots t_k (g_k V_0) t_{r_1} V_1 \dots t_{r_\beta} V_\beta$, and $\tau(U, \ell \to r, V, 1) = UrV = U_0 t_{i_1} U_1 \dots t_{i_\alpha} (U_\alpha h_0) t'_1 h_1 \dots t'_m (h_m V_0) t_{r_1} V_1 \dots t_{r_\beta} V_\beta$, that is, the atomic relative picture $(U, \ell \to r, V, 1)$ depicts the reduction step $U\ell V \Rightarrow_R UrV$, and $(U, \ell \to r, V, -1)$ depicts its converse.

The edge $\mathbb{E}$ is called *positive* if $\varepsilon = 1$. Paths in Γ are called *relative pictures*, and closed paths are called *spherical relative pictures*. For each vertex W, we use i_W to denote the empty path at that vertex.

If there is a path $\mathbb{P}$ from W_1 to W_2, then $\mathbb{P}$ represents a sequence of applications of rules of R and their inverses, that is, $W_1 \Leftrightarrow_R^* W_2$ holds.

Conversely, if $W_1 \Leftrightarrow_R^* W_2$, then there exists a path in Γ from W_1 to W_2 that corresponds to the sequence $W_1 \Leftrightarrow_R^* W_2$. Thus, the connected component of Γ that contains a given vertex W corresponds to the equivalence class $[W]_R$ of W mod $\Leftrightarrow_R^*$, and therewith it corresponds to the element of the monoid $\mathcal{M}$ that is presented by W.

There are left and right actions of F on Γ: if $Y, Z \in F$, then for any vertex W of Γ, $Y \cdot W \cdot Z := YWZ$ (product in F) and for any edge $\mathbb{E}$ as above, $Y \cdot \mathbb{E} \cdot Z := (YU, \ell \to r, VZ, \varepsilon)$. We extend these left and right actions to all paths in Γ: if $\mathbb{P} = \mathbb{E}_1\mathbb{E}_2 \ldots \mathbb{E}_m$ is a nonempty path in Γ with $\mathbb{E}_i$ an edge of Γ $(1 \leq i \leq m)$, then for any $Y, Z \in F$,

$$Y \cdot \mathbb{P} \cdot Z := (Y \cdot \mathbb{E}_1 \cdot Z)(Y \cdot \mathbb{E}_2 \cdot Z) \ldots (Y \cdot \mathbb{E}_m \cdot Z);$$

if $\mathbb{P}$ is the empty path at W, then $Y \cdot \mathbb{P} \cdot Z$ is the empty path at YWZ.

If $\mathbb{A}, \mathbb{B}$ are atomic relative pictures, then we use $[\mathbb{A}, \mathbb{B}]$ to denote the spherical relative picture

$$(\mathbb{A} \cdot \iota(\mathbb{B}))(\tau(\mathbb{A}) \cdot \mathbb{B})(\mathbb{A}^{-1} \cdot \tau(\mathbb{B}))(\iota(\mathbb{A}) \cdot \mathbb{B}^{-1}).$$

For all atomic pictures $\mathbb{A}, \mathbb{B}$, we adjoin a 2-cell with boundary $[\mathbb{A}, \mathbb{B}]$. In the following we will also denote this 2-cell by $[\mathbb{A}, \mathbb{B}]$. The resulting 2-complex $\mathcal{D} = \mathcal{D}(\mathcal{P})$ is called the *Squier complex* of the relative monoid presentation $\mathcal{P}$. The left and right actions of F on Γ extend to actions on $\mathcal{D}$, as $Y \cdot [\mathbb{A}, \mathbb{B}] \cdot Z = [Y \cdot \mathbb{A}, \mathbb{B} \cdot Z]$ holds.

By $\simeq_0$ we denote the homotopy relation on $\mathcal{D}$. It is easily seen that, for any two relative pictures $\mathbb{P}, \mathbb{Q}$,

$$(\mathbb{P} \cdot \iota(\mathbb{Q}))(\tau(\mathbb{P}) \cdot \mathbb{Q})(\mathbb{P}^{-1} \cdot \tau(\mathbb{Q}))(\iota(\mathbb{P}) \cdot \mathbb{Q}^{-1}) \simeq_0 i_{\iota(\mathbb{P})\iota(Q)}.$$

Let $\mathbf{X}$ be a set of spherical relative pictures of $\mathcal{D}(\mathcal{P})$. We form a new 2-complex $\mathcal{D}(\mathcal{P})^{\mathbf{X}}$ by adjoining to $\mathcal{D}(\mathcal{P})$ additional 2-cells with boundaries $Y \cdot \mathbb{X} \cdot Z$ $(Y, Z \in F, \mathbb{X} \in \mathbf{X})$. We say that $\mathbf{X}$ *trivializes* $\mathcal{D}(\mathcal{P})$, or that $\mathbf{X}$ is a *trivializer* of $\mathcal{D}(\mathcal{P})$, if $\mathcal{D}(\mathcal{P})^{\mathbf{X}}$ has trivial fundamental groups, that is, if $\mathbb{P}$ is any spherical relative picture in Γ, then $\mathbb{P}$ is homotopic in $\mathcal{D}(\mathcal{P})$ to a product of the form

$$(\mathbb{B}_1(U_1 \cdot \mathbb{X}_1^{\varepsilon_1} \cdot V_1)\mathbb{B}_1^{-1})(\mathbb{B}_2(U_2 \cdot \mathbb{X}_2^{\varepsilon_2} \cdot V_2)\mathbb{B}_2^{-1}) \ldots (\mathbb{B}_m(U_m \cdot \mathbb{X}_m^{\varepsilon_m} \cdot V_m)\mathbb{B}_m^{-1})$$

with $\mathbb{B}_i$ a relative picture, $U_i, V_i \in F$, $\mathbb{X}_i \in \mathbf{X}$, $\varepsilon_i \in \{\pm 1\}$ $(1 \leq i \leq m)$.

We will say that the relative monoid presentation $\mathcal{P}$ is of *finite derivation type* (FDT) if $\mathcal{P}$ is finite (that is, T and R are finite sets) and $\mathcal{D}(\mathcal{P})$ has a finite trivializer.

We want to analyse the relationship of the Squier complex $\mathcal{D}(\mathcal{P})$ to the Squier complex $\tilde{\mathcal{D}} := \mathcal{D}(\tilde{\mathcal{P}})$ of the monoid presentation $\tilde{\mathcal{P}}$ (see (6)) in the special case that the group $\mathcal{G}$ considered is finitely presented, and the presentation $\langle \Omega; S \rangle$ of $\mathcal{G}$ is finite.

The mapping $\varphi : (\underline{\Omega} \cup T)^* \to F$ of (5) extends to a mapping $\varphi : \tilde{\mathcal{D}} \to \mathcal{D}$ by sending a vertex $w \in (\underline{\Omega} \cup T)^*$ to $\varphi(w) \in F$, and an atomic picture $\mathbb{E} = (u, \ell' \to r', v, \varepsilon)$ to the empty path at $\varphi(\iota(\mathbb{E}))$, if $(\ell' \to r') \in S_0$, and to the atomic relative picture $\varphi(\mathbb{E}) := (\varphi(u), \ell \to r, \varphi(v), \varepsilon)$, if $(\ell' \to r') = (\psi(\ell) \to \psi(r)) \in \psi(R)$.

By an easy calculation it can be shown that φ is compatible with initial, terminal and inverse functions, and that it maps pictures in $\tilde{\mathcal{D}}$ onto relative pictures in $\mathcal{D}$. In particular, it maps spherical pictures of $\tilde{\mathcal{D}}$ onto spherical relative pictures of $\mathcal{D}$. Finally, if $[\mathbb{A}, \mathbb{B}]$ is a 2-cell of $\tilde{\mathcal{D}}$, then by φ its boundary is mapped to a spherical relative picture that is homotopic to the empty path at its initial vertex.

For establishing a correspondence between trivializers of $\mathcal{D}$ and trivializers of $\tilde{\mathcal{D}}$ we need the following technical result, which can be proved by induction on the length of the path $\mathbb{P}$.

Lemma 3.1. *(Lifting Lemma)*
For each relative picture $\mathbb{P}$ in $\mathcal{D}$, and all words $u, v \in (\underline{\Omega} \cup T)^*$ satisfying $\varphi(u) =_F \iota(\mathbb{P})$ and $\varphi(v) =_F \tau(\mathbb{P})$, there exists a picture $\tilde{\mathbb{P}}$ in $\tilde{\mathcal{D}}$ such that $\iota(\tilde{\mathbb{P}}) = u$, $\tau(\tilde{\mathbb{P}}) = v$, and $\varphi(\tilde{\mathbb{P}}) \simeq_0 \mathbb{P}$ in $\mathcal{D}$.

Based on this technical result we obtain the following result.

Theorem 3.1. *Let $\langle \Omega; S \rangle$ be a finite presentation of the group $\mathcal{G}$. If the complex $\tilde{\mathcal{D}} = \mathcal{D}(\tilde{\mathcal{P}})$ has a finite trivializer, then the group $\mathcal{G}$ is FDT and the complex $\mathcal{D} = \mathcal{D}(\mathcal{P})$ has a finite trivializer, too.*

Proof. Let $\mathbf{X}$ be a finite trivializer of $\tilde{\mathcal{D}} = \mathcal{D}(\tilde{\mathcal{P}})$. Then $\varphi(\mathbf{X})$ is a finite set of spherical relative pictures in $\mathcal{D} = \mathcal{D}(\mathcal{P})$, and based on Lemma 3.1 it can be shown that $\varphi(\mathbf{X})$ is a trivializer of $\mathcal{D}$.

The complex $\mathcal{D}_{\mathcal{G}} := \mathcal{D}((\underline{\Omega}; S_0))$ is a subcomplex of $\tilde{\mathcal{D}}$. Actually, $\mathcal{D}_{\mathcal{G}}$ is just the set of components of $\tilde{\mathcal{D}}$ for which the vertices do not contain any occurrences of symbols from T. Accordingly, the subset $\mathbf{X}_{\mathcal{G}} := \{\mathbb{X} \in \mathbf{X} \mid |\iota(\mathbb{X})|_T = 0\}$ of $\mathbf{X}$ is a finite set of spherical pictures in $\mathcal{D}_{\mathcal{G}}$. As $\mathbf{X}$ is a trivializer of $\tilde{\mathcal{D}}$, and as $|\ell|_T > 0$ and $|r|_T > 0$ for each rule $(\ell \to r) \in R$, it follows that $\mathbf{X}_{\mathcal{G}}$ is a trivializer of $\mathcal{D}_{\mathcal{G}}$. Thus, the group $\mathcal{G}$ is FDT. $\square$

In the following we will establish the converse of this theorem. For doing so we need the following technical result.

Lemma 3.2. *Let* $\tilde{\mathbb{P}}$ *and* $\tilde{\mathbb{Q}}$ *be pictures in* $\tilde{\mathcal{D}}$ *such that* $\iota(\tilde{\mathbb{P}}) = \iota(\tilde{\mathbb{Q}})$ *and* $\tau(\tilde{\mathbb{P}}) = \tau(\tilde{\mathbb{Q}})$. *If* $\varphi(\tilde{\mathbb{P}}) \simeq_0 \varphi(\tilde{\mathbb{Q}})$ *in* $\mathcal{D}$, *then* $\tilde{\mathbb{P}} \simeq_{\mathbf{X}_\mathcal{G}} \tilde{\mathbb{Q}}$, *where* $\mathbf{X}_\mathcal{G}$ *is a trivializer of* $\mathcal{D}_\mathcal{G}$.

Based on this lemma we obtain the following result.

Theorem 3.2. *Let* $\langle \Omega; S \rangle$ *be a finite presentation of the group* $\mathcal{G}$. *If the group* $\mathcal{G}$ *is FDT, and if the complex* $\mathcal{D} = \mathcal{D}(\mathcal{P})$ *has a finite trivializer, then the complex* $\tilde{\mathcal{D}} = \mathcal{D}(\tilde{\mathcal{P}})$ *has a finite trivializer, too.*

Proof. As in the proof of Theorem 3.1 we denote by $\mathcal{D}_\mathcal{G}$ the subcomplex of $\tilde{\mathcal{D}}$ that is associated to the monoid presentation $(\underline{\Omega}; S_0)$ of $\mathcal{G}$. By hypothesis $\mathcal{G}$ is FDT, and so $\mathcal{D}_\mathcal{G}$ has a finite trivializer $\mathbf{X}_\mathcal{G}$. Further, let $\mathbf{X}$ be a finite trivializer of the complex $\mathcal{D} = \mathcal{D}(\mathcal{P})$. Then each element $\mathbb{X}$ of $\mathbf{X}$ is a spherical relative picture of $\mathcal{D}$. Let $x := \psi(\iota(\mathbb{X}))$ $(= \psi(\tau(\mathbb{X})))$. By Lemma 3.1 there exists a spherical picture $\tilde{\mathbb{X}}$ of $\tilde{\mathcal{D}}$ such that $\iota(\tilde{\mathbb{X}}) = x = \tau(\tilde{\mathbb{X}})$ and $\varphi(\tilde{\mathbb{X}}) \simeq_0 \mathbb{X}$ in $\mathcal{D}$. By $\tilde{\mathbf{X}}$ we denote the set $\tilde{\mathbf{X}} := \{\, \tilde{\mathbb{X}} \mid \mathbb{X} \in \mathbf{X} \,\} \cup \mathbf{X}_\mathcal{G}$. Then $\tilde{\mathbf{X}}$ is a finite set of spherical pictures of $\tilde{\mathcal{D}}$, and it can be shown that $\tilde{\mathbf{X}}$ is a trivializer for $\tilde{\mathcal{D}}$. □

Theorems 3.1 and 3.2 yield the following characterization.

Corollary 3.1. *Let* $\langle \Omega; S \rangle$ *be a finite presentation of the group* $\mathcal{G}$, *let* $\mathcal{D} = \mathcal{D}(\mathcal{P})$ *be the complex associated to the relative monoid presentation* $\mathcal{P} = \langle \mathcal{G}, T; R \rangle$, *and let* $\tilde{\mathcal{D}} = \mathcal{D}(\tilde{\mathcal{P}})$ *be the complex that is associated to the monoid presentation* $\tilde{\mathcal{P}} = (\underline{\Omega} \cup T; S_0 \cup \psi(R))$. *Then the following two statements are equivalent:*

(1.) $\tilde{\mathcal{D}}$ *has a finite trivializer.*
(2.) $\mathcal{G}$ *is FDT, and* $\mathcal{D}$ *has a finite trivializer.*

In particular, if $\mathcal{G}$ is FDT, then $\tilde{\mathcal{D}}$ has a finite trivializer if and only if $\mathcal{D}$ has a finite trivializer.

4. $\mathcal{F}$-$\mathcal{G}$-sequences

Next we associate a set of sequences with two groups $\mathcal{F}$ and $\mathcal{G}$. We will see that this set is a groupoid, and we will establish a close correspondence between this groupoid and the fundamental groupoid of the complex $\mathcal{D}(\mathcal{P})$ for a certain relative monoid presentation $\mathcal{P}$ involving the group $\mathcal{G}$.

For the following consideration we fix two groups $\mathcal{F}$ and $\mathcal{G}$ and two group homomorphisms $\alpha, \beta : \mathcal{F} \to \mathcal{G}$. By λ we will denote the identity of $\mathcal{F}$, and by 1 we will denote the identity of $\mathcal{G}$.

By $\Sigma := \Sigma(\mathcal{F}, \mathcal{G}, \alpha, \beta)$ we denote the set of all sequences p of the form

$$p = (g_0, w_1, g_1, w_2, \ldots, w_m, g_m), \tag{7}$$

where $m \geq 0$, $g_0, g_1, \ldots, g_m \in \mathcal{G}$, and $w_1, \ldots, w_m \in \mathcal{F}$. A sequence with $w_1 = w_2 = \ldots = w_m = \lambda$ will be called a *trivial sequence*, and Λ denotes the set of all trivial sequences. We define two maps $\iota : \Sigma \to \Lambda$ and $\tau : \Sigma \to \Lambda$ as follows:

$$\begin{aligned}
\iota(p) &:= (g_0, \lambda, g_1, \lambda, \ldots, \lambda, g_m), \text{ and} \\
\tau(p) &:= (g_0 \cdot \alpha(w_1), \lambda, \beta(w_1)^{-1} \cdot g_1 \cdot \alpha(w_2), \lambda, \ldots, \lambda, \beta(w_m)^{-1} \cdot g_m).
\end{aligned}$$

A sequence $p \in \Sigma$ is called *closed* if $\iota(p) = \tau(p)$ holds. It is easily seen that p is closed if and only if the following equalities hold simultaneously in $\mathcal{G}$:

$$\begin{aligned}
g_0 &=_{\mathcal{G}} g_0 \cdot \alpha(w_1), \text{ that is, } \alpha(w_1) =_{\mathcal{G}} 1, \\
g_i &=_{\mathcal{G}} \beta(w_i)^{-1} \cdot g_i \cdot \alpha(w_{i+1}) \ (1 \leq i < m), \text{ and} \\
g_m &=_{\mathcal{G}} \beta(w_m)^{-1} \cdot g_m, \text{ that is, } \beta(w_m) =_{\mathcal{G}} 1.
\end{aligned}$$

In particular, each trivial sequence is closed.

Next we define a partial multiplication on Σ. For $p, q \in \Sigma$, if $\tau(p) = \iota(q)$, then pq is defined as

$$pq := (g_0, w_1 v_1, g_1, w_2 v_2, \ldots, w_m v_m, g_m), \tag{8}$$

where $p = (g_0, w_1, g_1, w_2, \ldots, w_m, g_m)$ and $q = (h_0, v_1, h_1, v_2, \ldots, v_m, h_m)$. Observe that $\tau(p) = \iota(q)$ implies in particular that $n = m$ holds.

Lemma 4.1. *Let $p, q, r \in \Sigma$.*

(a) If pq is defined, then $\iota(pq) = \iota(p)$ and $\tau(pq) = \tau(q)$.
(b) $\iota(p)p$ is defined, and $\iota(p)p = p$.
(c) $p\tau(p)$ is defined, and $p\tau(p) = p$.
(d) There exists a sequence $p^{-1} \in \Sigma$ such that

$$\iota(p^{-1}) = \tau(p), \ \tau(p^{-1}) = \iota(p), \ pp^{-1} = \iota(p), \text{ and } p^{-1}p = \tau(p).$$

(e) If pq and qr are defined, then $(pq)r$ and $p(qr)$ are defined and $(pq)r = p(qr)$, that is, the partial multiplication is associative, whenever it is defined.

This yields the following consequence.

Corollary 4.1. *Σ is a groupoid under the above partial multiplication.*

We define a second operation $\circ$ on Σ, which, however, is total. For $p = (g_0, w_1, g_1, \ldots, w_m, g_m)$ and $q = (h_0, v_1, h_1, \ldots, v_n, h_n)$, we define $p \circ q$ through

$$p \circ q := (g_0, w_1, g_1, \ldots, w_m, g_m h_0, v_1, h_1, \ldots, v_n, h_n). \tag{9}$$

Lemma 4.2. *Let $p, q, r \in \Sigma$.*

(a) $(p \circ q) \circ r = p \circ (q \circ r)$, that is, $\circ$ is an associative operation.

(b) $\iota(p \circ q) = \iota(p) \circ \iota(q)$ and $\tau(p \circ q) = \tau(p) \circ \tau(q)$, that is, $\circ$ is compatible with the maps ι and τ.

(c) $(p \circ \iota(q))(\tau(p) \circ q) = p \circ q = (\iota(p) \circ q)(p \circ \tau(q))$.

Finally we define some rewrite relations on Σ. For a set $B \subseteq \Sigma$ of closed sequences, the rewrite relation $\rightarrow_B \subseteq \Sigma \times \Sigma$ is defined as follows, where $B^{-1} := \{\, s^{-1} \mid s \in B \,\}$ (see Lemma 4.1(d)):

$$p \rightarrow_B q \text{ iff } \exists q_1, q_2 \in \Sigma\, \exists r \in (\Lambda \circ B \circ \Lambda) \cup (\Lambda \circ B^{-1} \circ \Lambda) \text{ such that}$$

$$p = q_1 r q_2 \quad \text{and} \quad q = q_1 q_2. \tag{10}$$

By $\leftrightarrow_B^*$ we denote the reflexive, symmetric, and transitive closure of $\rightarrow_B$.

From the fact that each trivial sequence is closed and from Lemma 4.2(b) we see that $(\Lambda \circ B \circ \Lambda) \cup (\Lambda \circ B^{-1} \circ \Lambda)$ only contains closed sequences. Thus, if $p = q_1 r q_2$ for some $r \in (\Lambda \circ B \circ \Lambda) \cup (\Lambda \circ B^{-1} \circ \Lambda)$, then $q_1 q_2$ is defined, and conversely, if $q = q_1 q_2$ and r is a sequence of the above form satisfying $\iota(r) = \tau(q_1)$, then $q_1 r q_2$ is defined. Hence, we obtain the following.

Lemma 4.3. *For $p, q \in \Sigma$, if $p \leftrightarrow_B^* q$, then $\iota(p) = \iota(q)$ and $\tau(p) = \tau(q)$. Further, if $s, r \in \Sigma$ satisfy $\tau(s) = \iota(p)$ and $\iota(r) = \tau(p)$, then $p \leftrightarrow_B^* q$ implies that $spr \leftrightarrow_B^* sqr$ holds.*

A set $B \subseteq \Sigma$ of closed sequences is called a *trivializer* of Σ if $p \leftrightarrow_B^* q$ holds for all $p, q \in \Sigma$ satisfying $\iota(p) = \iota(q)$ and $\tau(p) = \tau(q)$.

Lemma 4.4. *A set $B \subseteq \Sigma$ of closed sequences is a trivializer of Σ if and only if each closed sequence $p \in \Sigma$ satisfies $p \leftrightarrow_B^* \iota(p)$.*

Let $\mathcal{G}$ be a group, and let

$$\mathcal{P} = \langle \mathcal{G}, t; tb_i = a_i t \ (i = 1, \ldots, n) \rangle \tag{11}$$

be a relative monoid presentation, where t is a new letter, and $a_i, b_i \in \mathcal{G}$ $(1 \leq i \leq n)$. Let $X := \{x_1, \ldots, x_n\}$, let $\mathcal{F} := \mathcal{F}(X)$ denote the free group generated by X, and define $\alpha, \beta : \mathcal{F} \to \mathcal{G}$ through

$$\alpha(x_i) := a_i \ (1 \leq i \leq n) \quad \text{and} \quad \beta(x_i) := b_i \ (1 \leq i \leq n). \tag{12}$$

By $\Sigma := \Sigma(\mathcal{F}, \mathcal{G}, \alpha, \beta)$ we denote the set of all sequences that we obtain from $\mathcal{F}, \mathcal{G}, \alpha$, and β as described above. We will now establish a close correspondence between this groupoid Σ and the fundamental groupoid of the 2-complex $\mathcal{D}(\mathcal{P})$ that is associated with the relative monoid presentation $\mathcal{P}$.

We first define a mapping φ that maps pictures of $\mathcal{D} := \mathcal{D}(\mathcal{P})$ to sequences in Σ. Let $\mathbb{E} = (U, tb_i \to a_i t, V, \varepsilon)$ be an atomic picture in $\mathcal{D}$, that is, $U = U_0 t U_1 t \ldots t U_r$, $V = V_0 t V_1 t \ldots t V_s$ for some $U_j, V_k \in \mathcal{G}$, $r, s \geq 0$, some $i \in \{1, \ldots, n\}$, and $\varepsilon \in \{\pm 1\}$:

$$\varphi(\mathbb{E}) := \begin{cases} (U_0, \lambda, \ldots, \lambda, U_{r-1}, \lambda, U_r, x_i, b_i V_0, \lambda, V_1, \ldots, \lambda, V_s) & \text{if } \varepsilon = 1, \\ (U_0, \lambda, \ldots, \lambda, U_{r-1}, \lambda, U_r a_i, x_i^{-1}, V_0, \lambda, V_1, \ldots, \lambda, V_s) & \text{if } \varepsilon = -1. \end{cases}$$

Further, we define a vertex map $\varphi_v : \mathcal{G} * \{t\}^* \to \Lambda$ by taking

$$\varphi_v(g_0 t g_1 t \ldots t g_m) := (g_0, \lambda, g_1, \lambda, \ldots, \lambda, g_m).$$

An easy calculation shows the following.

Lemma 4.5. *$\varphi_v(\iota(\mathbb{E})) = \iota(\varphi(\mathbb{E}))$ and $\varphi_v(\tau(\mathbb{E})) = \tau(\varphi(\mathbb{E}))$.*

Thus, we can extend the mapping φ to pictures of $\mathcal{D}$ by taking $\varphi(\mathbb{P}) := \varphi(\mathbb{E}_1)\varphi(\mathbb{E}_2)\ldots\varphi(\mathbb{E}_m)$, if $\mathbb{P} = \mathbb{E}_1\mathbb{E}_2\ldots\mathbb{E}_m$, where $\mathbb{E}_i$ $(1 \leq i \leq m)$ are atomic pictures of $\mathcal{D}$. As $\mathbb{P}$ is a picture, we have $\tau(\mathbb{E}_i) = \iota(\mathbb{E}_{i+1})$ for all $i = 1, \ldots, m-1$. Hence, by Lemma 4.5 we have

$$\tau(\varphi(\mathbb{E}_i)) = \varphi_v(\tau(\mathbb{E}_i)) = \varphi_v(\iota(\mathbb{E}_{i+1})) = \iota(\varphi(\mathbb{E}_{i+1})) \quad (1 \leq i \leq m-1),$$

that is, the product $\varphi(\mathbb{E}_1)\varphi(\mathbb{E}_2)\ldots\varphi(\mathbb{E}_m)$ is defined in Σ. Because of Lemma 4.1(a) we see that the statement of Lemma 4.5 generalizes to arbitrary pictures in $\mathcal{D}$. This shows in particular that spherical pictures of $\mathcal{D}$ are mapped to closed sequences in Σ.

Lemma 4.6. *Let $\mathbb{A}, \mathbb{B}$ be atomic pictures in $\mathcal{D}$. Then*

$$\varphi((\mathbb{A} \cdot \iota(\mathbb{B}))(\tau(\mathbb{A}) \cdot \mathbb{B})) = \varphi((\iota(\mathbb{A}) \cdot \mathbb{B})(\mathbb{A} \cdot \tau(\mathbb{B}))).$$

Thus, the map φ is compatible with the basic homotopies of $\mathcal{D}$. If $\mathbb{E}$ is an atomic picture, then $\mathbb{E}\,\mathbb{E}^{-1} \simeq_0 i_{\iota(\mathbb{E})}$ and $\mathbb{E}^{-1}\mathbb{E} \simeq_0 i_{\tau(\mathbb{E})}$. It is easily verified that $\varphi(\mathbb{E}\,\mathbb{E}^{-1}) = \varphi(\mathbb{E})\varphi(\mathbb{E}^{-1}) = \iota(\varphi(\mathbb{E})) = \varphi_v(\iota(\mathbb{E}))$ and

$\varphi(\mathbb{E}^{-1}\mathbb{E}) = \varphi(\mathbb{E}^{-1})\varphi(\mathbb{E}) = \iota(\varphi(\mathbb{E}^{-1})) = \varphi_v(\tau(\mathbb{E}))$ hold. By induction this generalizes to arbitrary pictures $\mathbb{P}$.

Lemma 4.7. *For each picture $\mathbb{P}$ of $\mathcal{D}$, we have*

$$\varphi(\mathbb{P}\mathbb{P}^{-1}) = \iota(\varphi(\mathbb{P})) = \varphi_v(\iota(\mathbb{P})) \text{ and } \varphi(\mathbb{P}^{-1}\mathbb{P}) = \tau(\varphi(\mathbb{P})) = \varphi_v(\tau(\mathbb{P})).$$

These lemmas yield the following important fact.

Corollary 4.2. *If $\mathbb{P} \simeq_0 \mathbb{Q}$ for some pictures $\mathbb{P}, \mathbb{Q}$ in $\mathcal{D}$, then $\varphi(\mathbb{P}) = \varphi(\mathbb{Q})$.*

Hence, φ induces a map from the fundamental groupoid of $\mathcal{D}$ into Σ that is compatible with the operations of concatenation, ι, τ, and $^{-1}$ of the fundamental groupoid of $\mathcal{D}$, that is, φ is in fact a homomorphism from the fundamental groupoid of $\mathcal{D}$ into the groupoid Σ.

For defining a reverse map ψ from Σ into the fundamental groupoid of $\mathcal{D}$, we use the following technical results on Σ. Let $\underline{X} := X \cup \overline{X}$, where $\overline{X} := \{\bar{x}_1, \ldots, \bar{x}_n\}$ is a set of formal inverses of the generators X of $\mathcal{F}$.

Lemma 4.8. *For all $m \geq 1$, $g_0, g_1, \ldots, g_m \in \mathcal{G}$, $w_1, \ldots, w_m \in \underline{X}^*$, and $i \in \{1, \ldots, m\}$, let*

$$\sigma_i := (\tilde{g}_0, \lambda, \tilde{g}_1, \lambda, \ldots, \lambda, \tilde{g}_{i-2}, \lambda, \beta(w_{i-1})^{-1} \cdot g_{i-1}, w_i, g_i, \lambda, \ldots, \lambda, g_m),$$

where $\tilde{g}_0 := g_0 \cdot \alpha(w_1)$ and $\tilde{g}_i := \beta(w_i)^{-1} \cdot g_i \cdot \alpha(w_{i+1})$. Then

$$(g_0, w_1, g_1, w_2, \ldots, g_{m-1}, w_m, g_m) = \sigma_1\sigma_2 \ldots \sigma_m.$$

Thus, each element of Σ can be expressed as a product of elements that each only contain a single non-trivial factor from $\mathcal{F}$.

Lemma 4.9. *For all $m \geq 1$, $g_0, g_1, \ldots, g_m \in \mathcal{G}$, $i \in \{1, \ldots, m\}$, $k \geq 1$, and $x_1, \ldots, x_k \in \underline{X}$, let $y_j := x_1 \ldots x_{j-1}$ and*

$$\tau_{i,j} := (g_0, \lambda, \ldots, \lambda, g_{i-1} \cdot \alpha(y_j), x_j, \beta(y_j)^{-1} \cdot g_i, \lambda, g_{i+1}, \lambda, \ldots, \lambda, g_m),$$

where $1 \leq j \leq m$. Then

$$(g_0, \lambda, g_1, \lambda, \ldots, \lambda, g_{i-1}, x_1 \ldots x_k, g_i, \lambda, \ldots \lambda, g_m) = \tau_{i,1}\tau_{i,2} \ldots \tau_{i,m}.$$

Thus, each element of Σ can even be expressed as a product of elements that each only contain a single non-trivial factor from $\mathcal{F}$ of length one. Hence, it suffices to define ψ for sequences of this restricted form.

If g is the trivial sequence $(g_0, \lambda, g_1, \lambda, \ldots, \lambda, g_m)$, we take $\psi(g)$ to be the empty path at $g_0tg_1 \ldots tg_m$. If $g = (g_0, \lambda, \ldots, \lambda, g_i, x_j^\varepsilon, g_{i+1}, \ldots, \lambda, g_m)$ for some $x_j \in X$ and $\varepsilon \in \{\pm 1\}$, we take

$$\psi(g) := \begin{cases} (g_0tg_1 \ldots tg_i, tb_j \to a_jt, b_j^{-1}g_{i+1}tg_{i+2} \ldots tg_m, 1) & \text{if } \varepsilon = 1, \\ (g_0tg_1 \ldots tg_ia_j^{-1}, tb_j \to a_jt, g_{i+1}tg_{i+2} \ldots tg_m, -1) & \text{if } \varepsilon = -1, \end{cases}$$

that is, $\psi(g)$ is an atomic picture of $\mathcal{D}$. We can summarize the properties of ψ as follows.

Lemma 4.10. *For all sequences $g \in \Sigma$, $\psi(g)$ is a picture in $\mathcal{D}$ such that $\varphi(\psi(g)) = g$. If g is a closed sequence, then $\psi(g)$ is a spherical picture in $\mathcal{D}$.*

Proof. If g is a closed sequence, then $\psi(g)$ is a picture in $\mathcal{D}$ that has the following properties:

$$\varphi_v(\iota(\psi(g))) = \iota(g) = \tau(g) = \varphi_v(\tau(\psi(g))).$$

As $\varphi_v : \mathcal{G} * \{t\}^* \to \Lambda$ is an injective mapping, this shows that $\iota(\psi(g)) = \tau(\psi(g))$ holds, that is, $\psi(g)$ is a spherical picture. □

Lemma 4.10 implies in particular that φ is a surjective homomorphism from the fundamental groupoid of $\mathcal{D}$ onto the groupoid Σ, and the mapping ψ from Σ into the fundamental groupoid of $\mathcal{D}$ is injective. It remains to show that φ is also injective. This can be done by showing that $\psi(\varphi(\mathbb{P})) \simeq_0 \mathbb{P}$ holds for each picture $\mathbb{P}$. Hence, we have the following result.

Theorem 4.1. *The fundamental groupoid of the complex $\mathcal{D}(\mathcal{P})$ is isomorphic to the groupoid Σ.*

In addition, we want to establish a correspondence between the trivializers of $\mathcal{D}(\mathcal{P})$ and the trivializers of Σ.

Lemma 4.11. *If $\mathbf{X}$ is a trivializer of $\mathcal{D}(\mathcal{P})$, then $B := \varphi(\mathbf{X})$ is a trivializer of Σ.*

For establishing the converse implication the following technical result is useful.

Lemma 4.12. *Let B be a set of closed sequences in Σ, and let $p \in \Sigma$ be a closed sequence. If $p \leftrightarrow_B^* \iota(p)$, then*

$$p = q_1(u_1 \circ b_1^{\varepsilon_1} \circ v_1)q_1^{-1}q_2(u_2 \circ b_2^{\varepsilon_2} \circ v_2)q_2^{-1} \ldots q_m(u_m \circ b_m^{\varepsilon_m} \circ v_m)q_m^{-1}$$

for some $u_i, v_i \in \Lambda$, $b_i \in B$, $\varepsilon_i \in \{\pm 1\}$, and $q_i \in \Sigma$ $(1 \leq i \leq m)$.

Based on this lemma the following converse of Lemma 4.11 can be derived.

Lemma 4.13. *Let B be a trivializer of Σ, and let* $\mathbf{X}$ *be obtained by choosing one representative from each homotopy class $\psi(b)$ $(b \in B)$. Then* $\mathbf{X}$ *is a trivializer of $\mathcal{D}(\mathcal{P})$.*

Thus, we obtain the following extension of Theorem 4.1.

Corollary 4.3. *Under the isomorphism φ, the trivializers of the complex $\mathcal{D}(\mathcal{P})$ correspond to the trivializers of Σ.*

5. Application

Let $\mathcal{G}$ be a finitely presented group, and let $\mathcal{A}, \mathcal{B} \subseteq \mathcal{G}$ be subgroups of $\mathcal{G}$, where $\mathcal{A} = \langle a_i \mid i = 1, \ldots, n \rangle$ and $\mathcal{B} = \langle b_i \mid i = 1, \ldots, n \rangle$, such that the map $a_i \mapsto b_i$ $(1 \leq i \leq n)$ induces an isomorphism from $\mathcal{A}$ onto $\mathcal{B}$. Further, let $\mathcal{F} := \mathcal{F}(X)$, where $X = \{x_1, \ldots, x_n\}$, and let $\alpha : \mathcal{F} \to \mathcal{G}$ and $\beta : \mathcal{F} \to \mathcal{G}$ be defined through $\alpha(x_i) := a_i$ and $\beta(x_i) := b_i$ $(1 \leq i \leq n)$. Then $\alpha(w) =_{\mathcal{G}} 1$ if and only if $\beta(w) =_{\mathcal{G}} 1$, and so we see that $\mathrm{Ker}(\alpha) = \mathrm{Ker}(\beta)$. In the following we will denote $\mathrm{Ker}(\alpha)$ by $\mathcal{H}$.

Let $\mathcal{P} = \langle \mathcal{G}, t; tb_i = a_i t \ (i = 1, \ldots, n) \rangle$, where t is a new letter, let $\mathcal{D} := \mathcal{D}(\mathcal{P})$ be the 2-complex associated with $\mathcal{P}$, and let $\Sigma := \Sigma(\mathcal{F}, \mathcal{G}, \alpha, \beta)$. Below we will show that $\mathcal{D}$ has a finite trivializer if and only if the normal subgroup $\mathcal{H}$ of $\mathcal{F}$ is finitely generated as a normal subgroup. For deriving this result we need some preparations.

Lemma 5.1. $\pi_1(\mathcal{D}, t) \cong_{\varphi} \Sigma_{(1,\lambda,1)}$, *that is, the fundamental groupoid of $\mathcal{D}$ at the vertex t is isomorphic under φ to the groupoid of sequences p from Σ that satisfy $\iota(p) = (1, \lambda, 1) = \tau(p)$.*

Proof. Let $(g_0, w, g_1) \in \Sigma_{(1,\lambda,1)}$, that is, $g_0, g_1 \in \mathcal{G}$ and $w \in \underline{X}^*$ such that

$$\begin{aligned} \iota(g_0, w, g_1) &= (g_0, \lambda, g_1) = (1, \lambda, 1) \quad \text{and} \\ \tau(g_0, w, g_1) &= (g_0 \cdot \alpha(w), \lambda, \beta(w)^{-1} \cdot g_1) = (1, \lambda, 1). \end{aligned}$$

Thus, $g_0 =_{\mathcal{G}} 1 =_{\mathcal{G}} g_1$, and $\alpha(w) =_{\mathcal{G}} 1 =_{\mathcal{G}} \beta(w)$, that is, $w \in \mathcal{H}$. Now $\psi(g_0, w, g_1)$ is a spherical picture in $\mathcal{D}$ such that $\iota(\psi(g_0, w, g_1)) = g_0 t g_1 = t$.

Conversely, if $\mathbb{P}$ is a spherical picture such that $\iota(\mathbb{P}) = t$, then $\varphi(\mathbb{P})$ is a closed sequence of the form (g_0, w, g_1) such that $\iota(g_0, w, g_1) = \varphi_v(\iota(\mathbb{P})) = \varphi_v(t) = (1, \lambda, 1)$, that is, $\varphi(\mathbb{P}) \in \Sigma_{(1,\lambda,1)}$. □

Lemma 5.2. *Under the map $w \mapsto (1, w, 1)$, $\mathcal{H} = Ker(\alpha)$ is isomorphic to $\Sigma_{(1,\lambda,1)}$.*

Proof. Let $u, v \in \mathrm{Ker}(\alpha)$. Then $\tau(1, u, 1) = (\alpha(u), \lambda, \beta(u)^{-1}) = (1, \lambda, 1) = \iota(1, v, 1)$, that is, the product $(1, u, 1)(1, v, 1)$ is defined. Obviously, we have $(1, u, 1)(1, v, 1) = (1, uv, 1)$. Thus, $\mathcal{H} = \mathrm{Ker}(\alpha) \cong \Sigma_{(1,\lambda,1)}$. □

Lemma 5.3. $\Sigma_{(1,\lambda,1)}$ *is a trivializer of* Σ.

Proof. Let $p = (g_0, w_1, g_1, \ldots, w_m, g_m)$ be a closed sequence. Then $\iota(p) = \tau(p)$ implying inductively that $w_1, w_2, \ldots, w_m \in \mathrm{Ker}(\alpha) = \mathrm{Ker}(\beta) = \mathcal{H}$.

By Lemma 4.8 we have $(g_0, w_1, g_1, w_2, \ldots, g_{m-1}, w_m, g_m) = \sigma_1\sigma_2\ldots\sigma_m$, where $\tilde{g}_0 := g_0 \cdot \alpha(w_1)$, $\tilde{g}_i := \beta(w_i)^{-1} \cdot g_i \cdot \alpha(w_{i+1})$, and $\sigma_i := (\tilde{g}_0, \lambda, \tilde{g}_1, \lambda, \ldots, \lambda, \tilde{g}_{i-2}, \lambda, \beta(w_{i-1})^{-1} \cdot g_{i-1}, w_i, g_i, \lambda, \ldots, \lambda, g_m)$. As $w_i \in \mathcal{H}$, we see that $\sigma_i = (g_0, \lambda, \ldots, \lambda, g_{i-1}) \circ (1, w_i, 1) \circ (g_i, \lambda, \ldots, \lambda, g_m)$, that is, p is a product of elements from $\Lambda \circ \Sigma_{(1,\lambda,1)} \circ \Lambda$. This in turn shows that $p \to^*_{\Sigma_{(1,\lambda,1)}} \iota(p)$, that is, $\Sigma_{(1,\lambda,1)}$ is a trivializer of Σ by Lemma 4.4. □

Based on the above results on $\Sigma_{(1,\lambda,1)}$ we obtain the following.

Theorem 5.1. *Let* $\mathcal{P} = \langle \mathcal{G}, t; tb_i = a_i t \ (i = 1, \ldots, n)\rangle$, *where* $a_i \mapsto b_i$ $(1 \leq i \leq n)$ *induces an isomorphism from* $\mathcal{A} := \langle a_i \mid i = 1, \ldots, n \rangle$ *onto* $\mathcal{B} := \langle b_i \mid i = 1, \ldots, n \rangle$, *let* $\mathcal{F} = \mathcal{F}(X)$, *where* $X := \{x_1, \ldots, x_n\}$, *let* $\alpha : \mathcal{F} \to \mathcal{G}$ *and* $\beta : \mathcal{F} \to \mathcal{G}$ *be defined through* $\alpha(x_i) := a_i$ *and* $\beta(x_i) := b_i$ $(i = 1, \ldots, n)$, *and let* $\mathcal{H} := \mathit{Ker}(\alpha)$. *Then the complex* $\mathcal{D}(\mathcal{P})$ *has a finite trivializer if and only if* $\mathcal{H}$ *is finitely generated as a normal subgroup of* $\mathcal{F}$.

Proof. Let $\Sigma := \Sigma(\mathcal{F}, \mathcal{G}, \alpha, \beta)$ If $\mathcal{D}(\mathcal{P})$ has a finite trivializer, then Σ has a finite trivializer (Corollary 4.3), and hence, $\Sigma_{(1,\lambda,1)} = \{ (1, w, 1) \mid w \in \mathcal{H} \}$ contains a finite subset that is already a trivializer of Σ (Lemma 5.3), that is, there exists a finite subset $\{v_1, \ldots, v_s\} \subseteq \mathcal{H}$ such that

$$B := \{ (1, v_j, 1) \mid j = 1, \ldots, s \}$$

is a trivializer of Σ.

Now let $w \in \mathcal{H}$, that is, $(1, w, 1) \in \Sigma_{(1,\lambda,1)}$. As B is a trivializer of Σ, we have $(1, w, 1) \leftrightarrow^*_B \iota(1, w, 1) = (1, \lambda, 1)$, and so

$$(1, w, 1) = q_1(u_1 \circ (1, v_{j_1}^{\varepsilon_1}, 1) \circ w_1)q_1^{-1} \ldots q_m(u_m \circ (1, v_{j_m}^{\varepsilon_m}, 1) \circ w_m)q_m^{-1}$$

by Lemma 4.12, where $u_i, w_i \in \Lambda$, $j_i \in \{1, \ldots, s\}$, $\varepsilon_i \in \{\pm 1\}$, and $q_i \in \Sigma$ $(1 \leq i \leq m)$. Then $w =_{\mathcal{F}} z_1 v_{j_1}^{\varepsilon_1} z_1^{-1} \cdot \ldots \cdot z_m v_{j_m}^{\varepsilon_m} z_m^{-1}$, where $q_i = (h_{i,1}, z_i, h_{i,2})$ $(1 \leq i \leq m)$, which implies that $\{v_1, \ldots, v_s\}$ generates $\mathcal{H}$ as a normal subgroup.

Conversely, assume that there is a finite set $V := \{v_1, \ldots, v_s\} \subseteq \mathcal{H}$ that generates $\mathcal{H}$ as a normal subgroup of $\mathcal{F}$. Then, for each $w \in \mathcal{H}$, we have a representation of the form $w =_{\mathcal{F}} z_1 v_{j_1}^{\varepsilon_1} z_1^{-1} \cdot \ldots \cdot z_m v_{j_m}^{\varepsilon_m} z_m^{-1}$, where $z_i \in \underline{X}^*$, $j_i \in \{1, \ldots, s\}$, $\varepsilon_i \in \{\pm 1\}$, and $m \geq 0$. Then $B := \{ (1, v_j, 1) \mid j = 1, \ldots, s \} \subseteq \Sigma_{(1,\lambda,1)}$ is a trivializer of Σ. □

As an immediate consequence we obtain the following result.

Corollary 5.1. *Let $\mathcal{G}$ be a finitely presented group, let $\mathcal{A} := \langle a_i \mid i = 1, \ldots, n \rangle$ and $B := \langle b_i \mid i = 1, \ldots, n \rangle$ be subgroups of $\mathcal{G}$ that are isomorphic under the mapping $a_i \mapsto b_i$ $(1 \leq i \leq n)$, and let $\mathcal{M}$ be given through the relative monoid presentation $\mathcal{P} = \langle \mathcal{G}, t; tb_i = a_i t \ (i = 1, \ldots, n) \rangle$. Then $\mathcal{M}$ is FDT if and only if $\mathcal{G}$ is FDT and the subgroup $\mathcal{A}$ is finitely presented.*

Proof. If $\mathcal{M}$ is FDT, then by Theorem 3.1 the group $\mathcal{G}$ is FDT, and the complex $\mathcal{D} := \mathcal{D}(\mathcal{P})$ has a finite trivializer. By Theorem 5.1 this means that $\mathrm{Ker}(\alpha)$ is finitely generated as a normal subgroup of the free group $\mathcal{F} = \mathcal{F}(X)$, where $X := \{x_1, \ldots, x_n\}$ and $\alpha : \mathcal{F} \to \mathcal{G}$ is defined through $\alpha(x_i) := a_i$ $(1 \leq i \leq n)$. Hence, $\mathcal{A} \cong \mathcal{F}/\mathrm{Ker}(\alpha)$ is finitely presented.

Conversely, if $\mathcal{A}$ is finitely presented, then $\mathrm{Ker}(\alpha)$ is finitely generated as a normal subgroup, and hence, $\mathcal{D}$ has a finite trivializer by Theorem 5.1. Hence, if $\mathcal{G}$ is FDT, then also $\mathcal{M}$ is FDT by Theorem 3.2. □

The special case $a_i = b_i$ $(1 \leq i \leq n)$ yields Theorem 1(a) of [5].

References

1. S.I. Adjan, Defining relations and algorithmic problems for groups and semigroups, *Proc. Steklov Institute of Mathematics* **85**, Amer. Math. Soc., Providence, RI (1966).
2. M. Bestvina and N. Brady, Morse theory and finiteness properties of groups, *Invent. Math.* **129**, 445–470 (1997).
3. C.W. Kilgour, *Relative monoid presentations*, Preprint No. 96/44, University of Glasgow, Department of Mathematics (1996).
4. S.J. Pride and F. Otto, Relative monoid presentations and finite derivation type, *Mathematische Schriften Kassel*, 15/03 (2003).
5. S.J. Pride and F. Otto, For rewriting systems the topological finiteness conditions FDT and FHT are not equivalent, *J. London Math. Soc.* **69**, 363–382 (2004).
6. S.J. Pride and J. Wang, *Relative rewriting systems*, Preprint, University of Glasgow (2002).
7. C.C. Squier, F. Otto, and Y. Kobayashi, A finiteness condition for rewriting systems, *Theoret. Comput. Sci.* **131**, 271–294 (1994).
8. X. Wang and S.J. Pride, Second order Dehn functions of groups and monoids, *Int. J. Algebra and Comput.* **10**, 425–456 (2000).

LITERAL VARIETIES AND PSEUDOVARIETIES OF HOMOMORPHISMS ONTO ABELIAN GROUPS

L. POLÁK *

Department of Mathematics, Masaryk University
Janáčkovo nám 2a, 662 95 Brno, Czech Republic
E-mail: polak@math.muni.cz

We initiate a systematic study of the lattices of literal varieties and pseudovarieties of homomorphisms onto monoids. The case of abelian groups is completely solved.

1. Introduction

The study of varieties of languages and pseudovarieties of monoids is a well-established part of the algebraic language theory – see the book by Almeida [1] and the survey by Pin [5]. Recently, new horizons were opened by (Ésik and Ito)[2] by considering the so-called literal varieties of languages and more generally by Straubing [7] with his $\mathfrak{C}$-varieties, where $\mathfrak{C}$ is a category of finitely generated monoids with certain monoid homomorphisms. The classical Eilenberg correspondence was modified to relate the literal varieties of languages and the literal varieties of homomorphisms from finitely generated monoids onto finite monoids by (Ésik and Larsen) [3] and more generally by Straubing [7].

The equational characterizations of varieties and pseudovarieties of monoids are well-known results by Birkhoff and Reiterman, respectively. In case of homomorphisms they are due to the author [6] and Kunc [4].

The aim of this contribution is to start a systematic study of the lattices of literal varieties and pseudovarieties of homomorphisms from free monoids onto monoids. It is natural to consider the abelian groups first.

After this Section we continue with preliminaries and we recall a modified Birkhoff's theorem. In Section 3 we consider the literal varieties of

*Supported by the Ministry of Education of the Czech Republic under the Project MSM 0021622409

homomorphisms onto abelian groups and Section 4 deals with pseudovarieties. In Section 5 we describe the corresponding varieties of languages. We close the paper with two remarks.

2. Preliminaries

We use the following notation :
$\mathbb{N} = \{1, 2, \dots\}$, $\mathbb{N}_0 = \mathbb{N} \cup \{0\}$, $\mathbb{N}_0^\infty = \mathbb{N}_0 \cup \{\infty\}$,
$\mathbb{Z}$ is the class of all integers,
$\mathbb{P}$ is the class of all primes,
$X = \{x_1, x_2, \dots\}$ is the set of all variables, $X_n = \{x_1, \dots, x_n\}$ for $n \in \mathbb{N}$,
$\mathcal{M}$ is the class of all monoids.
For $\mathcal{V} \subseteq \mathcal{M}$, $\mathsf{Fin}\,\mathcal{V}$ is the class of all finite members of $\mathcal{V}$,
let

$$\boldsymbol{\mathcal{M}} = \{\, \phi : Y^* \twoheadrightarrow M \mid Y \text{ is a non-empty set, } M \in \mathcal{M} \,\}$$

be the class of all surjective homomorphisms from free monoids onto monoids,
for $\boldsymbol{\mathcal{V}} \subseteq \boldsymbol{\mathcal{M}}$, let

$$\mathbf{Fin}\,\boldsymbol{\mathcal{V}} = \{\, (\phi : Y^* \twoheadrightarrow M) \in \boldsymbol{\mathcal{V}} \mid \text{both of } Y,\, M \text{ are finite} \,\}$$

be the class of all *finite* homomorphisms from $\boldsymbol{\mathcal{V}}$,
for a monoid homomorphism $\alpha : M \to N$, let $\mathbf{im}\,\alpha$ denote the submonoid of N induced by the image of α,
for $u \in X^*$, $i \in \mathbb{N}$, let $|u|_i$ denote the number of occurrences of x_i in u; let $|u|$ be the length of u.
Let $\mathfrak{L}$ be the category having all Y^*'s as objects and

$$f \in \mathfrak{L}(Z^*, Y^*) \text{ if and only if } f(Z) \subseteq Y\ .$$

We speak about *literal* homomorphisms.

Recall that an n-ary *identity* is a pair $u = v$ where $u, v \in X_n^*$, $n \in \mathbb{N}$. A monoid $M \in \mathcal{M}$ *satisfies* $u = v$ if

$$(\ \forall \text{ homomorphism } \alpha : X_n^* \to M\ )\ \alpha(u) = \alpha(v)\ .$$

In fact, the choice of n is not significant and we write $M \models u = v$ in this case. For a class $\mathcal{V} \subseteq \mathcal{M}$, we put

$$\mathsf{Id}\,\mathcal{V} = \{\, (u, v) \in X^* \times X^* \mid (\ \forall\, M \in \mathcal{V}\ )\ M \models u = v \,\}\ .$$

Let $\Pi \subseteq X^* \times X^*$ be a set of identities. We put

$$\mathsf{Mod}\,\Pi = \{\, M \in \mathcal{M} \mid (\ \forall\, \pi \in \Pi\ )\ M \models \pi \,\}\ .$$

A class of monoids is a *variety* if it is closed with respect to the forming of homomorphic images, submonoids and products. Similarly, a class of finite monoids is a *pseudovariety* if it is closed with respect to the forming of homomorphic images, submonoids and products of finite families. A congruence ρ on Y^* is *fully invariant* if for each $u, v \in Y^*$ and each endomorphism $g : Y^* \to Y^*$, $u \,\rho\, v$ implies $g(u) \,\rho\, g(v)$.

The following theorem is classical :

Result 2.1. (Birkhoff) *The mappings*

$$\mathcal{V} \mapsto \mathsf{Ic}\, \mathcal{V} \text{ and } \Pi \mapsto \mathsf{Mod}\, \Pi$$

are mutually inverse bijections between the class of all varieties of monoids and the class of all fully invariant congruences on X^*.

For a class $\boldsymbol{\mathcal{V}} \subseteq \boldsymbol{\mathcal{M}}$, we define

$$\mathbf{H}\, \boldsymbol{\mathcal{V}} = \{\, \sigma\phi : Y^* \twoheadrightarrow N \mid (\phi : Y^* \twoheadrightarrow M) \in \boldsymbol{\mathcal{V}},\ \sigma : M \twoheadrightarrow N \text{ a surj. homom. } \},$$

$$\mathbf{S}_{\mathfrak{L}}\, \boldsymbol{\mathcal{V}} = \{\, \phi f : Z^* \twoheadrightarrow \mathbf{im}\,(\phi f) \mid Z \neq \emptyset,\ f \in \mathfrak{L}(Z^*, Y^*),\ (\phi : Y^* \twoheadrightarrow M) \in \boldsymbol{\mathcal{V}} \}\,,$$

$$\mathbf{P}\boldsymbol{\mathcal{V}} =$$

$$\{\, (\phi_\gamma)_{\gamma\in\Gamma} : Y^* \twoheadrightarrow \mathbf{im}\,((\phi_\gamma)_{\gamma\in\Gamma}) \mid \Gamma \text{ a set, } (\phi_\gamma : Y^* \twoheadrightarrow M_\gamma) \in \boldsymbol{\mathcal{V}} \text{ for } \gamma \in \Gamma \,\}$$

(here $(\phi_\gamma)_{\gamma\in\Gamma} : Y^* \to \prod_{\gamma\in\Gamma} M_\gamma,\ u \mapsto (\phi_\gamma(u))_{\gamma\in\Gamma}$).

A class $\boldsymbol{\mathcal{V}} \subseteq \boldsymbol{\mathcal{M}}$ is a $\mathfrak{L}$-*variety* if it is closed with respect to the operators $\mathbf{H}$, $\mathbf{S}_{\mathfrak{L}}$ and $\mathbf{P}$. Similarly, a class $\boldsymbol{\mathcal{X}} \subseteq \mathbf{Fin}\, \boldsymbol{\mathcal{M}}$ is an $\mathfrak{L}$-*pseudovariety* of finite homomorphisms onto monoids if it is closed with respect to $\mathbf{H}$, $\mathbf{S}_{\mathfrak{L}}$ and $\mathbf{P}_f$ (products of finite families).

Let $u, v \in X_n^*$. A homomorphism $(\phi : Y^* \twoheadrightarrow M) \in \boldsymbol{\mathcal{M}}$ $\mathfrak{L}$-*satisfies* $u = v$ if

$$(\,\forall\, f \in \mathfrak{L}(X_n^*, Y^*)\,)\ (\phi f)(u) = (\phi f)(v)\ .$$

We write $\phi \models_{\mathfrak{L}} u = v$. For a class $\boldsymbol{\mathcal{V}} \subseteq \boldsymbol{\mathcal{M}}$, we put

$$\mathsf{Id}_{\mathfrak{L}}\, \boldsymbol{\mathcal{V}} = \{\, (u, v) \in X^* \times X^* \mid (\,\forall\, \phi \in \boldsymbol{\mathcal{V}}\,)\ \phi \models_{\mathfrak{L}} u = v \}\ .$$

Let $\Pi \subseteq X^* \times X^*$ be a set of identities. We set

$$\phi \models_{\mathfrak{L}} \Pi \text{ if } (\forall\, \pi \in \Pi\,)\ \phi \models_{\mathfrak{L}} \pi\ , \text{ and}$$

$$\mathbf{Mod}_{\mathfrak{L}}\, \Pi = \{\, \phi \in \boldsymbol{\mathcal{M}} \mid \phi \models_{\mathfrak{L}} \Pi \}\ .$$

Further, for $\pi \in X^* \times X^*$, the meaning of $\Pi \models_{\mathfrak{L}} \pi$ is

$$(\forall \phi \in \mathcal{M}) (\phi \models_{\mathfrak{L}} \Pi \text{ implies } \phi \models_{\mathfrak{L}} \pi) ,$$

and Π is $\mathfrak{L}$-*closed* if

$$(\forall \pi \in X^* \times X^*) (\Pi \models_{\mathfrak{L}} \pi \text{ implies } \pi \in \Pi) .$$

A congruence ρ on Y^* is $\mathfrak{L}$-*invariant* if for each $u, v \in Y^*$ and each $g \in \mathfrak{L}(Y^*, Y^*)$, $u \,\rho\, v$ implies $g(u) \,\rho\, g(v)$.

Lemma 2.1. *Let* $\Pi \subseteq X^* \times X^*$. *Then* Π *is an* $\mathfrak{L}$-*closed set of identities if and only if it is an* $\mathfrak{L}$-*invariant congruence.*

Proof. $\Rightarrow$: Let $\phi \in \mathcal{M},\ u, v, w \in X_n,\ g \in \mathfrak{L}(X^*, X^*)$. Then

$$\phi \models_{\mathfrak{L}} u = u ,$$

$$\phi \models_{\mathfrak{L}} u = v \text{ implies } \phi \models_{\mathfrak{L}} v = u ,$$

$$\phi \models_{\mathfrak{L}} u = v,\ v = w \text{ implies } \phi \models_{\mathfrak{L}} v = w ,$$

$$\phi \models_{\mathfrak{L}} u = v \text{ implies } \phi \models_{\mathfrak{L}} uw = vw,\ wu = wv ,$$

$$\phi \models_{\mathfrak{L}} u = v \text{ implies } \phi \models_{\mathfrak{L}} g(u) = g(v) .$$

$\Leftarrow$: Conversely, let $\rho \subseteq X^* \times X^*$ be an $\mathfrak{L}$-invariant congruence, $u, v \in X_n^*,\ \rho \models_{\mathfrak{L}} u = v$. We will show that $(u, v) \in \rho$.

Realize first that

$$(\rho^\sharp : X^* \twoheadrightarrow X^*/\rho,\ w \mapsto w\rho) \models_{\mathfrak{L}} \rho .$$

Indeed, let $p, q \in X_m^*,\ (p, q) \in \rho,\ f \in \mathfrak{L}(X_m^*, X^*)$. Then $(f(p), f(q)) \in \rho$ and $(\rho^\sharp f)(p) = (\rho^\sharp f)(q)$.

Thus $\rho^\sharp \models_{\mathfrak{L}} u = v$ and for $f \in \mathfrak{L}(X_n^*, X^*)$ identical on X_n, we have $(\rho^\sharp f)(u) = (\rho^\sharp f)(v)$, that is, $(u, v) \in \rho$. $\square$

Theorem 2.1. *The mappings*

$$\mathcal{V} \mapsto \mathbf{Id}_{\mathfrak{L}}\, \mathcal{V} \text{ and } \Pi \mapsto \mathbf{Mod}_{\mathfrak{L}}\, \Pi$$

are mutually inverse bijections between the class of all $\mathfrak{L}$-*varieties of homomorphisms onto monoids and the class of all* $\mathfrak{L}$-*invariant congruences on* X^*.

Proof. Clearly, for each $\phi \in \mathcal{M}$, we have that $\mathbf{Id}_{\mathfrak{L}}\,\{\phi\}$ is an $\mathfrak{L}$-closed set of identities. Since the intersection of any family of $\mathfrak{L}$-closed sets is again $\mathfrak{L}$-closed, for each $\boldsymbol{\mathcal{V}} \subseteq \mathcal{M}$, the set $\mathbf{Id}_{\mathfrak{L}}\,\boldsymbol{\mathcal{V}}$ is also $\mathfrak{L}$-closed.

Further, if $\Pi \subseteq X^* \times X^*$ is $\mathfrak{L}$-closed, then $\mathbf{Id}_{\mathfrak{L}}\mathbf{Mod}_{\mathfrak{L}}\,\Pi \subseteq \Pi$. The opposite inclusion is valid for an arbitrary Π.

Let $\boldsymbol{\mathcal{V}}$ be an $\mathfrak{L}$-variety of homomorphisms onto monoids. By [6], Theorem 1, there exists $\Pi \subseteq X^* \times X^*$ such that $\boldsymbol{\mathcal{V}} = \mathbf{Mod}_{\mathfrak{L}}\,\Pi$. Clearly, $\mathbf{Id}_{\mathfrak{L}}\,\boldsymbol{\mathcal{V}} \supseteq \Pi$. Applying the operator $\mathbf{Mod}_{\mathfrak{L}}$, we get $\mathbf{Mod}_{\mathfrak{L}}\mathbf{Id}_{\mathfrak{L}}\,\boldsymbol{\mathcal{V}} \subseteq \mathbf{Mod}_{\mathfrak{L}}\,\Pi = \boldsymbol{\mathcal{V}}$. The opposite inclusion holds for an arbitrary class $\boldsymbol{\mathcal{V}}$.

Using Lemma 2.1. we get the statement. □

We write $\rho_{\mathcal{V}}$ and $\boldsymbol{\rho}_{\mathfrak{L},\boldsymbol{\mathcal{V}}}$ instead of $\mathsf{Mod}\,\mathcal{V}$ and $\mathbf{Mod}_{\mathfrak{L}}\,\boldsymbol{\mathcal{V}}$. Further, for $\rho \subseteq X^* \times X^*$, we write $\rho_n = \rho \cap (X_n^* \times X_n^*)$.

3. Varieties of homomorphisms onto abelian groups

Lemma 3.1. *Let $\mathcal{V}$ be a variety of monoids such that $\mathsf{Fin}\,\mathcal{V}$ consists of groups. Then there exists $l \in \mathbb{N}$ such that $\mathcal{V}$ satisfies the identity $x^l = 1$.*

Proof. Let F be the (relatively) free monoid in $\mathcal{V}$ over the set $\{a\}$. Suppose that $F = \{1, a, a^2, \dots\}$ is infinite. Then $k, l \in \mathbb{N}_0$, $a^k = a^l$ implies $k = l$. Putting $a, a^2, \dots$ into a single class we get a congruence τ such that F/τ is finite and it is not a group, a contradiction. Consequently, F is of the form

$$1, a, a^2, \dots, a^{k+l-1} \text{ pairwise different, } a^{k+l} = a^k \ ,$$

for some $k \in \mathbb{N}_0$, $l \in \mathbb{N}$. Since F is a group, we have $k = 0$. □

The following is easy and well-known :

Result 3.1. *The varieties of monoids consisting of abelian groups are exactly*

$$\mathcal{A}(l) = \mathsf{Mod}\,(\,xy = yx,\ x^l = 1\,),\ \text{where } l \in \mathbb{N}\ .$$

Moreover,

$$\mathcal{A}(l) \subseteq \mathcal{A}(l') \text{ if and only if } l|l'\ .$$

The corresponding fully invariant congruences are

$$\rho(l) = \{\,(u,v) \in X^* \times X^* \mid (\ \forall\, i \in \mathbb{N}\ )\ |u|_i \equiv |v|_i \bmod l\,\}\ .$$

We present a proof as a preparation for the proof of the next theorem.
Proof. Notice first that a group M satisfying the identities $x^k = 1,\ x^l = 1$ satisfies also $x^{\gcd(k,l)} = 1$. Indeed, by Bezout's Theorem there exist $r, s \in \mathbb{Z}$ such that $rk + sl = \gcd(k,l)$. Then for $a \in M$, $a^{rk+sl} = (a^k)^r (a^l)^s = 1$. Let $\mathcal{V}$ be a variety of monoids consisting of abelian groups. By Lemma 3.1, there is $l \in \mathbb{N}$ such that $\mathcal{V} \models x^l = 1$; take the smallest one with respect to the divisibility. Clearly, $\rho(l) \subseteq \rho_{\mathcal{V}}$. Let

$$(x_1^{k_1} \dots x_n^{k_n}, x_1^{l_1} \dots x_n^{l_n}) \in \rho_{\mathcal{V}} \setminus \rho(l) .$$

There exists i such that $l \not| (k_i - l_i)$ and we have $x^{\gcd(l, k_i - l_i)} = 1$, a contradiction. □

Theorem 3.1. *The $\mathfrak{L}$-varieties of homomorphisms onto abelian groups are exactly*

$$\boldsymbol{\mathcal{A}}(k,l) = \mathbf{Mod}_{\mathfrak{L}}\,(\, xy = yx,\ x^k = y^k,\ x^l = 1\,),\ \text{where } k, l \in \mathbb{N},\ k \mid l\ .$$

Moreover,

$$\boldsymbol{\mathcal{A}}(k,l) \subseteq \boldsymbol{\mathcal{A}}(k',l') \text{ if and only if } k|k' \text{ and } l|l' .$$

The corresponding $\mathfrak{L}$-invariant congruences are $\boldsymbol{\rho}(k,l) =$

$$= \{\, (u,v) \in X^* \times X^* \mid (\ \forall\, i \in \mathbb{N}\ )\ |u|_i \equiv |v|_i \bmod k,\ \text{and } |u| \equiv |v| \bmod l\,\} .$$

Proof. Similarly as above, if a homomorphism $\phi : Y^* \twoheadrightarrow M$ onto an abelian group $\mathfrak{L}$-satisfies the identities $x^k = y^k,\ x^l = y^l$, then it also $\mathfrak{L}$-satisfies $x^{\gcd(k,l)} = y^{\gcd(k,l)}$.

Notice that $(\phi : Y^* \twoheadrightarrow M) \models_{\mathfrak{L}} xy = yx$ if and only if $M \models xy = yx$. In the presence of $xy = yx$ the same is true for the identity $x^l = 1$. Let $\boldsymbol{\mathcal{V}}$ be an $\mathfrak{L}$-variety of homomorphisms onto abelian groups. By Lemma 3.1, there is $l \in \mathbb{N}$ such that $\mathcal{V} \models x^l = 1$; take the smallest one with respect to the divisibility. Of course, $\boldsymbol{\mathcal{V}} \models_{\mathfrak{L}} x^l = y^l$. Let k be the smallest l with the last property. Clearly, $\boldsymbol{\rho}(k,l) \subseteq \boldsymbol{\rho}_{\boldsymbol{\mathcal{V}}}$. Let $(u,v) \in \boldsymbol{\rho}_{\boldsymbol{\mathcal{V}}} \setminus \boldsymbol{\rho}(k,l)$. If $|u| \not\equiv |v| \bmod l$, substitute a variable x for all variables. This would lead to a contradiction with the choice of l.

Using the identities $xy = yx,\ x^l = 1$, there is an $\mathfrak{L}$-equivalent identity to $u = v$, say $x_1^{k_1} \dots x_n^{k_n} = 1$ with some i such that $k \not| k_i$. Using consecutively the substitutions $x_i \mapsto z,\ x_i \mapsto t$ we would get $z^{k_i} = t^{k_i},\ z^{\gcd(k,k_i)} = t^{\gcd(k,k_i)}$, a contradiction. □

4. Pseudovarieties of finite homomorphisms onto abelian groups

Result 4.1. *(i)* (Eilenberg and Schützenberger, Baldwin and Berman, Ash, see Almeida [1] Proposition 3.2.6) *Pseudovarieties of finite monoids are exactly classes of the form* $\mathsf{Fin}\,\mathcal{V}$ *where* $\mathcal{V}$ *is a union of a chain of varieties of monoids.*
(ii) (Polák [6] Theorem 3) $\mathfrak{L}$*-pseudovarieties of finite homomorphisms onto monoids are exactly classes of the form* $\mathbf{Fin}\,\boldsymbol{\mathcal{V}}$ *where* $\boldsymbol{\mathcal{V}}$ *is a union of a chain of* $\mathfrak{L}$*-varieties of homomorphisms onto monoids.*

Notice that the class $\mathcal{A}$ of all abelian groups does not form a variety of monoids and that the class $\mathsf{Fin}\,\mathcal{A}$ is a pseudovariety of finite monoids. Consequently, we can speak about pseudovarieties of finite abelian groups instead of pseudovarieties of finite monoids consisting of abelian groups. Moreover, the following is true :

Corollary 4.1. *(i) Pseudovarieties of finite abelian groups are exactly classes of the form* $\mathsf{Fin}\,\mathcal{V}$ *where* $\mathcal{V}$ *is a union of a chain of varieties of monoids consisting of abelian groups.*
(ii) $\mathfrak{L}$*-pseudovarieties of finite homomorphisms onto abelian groups are exactly classes of the form* $\mathbf{Fin}\,\boldsymbol{\mathcal{V}}$ *where* $\boldsymbol{\mathcal{V}}$ *is a union of a chain of* $\mathfrak{L}$*-varieties of homomorphisms onto abelian groups.*

Proof. (i) Let $\mathcal{X}$ be a pseudovariety of finite groups. By Result 4.1 there exists a chain $(\mathcal{V}_i)_{i\in I}$ of varieties of monoids such that $\mathcal{X} = \mathsf{Fin}\,(\ \bigcup_{i\in I} \mathcal{V}_i\ )$. The last class could be written as $\bigcup_{i\in I} \mathsf{Fin}\,\mathcal{V}_i$. By Lemma 3.1, for each $i \in I$, there is $l_i \in \mathbb{N}$ such that all members of $\mathcal{V}_i$ satisfy the identity $x^{l_i} = 1$. Now $\mathcal{V}_i \cap \mathcal{A}$ is a variety of monoids consisting of abelian groups, for $i \in I$, and $\mathcal{X} = \mathsf{Fin}\,(\ \bigcup_{i\in I}\ (\ \mathcal{V}_i \cap \mathcal{A}\ )\ )$.
The proof of (ii) is similar. □

For $f \in \prod_{p\in\mathbb{P}} \mathbb{N}_0^\infty$, we put $\langle f\rangle =$

$$\{\, p_1^{g_1} \dots p_r^{g_r} \mid r \in \mathbb{N},\ p_1,\dots,p_r \in \mathbb{P} \text{ pairwise diff.},\ g_1 \le f_{p_1},\dots,g_r \le f_{p_r} \,\}\ .$$

Theorem 4.1. *(i) Pseudovarieties of abelian groups correspond to the elements of* $\prod_{p\in\mathbb{P}} \mathbb{N}_0^\infty$*. More precisely,*

$$\mathcal{A}[f] = \{\, M \in \mathsf{Fin}\,\mathcal{M} \mid (\ \forall\, a \in M\ )\ (\ \exists\, l \in \langle f\rangle\ ) \text{ such that } a^l = 1 \,\}\ .$$

Moreover,

$$\mathcal{A}[f] \subseteq \mathcal{A}[f'] \text{ if and only if } f \le f'\ .$$

(ii) $\mathfrak{L}$-pseudovarieties of finite homomorphisms onto abelian groups correspond to the elements of $(e, f) \in \prod_{p\in\mathbb{P}} \mathbb{N}_0^\infty \times \prod_{p\in\mathbb{P}} \mathbb{N}_0^\infty$ *with* $e \leq f$. *More precisely,*

$$\boldsymbol{\mathcal{A}}[e,f] = \{ (\phi : Y^* \twoheadrightarrow M) \in \textsf{\textbf{Fin}}\,\boldsymbol{\mathcal{M}} \mid$$

$$M \in \mathcal{A}[f] \text{ and } (\ \forall\ y,z \in Y\ )\ (\ \exists\ k \in \langle e \rangle\ ) \text{ such that } (\phi(y))^k = (\phi(z))^k \}\ .$$

Moreover,

$$\boldsymbol{\mathcal{A}}[e,f] \subseteq \boldsymbol{\mathcal{A}}[e',f'] \text{ if and only if } e \leq e',\ f \leq f'\ .$$

Proof. (i) Let us consider the union $\mathcal{V}$ of a chain of varieties of monoids consisting of abelian groups. We can suppose that our chain starts at $\mathcal{A}(1)$ and each pair of consecutive members forms a covering in the lattice of all varieties of monoids. Notice that each $\mathcal{A}(l)$ is covered exactly by $\mathcal{A}(pl)$, $p \in \mathbb{P}$. The union of our chain is $\mathcal{A}[f]$ where, for $p \in \mathbb{P}$, the number f_p says how many times we used a covering corresponding to p.

Let $\mathcal{V}$ and $\mathcal{V}'$ be two different unions as above. Using finite abelian groups $(\mathbb{Z}/n\mathbb{Z}, +)$, $n \in \mathbb{N}$, we see that $\mathsf{Fin}\,\mathcal{V} \neq \mathsf{Fin}\,\mathcal{V}'$. The rest of the statement is clear.

(ii) The proof is analogous to that of (i). When showing that **Fin** is injective, use the homomorphisms

$$\phi(k,l) : \{y,z\}^* \twoheadrightarrow (\mathbb{Z}/k\mathbb{Z}, +) \times (\mathbb{Z}/l\mathbb{Z}, +),\ y \mapsto (0,1),\ z \mapsto (1,1)\ ,$$

for $k, l \in \mathbb{N}$, $k|l$. □

5. Corresponding languages

Let Y be non-empty finite alphabet. A regular language $L \subseteq Y^*$ defines the *syntactic congruence* $\sim_L$ on Y^* by

$$u \sim_L v \text{ if and only if } (\ \forall\ p,q \in Y^*\ )\ (\ puq \in L \Leftrightarrow pvq \in L\ )\ .$$

The factor-structure $M(L) = Y^*/\sim_L$ is called the *syntactic monoid* of L and $\phi_L : Y^* \twoheadrightarrow M(L)$, $u \mapsto u \sim_L$ is the *syntactic homomorphism* of L.

A *class* of (regular) languages is an operator $\mathscr{L}$ assigning to every non-empty finite set Y a set $\mathscr{L}(Y)$ of regular languages over the alphabet Y. Such a class is a *variety* if

(i) each $\mathscr{L}(Y)$ is closed with respect to finite intersections, finite unions, complements, and quotients, and

(ii) for each finite sets Y and Z and $f : Z^* \to Y^*$, $K \in \mathscr{L}(Y)$ implies

$$f^{-1}(K) = \{ v \in Z^* \mid f(v) \in K \} \in \mathscr{L}(Z)\ .$$

It is a $\mathfrak{L}$-*variety* if (i) is true and (ii) is satisfied for all $f \in \mathfrak{L}(Z^*, Y^*)$.

We assign to a class $\mathscr{L}$ of regular languages the pseudovariety

$$\mathsf{M}(\mathscr{L}) = \langle \{ M(L) \mid Y \text{ is a non-empty finite set and } L \in \mathscr{L}(Y) \} \rangle_{\mathcal{M}}$$

of finite monoids generated by all syntactic monoids of members of $\mathscr{L}$, and the $\mathfrak{L}$-pseudovariety

$$\mathbf{M}(\mathscr{L}) = \langle \{ \phi_L \mid Y \text{ is a non-empty finite set and } L \in \mathscr{L}(Y) \} \rangle_{\boldsymbol{\mathcal{M}}}$$

of finite homomorphisms onto monoids generated by all syntactic homomorphisms of members of $\mathscr{L}$. Conversely, for a pseudovariety $\mathcal{X}$ of finite monoids, an $\mathfrak{L}$-pseudovariety $\boldsymbol{\mathcal{X}}$ of finite homomorphisms onto monoids, and a non-empty finite set Y we put

$$(\mathsf{L}(\mathcal{X}))(Y) = \{ L \subseteq Y^* \mid M(L) \in \mathcal{X} \} \text{ and}$$

$$(\mathbf{L}(\boldsymbol{\mathcal{X}}))(Y) = \{ L \subseteq Y^* \mid \phi_L \in \boldsymbol{\mathcal{X}} \} .$$

Result 5.1. *(i)* (Eilenberg, see Almeida [1] Corollary 3.3.7, Pin [5] Corollary 1.4.8) *The operators* M *and* L *are mutually inverse bijections between the class of all varieties of languages and the class of all pseudovarieties of finite monoids.*
(ii) (Straubing [7] Theorems 1 and 2, (Ésik and Larsen) [3] Theorem 5.8) *The operators* $\mathbf{M}$ *and* $\mathbf{L}$ *are mutually inverse bijections between the class of all* $\mathfrak{L}$*-varieties of languages and the class of all* $\mathfrak{L}$*-pseudovarieties of finite homomorphisms onto monoids.*

Theorem 5.1. *(i) Let* $\mathcal{V}$ *be a variety of monoids, and let* $n \in \mathbb{N}$*. Then* $L \in (\mathsf{L}(\mathsf{Fin}\,\mathcal{V}))(X_n)$ *if and only if* L *is a regular language which is a union of classes of* $X_n^*/\rho_{\mathcal{V},n}$.
(ii) Let $\boldsymbol{\mathcal{V}}$ *be an* $\mathfrak{L}$*-variety of monoids, and let* $n \in \mathbb{N}$*. Then* $L \in (\mathbf{L}(\mathbf{Fin}\,\boldsymbol{\mathcal{V}}))(X_n)$ *if and only if* L *is a regular language which is a union of classes of* $X_n^*/\boldsymbol{\rho}_{\mathfrak{L},\boldsymbol{\mathcal{V}},n}$.

Proof. (ii) Let $L \subseteq X_n^*,\ u, v \in X_m^*$. Then

$$(\phi_L : X_n^* \twoheadrightarrow M(L)) \models_{\mathfrak{L}} u = v$$

if and only if

$$(\forall\, x_{i_1}, \ldots, x_{i_m} \in X_n\,)\ \phi_L u(x_{i_1}, \ldots, x_{i_m}) = \phi_L v(x_{i_1}, \ldots, x_{i_m}) ,$$

which is equivalent to $(\forall\, x_{i_1}, \ldots, x_{i_m} \in X_n,\ p, q \in X_n^*\,)$

$$pu(x_{i_1}, \ldots, x_{i_m})q \in L \iff pv(x_{i_1}, \ldots, x_{i_m})q \in L \qquad (\dagger) .$$

$\Rightarrow$ of the statement : Let $u \in L$, $v \in X_n^*$, $\boldsymbol{\mathcal{V}} \models_{\mathfrak{L}} u = v$. Put $p = q = 1$, $x_{i_1} = x_1, \ldots, x_{i_n} = x_n$ (we have $m = n$) in (†).
$\Leftarrow$: Let $u, v \in X_n^*$, $\boldsymbol{\mathcal{V}} \models_{\mathfrak{L}} u = v$. Let $x_{i_1}, \ldots, x_{i_m} \in X_n$, $p, q \in X_n^*$. Then $\boldsymbol{\mathcal{V}} \models_{\mathfrak{L}} pu(x_{i_1}, \ldots, x_{i_m})q = pv(x_{i_1}, \ldots, x_{i_m})q$. Using that L is a union of classses of $X_n^*/\rho_{\mathfrak{L},\boldsymbol{\mathcal{V}},n}$, we get (†) and $\phi_L \models_{\mathfrak{L}} u = v$.

The proof of (i) is very similar; we write only $\mathcal{V}$, $\models$, $w_1, \ldots, w_m \in X_n^*$ instead of $\boldsymbol{\mathcal{V}}$, $\models_{\mathfrak{L}}$, $x_{i_1}, \ldots, x_{i_m} \in X_n$. □

Remark. If $X_n^*/\rho_{\mathcal{V},n}$ in Th. 5.1. (i) is finite then each union of its classes is a regular language; similarly in the item (ii). Notice that this is the case for all varieties from Result 3.1. and Theorem 3.1.

For $n, k, l \in \mathbb{N}$, $m_1, \ldots, m_n, m \in \mathbb{N}_0$ with $m_1 + \cdots + m_n \equiv m \bmod l$, we put

$$L(n, l; m_1, \ldots, m_n) = \{\, u \in X_n^* \mid |u|_1 \equiv m_1, \ldots, |u|_n \equiv m_n \bmod l \,\} ,$$

$$L(n, k, l; m_1, \ldots, m_n, m) =$$

$$= \{\, u \in X_n^* \mid |u|_1 \equiv m_1, \ldots, |u|_n \equiv m_n \bmod k,\ |u| \equiv m \bmod l \,\} .$$

Theorem 5.2. *(i)* (see also Pin [5] Proposition 5.8) *For $n, l \in \mathbb{N}$, the class $(\mathsf{L}(\mathsf{Fin}\,\mathcal{A}(l)))(X_n)$ consists of unions of the languages $L(n, l; m_1, \ldots, m_n)$.*
(ii) For $n, k, l \in \mathbb{N}$, the class $(\mathbf{L}(\mathbf{Fin}\,\boldsymbol{\mathcal{A}}(k, l)))(X_n)$ consists of unions of the languages $L(n, k, l; m_1, \ldots, m_n, m)$.

Proof. It follows immediately from Result 3.1, Theorems 3.1 and 5.1. and the Remark above. □

To get the classes of languages corresponding pseudovarieties which do not arise as classes of finite members of varieties, we can use :

Theorem 5.3. *(i) Let $\mathcal{V}$ be a union of a family of varieties of monoids, $\mathcal{V} = \bigcup_{i \in I} \mathcal{V}_i$. Then, for $n \in \mathbb{N}$,*

$$(\mathsf{L}(\mathsf{Fin}\,\mathcal{V}))(X_n) = \bigcup_{i \in I} (\mathsf{L}(\mathsf{Fin}\,\mathcal{V}_i))(X_n) .$$

(ii) Let $\boldsymbol{\mathcal{V}}$ be a union of a family of $\mathfrak{L}$-varieties of homomorphisms onto monoids, $\boldsymbol{\mathcal{V}} = \bigcup_{i \in I} \boldsymbol{\mathcal{V}}_i$. Then, for $n \in \mathbb{N}$,

$$(\mathbf{L}(\mathbf{Fin}\,\boldsymbol{\mathcal{V}}))(X_n) = \bigcup_{i \in I} (\mathbf{L}(\mathbf{Fin}\,\boldsymbol{\mathcal{V}}_i))(X_n) .$$

Proof. Both statement are obvious. □

6. Final Remarks

(1) We can present equational characterizations of classes from Theorem 4.1. In fact, it is quite similar to Theorems 8 and 11 of [6].

(2) Instead of the category $\mathfrak{L}$ of free monoids and literal homomorphisms we can consider the category $\mathfrak{M}$ of free monoids and length multiplying homomorphisms : $f \in \mathfrak{M}(Z^*, Y^*)$ if and only if there exists $m \in \mathbb{N}_0$ such that for all $z \in Z$ we have $|f(z)| = m$. One can show that the statements of Theorem 4.1 remain true.

References

1. J. Almeida, *Finite Semigroups and Universal Algebra*, World Scientific, 1994.
2. Z. Ésik and M. Ito, Temporal logic with cyclic counting and the degree of aperiodicity of finite automata, *Acta Cybernetica* **16**, 1–28 (2003) a preprint BRICS 2001.
3. Z. Ésik and K.G. Larsen, Regular languages defined by Lindström quantifiers, *Theoretical Informatics and Applications* **37**, 197–242 (2003), preprint BRICS 2002.
4. M. Kunc, Equational description of pseudovarieties of homomorphisms, *Theoretical Informatics and Applications* **37**, 243–254 (2003).
5. J.-E. Pin, Syntactic semigroups, Chapter 10 in *Handbook of Formal Languages*, G. Rozenberg and A. Salomaa eds, Springer (1997).
6. L. Polák, On varieties, generalized varieties and pseudovarieties of homomorphisms, *Contributions to General Algebra* **16**, Verlag Johannes Heyn, Klagenfurt, 173-187 (2005).
7. H. Straubing, On logical descriptions of regular languages, *Proc. LATIN 2002, Springer Lecture Notes in Computer Science* **2286**, 528–538 (2002).

FACTORIZABILITY IN CERTAIN CLASSES OVER INVERSE SEMIGROUPS

MÁRIA B. SZENDREI*
Bolyai Institute, University of Szeged,
Aradi vértanúk tere 1,
H-6720 Szeged, Hungary
E-mail: M.Szendrei@math.u-szeged.hu

McAlister's P-theorem on E-unitary inverse semigroups and his theorem on E-unitary covers are among the most beautiful results in semigroup theory. They have motivated research in a lot of directions. In this paper, we present the recent attempts in generalizing the notion of almost factorizability for orthodox and for weakly ample semigroups, respectively, and summarize the results obtained.

1. Preliminaries

For the undefined notions and notation, we refer to Howie [15], Lawson [17] and Petrich [25].

Given a monoid M, its identity is usually denoted by 1 and its group of units (i. e. the $\mathcal{H}$-class of 1) by $U(M)$. For example, the group of units of the symmetric inverse monoid $\mathcal{I}_A$ is the symmetric group $\mathcal{S}_A$. A monoid is called *unipotent* if the identity is the only idempotent element in it.

Let S be a semigroup. The relation $\widetilde{\mathcal{R}}$ on S is defined as follows: for all $a, b \in S$, we mean by $a\widetilde{\mathcal{R}}b$ that, for arbitrary $e \in E(S)$, the equality $ea = a$ holds if and only if $eb = b$ holds.

A semigroup S is said to be *weakly left ample* if the following conditions are fulfilled: the idempotents of S form a semilattice; the relation $\widetilde{\mathcal{R}}$ is a left congruence on S; for each $a \in S$, the $\widetilde{\mathcal{R}}$-class $R_a(S)$ of S containing a has an idempotent, necessarily unique, denoted by a^+; and for all $a \in S$ and $e \in E(S)$, we have $ae = (ae)^+a$. Note that, in a weakly left ample

Key words: inverse semigroup, orthodox semigroup, weakly ample semigroup, semidirect product, permissible set, translation, almost factorizability.
*Research partially supported by the Hungarian National Foundation for Scientific Research, grants no. T37877 and T48809.

monoid M, the $\widetilde{\mathcal{R}}$-class $\widetilde{R}_1(M)$ of 1 constitutes a unipotent submonoid in M.

The weakly left ample semigroups are considered as algebras of type $(2,1)$, so we refer, for example, to $(2,1)$-subsemigroups, $(2,1)$-homomorphism, etc. *Weakly right ample semigroups* are defined dually, in which case, we define $\widetilde{\mathcal{L}}$, and denote the respective unary operation by *. By a *weakly ample semigroup* we mean a semigroup that is both weakly left and weakly right ample. Therefore a weakly ample semigroup is an algebra of type $(2,1,1)$.

On every weakly left ample semigroup S, there exists a *least unipotent congruence* σ, and, for any $a, b \in S$, we have $a\sigma b$ if and only if $ea = eb$ for some $e \in E(S)$. In a weakly right ample semigroup, the least unipotent congruence can be dually given, and in a weakly ample semigroup, both descriptions are valid.

A *left ample semigroup* S is a weakly left ample semigroup in which the relation $\widetilde{\mathcal{R}}$ coincides with the relation $\mathcal{R}^*$ defined in the following way: for every $a, b \in S$, we mean by $a\mathcal{R}^*b$ that, for arbitrary elements $x, y \in S^1$, the equality $xa = ya$ holds if and only if $xb = yb$ holds. Thus, in a left ample monoid M, the submonoid $\widetilde{R}_1(M)$ is right cancellative. On a left ample semigroup, the relation σ is the *least right cancellative congruence*. Dually to a left ample semigroup, we may define a *right ample semigroup*, and by an *ample semigroup* we mean a semigroup that is both left and right ample.

In particular, an inverse semigroup S can be considered as a (weakly) ample semigroup with respect to the unary operations $^+$ and * defined by $a^+ = aa^{-1}$ and $a^* = a^{-1}a$, respectively, for any $a \in S$. In this case $\mathcal{R} = \mathcal{R}^* = \widetilde{\mathcal{R}}$, $\mathcal{L} = \mathcal{L}^* = \widetilde{\mathcal{L}}$, and σ is the least group congruence on S.

The notion of a semidirect product is needed in a somewhat different setting than that presented in Lawson [17].

Let X be an arbitrary semigroup and M a monoid. Suppose that M acts on X on the left by endomorphisms such that the endomorphism corresponding to 1 is the identity automorphism. In this case, we briefly say that M acts on X on the left. The image of an element $x \in X$ under the endomorphism corresponding to an element $a \in M$ is denoted by ${}^a x$. The *semidirect product* $X * M$ of X by M is defined to be the set $X \times M$ with the multiplication

$$(x, a)(y, b) = (x\, {}^a y, ab)\,.$$

The left-right dual of this construction is called a *reverse semidirect product* and is denoted by $M *_{\mathrm{rev}} X$. In particular, if M is a group then each reverse semidirect product of X by M is isomorphic to a semidirect product of X by M, and vice versa. If, moreover, X is a semilattice [band] then $X * M$ is an inverse [orthodox] semigroup. Furthermore, if X is a semilattice and M is a unipotent monoid [right cancellative monoid] then $X * M$ is a weakly left ample [left ample] semigroup.

An alternative construction stemming from a reverse semidirect product of a semilattice by a right cancellative monoid was introduced in the theory of left ample semigroups by Fountain and Gomes [10]. For our later convenience, we present the definition in a more general setting.

Let M be a unipotent monoid acting on a semilattice X on the right by injective endomorphisms whose images are order ideals of X. It is easy to see that

$$W(M,X) = \{(a, x^a) \in M *_{\mathrm{rev}} X : a \in M,\ x \in X\}$$

is a full subsemigroup in the reverse semidirect product $M *_{\mathrm{rev}} X$. Moreover, it can be checked to be weakly left ample with respect to the unary operation defined by $(a, x^a)^+ = (1, x)$ for any $(a, x^a) \in W(M, X)$. We refer to $W(M, X)$ as a *W-product* of X by M. In particular, if M is a right cancellative monoid then $W(M, X)$ is left ample.

2. Inverse semigroups

It is a general idea in the structure theory of inverse semigroups to build up inverse semigroups from semilattices and groups. There are two main approaches to obtain all inverse semigroups from semidirect products of semilattices by groups. The first approach produces any inverse semigroup as an idempotent separating homomorphic image of an inverse subsemigroup of such a semidirect product. The second one goes the other way round, that is, it produces any inverse semigroup as an inverse subsemigroup of an idempotent separating homomorphic image of such a semidirect product.

2.1. *First approach*

The fundamental step in the first approach is due to McAlister [18, 19] (see also Lawson [17] and Petrich [25]) who introduced the notion of an E-unitary inverse semigroup, described their structure — generally referred to as the P-theorem — by means of partially ordered sets and groups acting on them, and proved that each inverse semigroup has an E-unitary cover. The fact that the E-unitary inverse semigroups are just the inverse subsemigroups of semidirect products of semilattices by groups was proved by O'Carroll [23], by applying the P-theorem.

2.2. *Second approach*

McAlister's role was fundamental also in the second approach, see McAlister [20] and McAlister and Reilly [21]. A crucial observation of the latter paper (see also Lawson [17]) is that, to each inverse semigroup S, an E-unitary cover T can be constructed in the following way: we consider an embedding ι of S in a symmetric inverse monoid $\mathcal{I}_A$ such that each element $s\iota$ $(s \in S)$ can be extended to a permutation of A, and define T to be the subsemigroup

$$T = \{(s, \pi) : s\iota \text{ is a restriction of } \pi\}$$

in the direct product $S \times \mathcal{S}_A$. The fact that such an embedding exists is due to Chen and Hsieh [5]: if S is finite then, for example, ι can be chosen to be the Wagner–Preston representation, and if S is infinite then the Wagner–Preston representation can be modified by doubling the base set. The property important in this embedding led them to the following notion. An inverse monoid is called *factorizable* if $M = E(M) \cdot U(M)$, or, equivalently, if for each element $a \in M$, there is a unit $u \in U(M)$ with $a \leq u$. So they proved the following embedding theorem:

Theorem 2.1. *Each inverse semigroup (monoid) is embeddable in a factorizable inverse monoid.*

The connection of factorizable inverse monoids and semidirect products was cleared up by McAlister [20]:

Theorem 2.2. *For any inverse monoid M, the following statements are equivalent:*

(i) M is factorizable.
(ii) M is a homomorphic image of a semidirect product of a semilattice monoid by a group.
(iii) M is an idempotent separating homomorphic image of a semidirect product of a semilattice monoid by a group.

Theorem 2.3. *An inverse monoid is isomorphic to a semidirect product of a semilattice monoid by a group if and only if it is E-unitary and factorizable.*

The motivating observation on how to construct an E-unitary cover of an inverse semigroup from a certain embedding can be easily generalized for any embedding in a factorizable inverse monoid. It was proved by McAlister and Reilly [21] that all E-unitary covers are obtained in this way:

Theorem 2.4. *Given an embedding ι of an inverse semigroup S in a factorizable inverse monoid M, the subsemigroup $T = \{(s, u) : s\iota \leq u\}$ of the direct product $S \times U(M)$ is an E-unitary cover of S. Conversely, each E-unitary cover of S is, up to isomorphism, of this form.*

This theorem establishes an intimate connection between the two approaches. However, there is an asymmetry in the results cited so far: in the second approach, that is, in Theorems 2.2 and 2.3, we dealt only with semidirect products of semilattice monoids by groups. It is natural to ask what about the semigroup case. Although this question was addressed only much later by Lawson [16] (see also Lawson [17]), most of his main ideas and some of the results were implicit in McAlister [20]. An appropriate semigroup analogue of a factorizable inverse monoid was introduced and the analogues of Theorems 2.2 and 2.3 were proved.

Given an inverse semigroup S, we denote by $\hat{S}$ the monoid of the injective partial right translations of S, and by $C(S)$ the monoid of the permissible subsets of S. It is well known (see Petrich [25]) that $\hat{S}$ is isomorphic to $C(S)$, and there exists a natural embedding $\nu\colon S \to \hat{S}$ [similarly, $\nu\colon S \to C(S)$]. The inverse semigroup S is called *almost factorizable* if $S\nu \subseteq E(S\nu) \cdot U(\hat{S})$ [similarly, $S\nu \subseteq E(S\nu) \cdot U(C(S))$].

Theorem 2.5. *An inverse monoid is almost factorizable if and only if it is factorizable.*

Theorem 2.6. *If M is a factorizable inverse monoid then $M \setminus U(M)$ is an almost factorizable inverse semigroup. Conversely, each almost factorizable inverse semigroup is, up to isomorphism, of this form.*

Theorem 2.7. *For any inverse semigroup S, the following statements are equivalent:*

- *(i) S is almost factorizable.*
- *(ii) S is a homomorphic image of a semidirect product of a semilattice by a group.*
- *(iii) S is an idempotent separating homomorphic image of a semidirect product of a semilattice by a group.*

Theorem 2.8. *An inverse semigroup is isomorphic to a semidirect product of a semilattice by a group if and only if it is E-unitary and almost factorizable.*

Finally, let us mention that Theorems 2.1 and 2.5 easily imply

Corollary 2.1. *Each inverse semigroup is embeddable in an almost factorizable inverse semigroup.*

3. More general classes of semigroups

3.1. *First approach*

The first approach has been generalized for a number of classes which contain all inverse semigroups. Let us mention several of them.

An *E-unitary regular semigroup* is defined to be a regular semigroup S in which $E(S)$ is a unitary subset, or, equivalently, in which $E(S)$ is the kernel of the least group congruence. Thus an E-unitary regular semigroup is necessarily orthodox. An E-unitary regular semigroup T is called an *E-unitary cover* of an orthodox semigroup S if there exists an idempotent separating homomorphism from T onto S. Takizawa [28] and Szendrei [26] independently proved that each orthodox semigroup has an E-unitary cover. The question whether each E-unitary regular semigroup is embeddable in a semidirect product of a band by a group was answered in the negative by Billhardt [2]. However, there are plenty of E-unitary regular semigroups embeddable in a semidirect product of a band by a group in the sense that each orthodox semigroup has such an E-unitary cover, see Szendrei [27].

McAlister's result on the existence of E-unitary covers for inverse semigroups and his P-theorem were generalized for locally inverse semigroups by Pastijn [24]. An important role was played there by the locally inverse semigroups, called *straight locally inverse semigroups*, whose set of idempotents was a disjoint union of subsemilattices. For any locally inverse semigroup S, a special straight locally inverse semigroup T was constructed, such that there exists a homomorphism of a special kind from T onto S. Moreover, the structure of such straight locally inverse semigroups T was described by means of a construction reminiscent to a P-semigroup. A type of product, similar to a semidirect product, of a semilattice by a completely simple semigroup was introduced that was later referred to as a Pastijn product. A locally inverse semigroup S is called *weakly E-unitary* if, for every $e \in E(S)$ and $s \in S$ with $e \leq s$, we have $s \in E(S)$, or, equivalently, if the idempotent classes of the least completely simple congruence of S consist of idempotents only. The straight locally inverse semigroups T mentioned and the Pastijn products of semilattices — more generally, of normal bands — by completely simple semigroups are easily seen to be weakly E-unitary locally inverse semigroups. The structure theorem on straight locally inverse semigroups T mentioned was extended to any weakly E-unitary locally inverse semigroup by Billhardt and Szendrei [3].

Within the class of weakly left ample semigroups, the appropriate generalization of the notion of an E-unitary inverse semigroup is as follows: a weakly left ample semigroup S is called *proper* if $\widetilde{\mathcal{R}} \cap \sigma$ is the equality relation of S. In particular, an inverse semigroup, considered as a weakly left ample semigroup, is proper if and only if it is E-unitary. A proper weakly left ample semigroup T is said to be a *proper cover* of a weakly left ample semigroup S if there exists an idempotent separating $(2, 1)$-homomorphism from T onto S. The first results concerned left ample semigroups. Fountain [8] proved that each left ample semigroup has a proper cover, and Fountain and Gomes [10] showed that every proper left ample semigroup is embeddable in a W-product of a semilattice by a right cancellative monoid. Billhardt [1] proved an alternative embedding for proper left ample semigroups where a usual semidirect product appeared instead of a W-product. Gomes and Gould [11] established that, more generally, each weakly left ample semigroup has a proper cover, and every proper weakly left ample semigroup is embeddable in a semidirect product of a semilattice by a unipotent monoid.

3.2. *Second approach*

As far as the second approach is concerned, much less is known for more general classes than that of inverse semigroups.

Theorems 2.5, 2.6, 2.7 and Corollary 2.1 were generalized by Dombi [6] for straight locally inverse semigroups. She introduced notions of factorizability and of almost factorizability for this class. Furthermore, she proved, among others, that the almost factorizable straight locally inverse semigroups are just the idempotent separating homomorphic images of Pastijn products of semilattices by completely simple semigroups. Moreover, the structural desription she found for the almost factorizable straight locally inverse semigroups helped her to establish, by making use of a result due to Pastijn [24], that each straight locally inverse semigroup is embeddable in a(n almost) factorizable straight locally inverse semigroup. Although not stated explicitely, the analogue of Theorem 2.8 also holds.

A notion of, necessarily one-sided, factorizability was introduced for left ample monoids by El Qallali [7] as follows: a left ample monoid M is said to be *factorizable* if $M = E(M) \cdot R_1^*(M)$. Fountain and El Qallali [9] generalized Theorem 2.4 for E-dense left ample semigroups. Recall that a semigroup S is termed *E-dense* if, for every $a \in S$, there exists $b \in S$ with $bab = b$. They notice that, in particular, an E-dense left ample monoid is factorizable if and only if it is a factorizable inverse monoid. One of their main results characterizes, for any E-dense left ample semigroup S, its proper covers arising, in a very similar way as in the inverse case, from embeddings of S in factorizable inverse monoids.

Now we turn to summarizing the new results achieved recently in the classes of orthodox semigroups and of weakly (left) ample semigroups.

It is natural to generalize the notion of factorizability for orthodox monoids as follows: let us say that an orthodox monoid M is *factorizable* if $M = E(M) \cdot U(M)$. Similarly to the inverse case, one can prove the following generalization of Theorem 2.2:

Theorem 3.1. *For any orthodox monoid M, the following statements are equivalent:*

(i) M is factorizable.
(ii) M is a homomorphic image of a semidirect product of a band monoid by a group.
(iii) M is an idempotent separating homomorphic image of a semidirect product of a band monoid by a group.

Note that factorizable orthodox monoids have been already investigated by Blyth and McFadden [4], and they were called unit orthodox semigroups. Theorem 3.1 partly appears in McFadden [22].

When looking for an appropriate notion of almost factorizability within the class of orthodox semigroups, the ideas applied in the inverse case give less help. There is no hope to generalize permissible sets since the natural partial order is not compatible with the multiplication in general. There is no hope either to generalize $\hat{S}$ for any orthodox semigroup S. However, notice that if S is an inverse semigroup then the natural embedding $\nu\colon S \to \hat{S}$ can be factorized through the translational hull $\Omega(S)$ of S, that is, $\nu = \nu_0\psi$ where $\nu_0\colon S \to \Omega(S)$ is the natural embedding of S in $\Omega(S)$ and the definition of $\psi\colon \Omega(S) \to \hat{S}$ can be found in Petrich [25]. Furthermore, ψ is a monoid homomorphism whose restriction to the groups of units is an isomorphism. This gives the idea of the following definition. Let us call an orthodox semigroup *almost factorizable* if $S\nu_0 \subseteq E(S\nu_0) \cdot U(\Omega(S))$.

It is proved by Hartmann [14] that a part of Theorems 2.5–2.7 remains valid within the class of orthodox semigroups:

Theorem 3.2. *An orthodox monoid is almost factorizable if and only if it* *is factorizable.*

Theorem 3.3. *If M is a factorizable orthodox monoid then $M \setminus U(M)$ is an almost factorizable orthodox semigroup. Conversely, each almost factorizable orthodox semigroup is, up to isomorphism, of this form.*

Theorem 3.4. *For any orthodox semigroup S, the following statements are equivalent:*

(i) *S is almost factorizable.*
(ii) *S is an idempotent separating homomorphic image of a semidirect product of a band by a group.*

He notices that the class of all almost factorizable generalized inverse semigroups constitutes a proper subclass within the class of all homomorphic images of semidirect products of normal bands by groups. The latter class is characterized as follows:

Theorem 3.5. *A generalized inverse semigroup is a homomorphic image of a semidirect product of a normal band by a group if and only if its greatest inverse semigroup homomorphic image is almost factorizable.*

In general, the situation is much more complicated. He establishes a general characterization of all homomorphic images of semidirect products of bands by group.

Now we turn our attention to the class of weakly (left) ample semigroups. As we have mentioned, a notion of factorizability was introduced by Fountain and El Qallali [9] for left ample monoids. They also defined a notion of permissible sets, and generalized for left ample semigroups the result by McAlister and Reilly [21] on the construction of E-unitary covers of inverse semigroups from permissible sets. In a joint work with Gomes [12], we intended to find notions of factorizability and of almost factorizability that have closer link to the results on (weakly) left ample semigroups which belong to the first approach and are cited above. The simple but crucial initial observation is the following. If X is a semilattice and M is a unipotent monoid acting on X on the right then the reversed semidirect product $M *_{\mathrm{rev}} X$ is a weakly right ample semigroup with respect to the unary operation defined by $(a, x)^* = (1, x)$ for every $(a, x) \in M *_{\mathrm{rev}} X$. Since $W(M, X)$ is a full subsemigroup in $M *_{\mathrm{rev}} X$, it is also a $(2,1)$-subsemigroup, and so the weakly left ample semigroup $W(M, X)$ is also weakly ample.

Modifying the above definition due to El Qallali [7], we say that a weakly ample monoid M is *left factorizable* if $M = E(M) \cdot \widetilde{R}_1(M)$. It turns out that Theorem 2.2 can be generalized for weakly ample monoids as follows:

Theorem 3.6. *For any weakly ample monoid M, the following statements are equivalent:*

(i) *M is left factorizable.*
(ii) *M is a $(2,1,1)$-homomorphic image of a W-product of a semilattice monoid by a unipotent monoid.*
(iii) *M is an idempotent separating $(2,1,1)$-homomorphic image of a W-product of a semilattice monoid by a unipotent monoid.*

Since a W-product of a semilattice by a unipotent monoid is proper both as a weakly left and as a weakly right ample semigroup, it is natural to define a weakly left ample semigroup to be *proper*, if each of the equivalences $\widetilde{\mathcal{R}} \cap \sigma$ and $\widetilde{\mathcal{L}} \cap \sigma$ is the equality relation. The following analogue of Theorem 2.3 holds.

Theorem 3.7. *A weakly ample monoid is isomorphic to a W-product of a semilattice monoid by a unipotent monoid if and only if it is proper and left factorizable.*

An appropriate notion of almost factorizability is introduced by adapting for weakly ample semigroups the definition of a permissible set due to Fountain and El Qallali [9]. Let S be a weakly ample semigroup. A non-empty subset $A \subseteq S$ is said to be *permissible* if, for every $a \in A$ and $s \in S$ with $s^+a = s$ [or, equivalently, $as^* = s$], we have $s \in A$, and, for every $a, b \in S$, we have $a^+b = b^+a$ and $ab^* = ba^*$. One can check that the set $C(S)$ of all permissible sets in S forms a weakly ample monoid with respect to the usual set multiplication and to the following unary operations: $A^+ = \{a^+ : a \in A\}$ and $A^* = \{a^* : a \in A\}$. Moreover, S is naturally embedded in $C(S)$ by $\nu\colon S \to C(S)$, $s\nu = (s]$ where $(s] = \{z \in S : z^+s = z\}$. Note that if S is an inverse semigroup then a subset is permissible in it in this sense if and only if it is permissible in the usual sense. A weakly left ample semigroup S is termed *almost left factorizable* if $S\nu \subseteq E(S\nu) \cdot \widetilde{R}_1(C(S))$.

Theorems 2.5, 2.7 and 2.8 can be generalized for weakly ample semigroups which justifies the latter definition.

Theorem 3.8. *A weakly ample monoid is almost left factorizable if and only if it is left factorizable.*

Theorem 3.9. *For any weakly ample semigroup S, the following statements are equivalent:*

(i) *S is almost left factorizable.*
(ii) *S is a $(2,1,1)$-homomorphic image of a W-product of a semilattice by a unipotent monoid.*
(iii) *S is an idempotent separating $(2,1,1)$-homomorphic image of a W-product of a semilattice by a unipotent monoid.*

Theorem 3.10. *A weakly ample semigroup is isomorphic to a W-product of a semilattice by a unipotent monoid if and only if it is proper and almost left factorizable.*

However, the analogue of Theorem 2.6 fails for weakly ample semigroups in general. Although $M \setminus \widetilde{R}_1(M)$ is an almost left factorizable weakly ample semigroup for any left factorizable weakly ample monoid M, the converse is not valid even for W-products of semilattices by commutative cancellative monoids.

Finally, Gomes and the author [13] have investigated which proper covers of left ample semigroups arise from embeddings in factorizable left ample monoids. The characterization of such covers of E-dense left ample semigroups is generalized for any left ample semigroup. Since the characterization is rather complicated, let us mention only that it turns out to be equivalent whether a proper cover of a left ample semigroup arises from an embedding in a factorizable left ample monoid or it arises from an embedding in a factorizable left ample monoid which is a $(2,1)$-submonoid of an inverse monoid.

Acknowledgments

The author is greatful to the referee for calling her attention to references [1], [4] and [22].

References

1. B. Billhardt, Extensions of semilattices by left type-A semigroups, *Glasgow Math. J.* **39**, 7–16 (1997).
2. B. Billhardt, On embeddability into a semidirect product of a band by a group, *J. Algebra* **206**, 40–50 (1998).
3. B. Billhardt and M. B. Szendrei, Weakly E-unitary locally inverse semigroups, *J. Algebra* **267**, 559–576 (2003).
4. T. S. Blyth and R. McFadden, Unit orthodox semigroups, *Glasgow Math. J.* **24**, 39–42 (1983).

5. S. Y. Chen and S. C. Hsieh, Factorisable inverse semigroups, *Semigroup Forum* **8**, 283–297 (1974).
6. E. Dombi, Almost factorisable straight locally inverse semigroups, *Acta Sci. Math.* **69**, 569–589 (2003).
7. A. El Qallali, Factorisable right adequate semigroups, *Proc. Edinburgh Math. Soc.* **24**, 171–178 (1981).
8. J. Fountain, A class of right PP monoids, *Quart. J. Math. Oxford* **28**, 285–300 (1977).
9. J. Fountain and A. El Qallali, Proper covers for left ample semigroups, to appear in *Semigroup Forum*.
10. J. Fountain and G. M. S. Gomes, Proper left type-A monoids revisited, *Glasgow Math. J.* **35**, 293–306 (1993).
11. G. M. S. Gomes and V. Gould, Proper weakly left ample semigroups, *Internat. J. Algebra Comput.* **9**, 721–739 (1999).
12. G. M. S. Gomes and M. B. Szendrei, Almost factorizable weakly ample semigroups, preprint.
13. G. M. S. Gomes and M. B. Szendrei, Even covers for left ample semigroups, preprint.
14. M. Hartmann, Almost factorizable orthodox semigroups, preprint.
15. J. M. Howie, *Fundamentals of Semigroup Theory*, London Math. Soc. Monographs, New Series 12, Oxford Science Publications, 1995.
16. M. V. Lawson, Almost factorisable inverse semigroups, *Glasgow Math. J.* **36**, 97–111 (1994).
17. M. V. Lawson, *Inverse semigroups: The Theory of Partial Symmetries*, World Scientific, Singapore, 1998.
18. D. B. McAlister, Groups, semilattices and inverse semigroups, *Trans. Amer. Math. Soc.* **192**, 227–244 (1974).
19. D. B. McAlister, Groups, semilattices and inverse semigroups II, *Trans. Amer. Math. Soc.* **196**, 351–370 (1974).
20. D. B. McAlister, Some covering and embedding theorems for inverse semigroups, *J. Austral. Math. Soc.* **22**, 188–211 (1976).
21. D. B. McAlister and N. R. Reilly, E-unitary covers for inverse semigroups, *Pacific J. Math.* **68**, 161–174 (1977).
22. R. B. McFadden, Unit-regular orthodox semigroups, *Glasgow Math. J.* **25**, 229–240 (1984).
23. L. O'Carroll, Embedding theorems for proper inverse semigroups, *J. Algebra* **42**, 26–40 (1976).
24. F. Pastijn, The structure of pseudo-inverse semigroups, *Trans. Amer. Math. Soc.* **273**, 631–655, (1982).
25. M. Petrich, *Inverse Semigroups*, Wiley & Sons, New York, 1984.
26. M. B. Szendrei, On a pull-back diagram for orthodox semigroups, *Semigroup Forum* **20**, 1–10 (1980); corr. **25**, 311–324 (1982).
27. M. B. Szendrei, On E-unitary covers of orthodox semigroups, *Internat. J. Algebra Comput.* **3**, 317–333 (1993).
28. K. Takizawa, Orthodox semigroups and E-unitary regular semigroups, *Bull. Tokyo Gakugei Univ., Ser. IV* **31**, 41–43 (1979).